ENVIRONMENTAL CHEMISTRY

New Techniques and Data

ENVIRONMENTAL CHEMISTRY

New Techniques and Data

Edited By

Harold H. Trimm, PhD, RSO

Chairman, Chemistry Department, Broome Community College;
Adjunct Analytical Professor, Binghamton University,
State University of New York, Binghamton, U.S.A.

William Hunter III

Researcher, National Science Foundation, U.S.A.

Apple Academic Press

TORONTO NEW YORK

Apple Academic Press Inc.
3333 Mistwell Crescent
Oakville, ON L6L 0A2
Canada

Apple Academic Press Inc.
1613 Beaver Dam Road, Suite # 104
Point Pleasant, NJ 08742
USA

Exclusive worldwide distribution by CRC Press, a Taylor & Francis Group

International Standard Book Number: 978-1-926692-77-7 (Hardback)

Library and Archives Canada Cataloguing in Publication

Environmental chemistry: new techniques and data/[edited by] Harold H. Trimm, William Hunter.

Includes index.
ISBN 978-1-926692-77-7
1. Environmental chemistry. I. Trimm, Harold H. II. Hunter, William, 1971-
(Harold Henry), 1955- III. Series: Research progress in chemistry

TD193.E544 2011 577›.14 C2011-901373-8

Preface

Just as the environment is a large and diverse place, environmental chemistry is a large and diverse field, incorporating atmospheric, aquatic and soil chemistry, as well as toxicology and biochemistry. Environmental chemistry is used to discover the effects of pollutants on human health and to hypothesize the evolutionary origins of particular traits in plants and animals. It is used to understand the ecological processes determining water quality or soil fertility and to develop effective crop management systems. The applications of environmental chemistry are far-reaching and are of particular importance in today's changing environment. The articles collected in this book will familiarize you with new technologies, recently developed methods of research, and relevant data that are providing new insight into the chemical workings of our environment.

The linkages between human health and environmental contaminants represent a growing body of research in environmental chemistry. Numerous studies in this book address this topic. Rushton and Mann examine the extent and nature of pesticide-related illness presented and diagnosed in primary care across Great Britain, while investigating the feasibility of implementing a routine monitoring system for pesticide-related illness. Later, in Chapter 9, Atari and Luginaah analyze the distribution and origins of outdoor toxins known as volatile organic compounds (VOCs) in Sarnia, Ontario, Canada. These outdoor toxins have been associated with increased health problems in exposed populations, motivating research on successful methods for modeling their distribution.

Addressing a well-known vector of pathogens causing human diseases, Leal et al.'s investigation into chemical ecological approaches for mosquito surveillance and control programs, presented in Chapter 12, makes innovative contributions to the effective monitoring of this important environmental risk. These studies have potential applications in the field of public health in the form of implementing successful monitoring and exposure prevention programs. Findings in these studies may also contribute to valuable government and corporate policy.

As environmental contaminants have the ability to impact human health, the methods by which contaminants affect humans, particularly the specific chemical channels employed, create another important arena of research in environmental Chemistry. Li et al. contribute to this area of research in Chapter 11 with their discovery of a previously unrecognized regulatory pathway through which exposure to chemically diverse environmental toxicants can disrupt normal development of the central nervous system. Pages et al. use the bacterium *Strenotrophomonas maltophilia* to explore mechanisms of bacterial resistance to antibiotics and metal toxicity. Later, Garritano et al. look at the estrogenic activity in seafood from the Meditteranean Sea, studying the effects of environmental pollutants on the chemical makeup of marine organisms destined for human consumption. The last article in this compilation looks at the exposure of rat embryos to the AhR pathway, a pathway known to frequently mediate the effects of environmental toxicants on the health of exposed species. These studies provide insight into the processes by which environmental toxins affect our health and the health of our surrounding ecosystems. They impart key knowledge for ongoing efforts to remedy and prevent damaging consequences of environmental pollutants.

Environmental chemistry is also an essential component in understanding the condition of natural resources on which human livelihoods and all ecosystems rely. Water is one such resource, and several studies in this collection address the quality of important water supplies. In Chapter 4, Dalvie et al. assess pesticide contamination of ground, surface, and drinking water in three agricultural intensive regions in the Western Cape of South Africa. Their findings lead them to recommend monitoring of surface and ground water for pesticide contamination and the development of drinking water quality standards for specific pesticides in South Africa. In the following chapter, Mendosa et al. look at water quality in the Rio Grande River, an important water source for the border region between the U.S. and Canada. The water supplies of the area are subjected to contamination by agricultural, industrial, and domestic activities of the region. In addition contamination results from inconsistencies in regulations and their enforcement, a unique obstacle in the border region.

Soil is another environmental resource vital to human livelihoods. The agricultural sector of our economy relies on it for its income and we all depend on soil for the growing of food for our consumption. Furthermore, soil, and the plants growing from it, serves as an important sink for many molecules. Carbon and nitrogen are two such molecules. Nitrogen, a critical factor in plant growth, is the main component in most fertilizers. Nitrogen compounds released from inefficient plant uptake of these fertilizers can have detrimental effects on waterways and the atmosphere. Prinsi et al. contribute to the understanding of how cereal crops use nitrate applications using a recently developed proteomic approach. Sayer et al. look at changes in soil carbon levels resulting from increased litterfall in the tropics. Their study is particularly relevant given the expected increase in litterfall resulting from global changes in rainfall patterns, temperatures, and atmospheric dioxide carbon concentrations. They find that an increase in litterfall, which represents a large flux of carbon from vegetation to soil, can cause a substantial release of soil carbon into the atmosphere. Both studies contribute to the understanding of the role of soil and vegetation in nutrient cycling in the larger environment.

In addition to evaluating and monitoring current environmental phenomena, environmental chemistry is useful in explaining the evolutionary history of many traits exhibited by species present today. Gohli and Högstedt present a novel explanation for the evolution of warning coloration, a process that has troubled evolutionary biologists for some time. Next, Matz et al. explore bacterial defense mechanisms that may offer the ecological and evolutionary context for particular defense mechanisms used by many plants and animals to protect themselves from predation. Environmental chemistry can also shed light on the history of our environment at large and the evolution undergone by the natural landscapes that still surround us. In Chapter 10, Bowman and Sachs explain how saline lakes in the Canadian Northern Great Plains provide an ideal testing ground for new tools used in determining past salinity from the sediment record.

Methods for measuring the natural processes that govern our environment are continuously evolving. Collectively, the chapters in this book present innovative techniques for measuring important phenomena in our changing environment and provide new data on relevant topics in environmental chemistry.

— **Harold H. Trimm, PhD, RSO**

List of Contributors

Wafa Achouak
Commissariat à l'Energie Atomique (CEA), Cadarache, Direction des Sciences du Vivant (DSV), Institut de Biologie Environnementale et Biotechnologie (IBEB), Lab Ecol Microb Rhizosphere ans Environ Extrem (LEMiRE), Saint-Paul-lez-Durance, France.
Centre National de la Recherche Scientifique (CNRS), UMR Biol Veget and Microbiol Enviro, Saint-Paul-lez-Durance, France.
Aix-Marseille Université, Saint-Paul-lez-Durance, France.

David F. Albertini
The Center for Reproductive Sciences, Department of Molecular and Integrative Physiology, University of Kansas Medical Center, 3901 Rainbow Boulevard, Kansas City, KS 66160, USA.
Marine Biological Laboratory, Woods Hole, MA 02543, USA.

Maria Alvarez
Biology Department, El Paso Community College, P.O. Box 20500, El Paso, TX 79998, USA.

Renata Amodio-Cocchieri
Department of Food Science, University of Naples Federico II, Via Università 100, 80055 Portici, Italy.

Dominic Odwa Atari
Department of Geography, University of Western Ontario, London, Ontario, Canada.

Rosângela M. R. Barbosa
Departamento de Entomologia, Centro de Pesquisas Ageu Magalhaes-Fiocruz, Recife, Brazil.

James Botsford
Biology Department, New Mexico State University, MSC 3AF, Las Cruces, NM, 88003, USA.

Jeff S. Bowman
School of Oceanography, University of Washington, Seattle 98195-5351, USA.

Eugene Cairncross
Department of Physical Science, Peninsula Technicon, P.O. Box 1906, Bellville 7535, South Africa.

Marco Calderisi
Department of Experimental Pathology, Medical Biotechnology, Infectivology and Epidemiology, University of Pisa, Via San Zeno 37, 56127 Pisa, Italy.

Patrick Carrier
Commissariat à l'Energie Atomique (CEA), Cadarache, Direction des Sciences du Vivant (DSV), Institut de Biologie Environnementale et Biotechnologie (IBEB), Service de Biologie Végétale et de Microbiologie Environnementale (SBVME), Lab Bioenerget Biotechnol Bacteries and Microalgues (LB3M), Saint-Paul-lez-Durance, France.

Angela M. Chen
Honorary Maeda–Duffey Laboratory, Department of Entomology, University of California Davis, Davis, CA, USA.

Teresa Cirillo
Department of Food Science, University of Naples Federico II, Via Università 100, 80055 Portici, Italy.

Maurizio Cocucci
Dipartimento di Produzione Vegetale, University of Milan, via Celoria 2, I-20133 Milano, Italy.

Sandrine Conrod
Commissariat à l'Energie Atomique (CEA), Cadarache, Direction des Sciences du Vivant (DSV), Institut de Biologie Environnementale et Biotechnologie (IBEB), Lab Ecol Microb Rhizosphere and Environ Extrem (LEMiRE), Saint-Paul-lez-Durance, France.
Centre National de la Recherche Scientifique (CNRS), UMR Biol Veget and Microbiol Enviro, Saint-Paul-lez-Durance, France.
Aix-Marseille Université, Saint-Paul-lez-Durance, France.

Anthony J. Cornel
Mosquito Control Research Laboratory, Department of Entomology, University of California Davis, Davis, CA, USA.

Stephane Cuine
Commissariat à l'Energie Atomique (CEA), Cadarache, Direction des Sciences du Vivant (DSV), Institut de Biologie Environnementale et Biotechnologie (IBEB), Service de Biologie Végétale et de Microbiologie Environnementale (SBVME), Lab Bioenerget Biotechnol Bacteries and Microalgues (LB3M), Saint-Paul-lez-Durance, France.

Mohamed A. Dalvie
Occupational and Environmental Health Research Unit, Department of Public Health, Medical School, University of Cape Town, Anzio Road, Observatory 7925, Cape Town, South Africa.

Tiefei Dong
Department of Biomedical Genetics, University of Rochester Medical Center, Rochester, New York, USA.

Suhelen Egan
School of Biotechnology and Biomolecular Sciences and Centre for Marine Bio-innovation, University of New South Wales, Sydney, Australia.

Luca Espen
Dipartimento di Produzione Vegetale, University of Milan, via Celoria 2, I-20133 Milano, Italy.

André Furtado
Departamento de Entomologia, Centro de Pesquisas Ageu Magalhaes-Fiocruz, Recife, Brazil.

Sonia Garritano
IARC, 150 Cours Albert Thomas, 69372 Lyon Cedex 08, France.

Jostein Gohli
Department of Biology, University of Bergen, Bergen, Norway.

Jose Hernandez
Biology Department, El Paso Community College, P.O. Box 20500, El Paso, TX 79998, USA.

Thierry Heulin
Commissariat à l'Energie Atomique (CEA), Cadarache, Direction des Sciences du Vivant (DSV), Institut de Biologie Environnementale et Biotechnologie (IBEB), Lab Ecol Microb Rhizosphere and Environ Extrem (LEMiRE), Saint-Paul-lez-Durance, France.
Centre National de la Recherche Scientifique (CNRS), UMR Biol Veget and Microbiol Enviro, Saint-Paul-lez-Durance, France.
Aix-Marseille Université, Saint-Paul-lez-Durance, France.

Göran Högstedt
Department of Biology, University of Bergen, Bergen, Norway.

Karla J. Hutt
The Center for Reproductive Sciences, Department of Molecular and Integrative Physiology, University of Kansas Medical Center, 3901 Rainbow Boulevard, Kansas City, KS 66160, USA.

Yuko Ishida
Honorary Maeda–Duffey Laboratory, Department of Entomology, University of California Davis, Davis, CA, USA.

Staffan Kjelleberg
School of Biotechnology and Biomolecular Sciences and Centre for Marine Bio-innovation, University of New South Wales, Sydney, Australia.

Nicolas Latte
Honorary Maeda–Duffey Laboratory, Department of Entomology, University of California Davis, Davis, CA, USA.

Walter S. Leal
Honorary Maeda–Duffey Laboratory, Department of Entomology, University of California Davis, Davis, CA, USA.

Zaibo Li
Department of Biomedical Genetics, University of Rochester Medical Center, Rochester, New York, USA.

Leslie London
Occupational and Environmental Health Research Unit, Department of Public Health, Medical School, University of Cape Town, Anzio Road, Observatory 7925, Cape Town, South Africa.

Isaac N. Luginaah
Department of Geography, University of Western Ontario, London, Ontario, Canada.

Vera Mann
Medical Statistics Unit, London School of Hygiene and Tropical Medicine, London, WC1E 7HT, UK.

Carsten Matz
School of Biotechnology and Biomolecular Sciences and Centre for Marine Bio-innovation, University of New South Wales, Sydney, Australia.
Division of Cell and Immune Biology, Helmholtz Centre for Infection Research, Braunschweig, Germany.

Jose Mendoza
Biology Department, El Paso Community College, P.O. Box 20500, El Paso, TX 79998, USA.

Anna Montoya
Biology Department, El Paso Community College, P.O. Box 20500, El Paso, TX 79998, USA.

Tania I. Morgan
Honorary Maeda–Duffey Laboratory, Department of Entomology, University of California Davis, Davis, CA, USA.

Alfredo S. Negri
Dipartimento di Produzione Vegetale, University of Milan, via Celoria 2, I-20133 Milano, Italy.

Mark Noble
Department of Biomedical Genetics, University of Rochester Medical Center, Rochester, New York, USA.

Delphine Pages
Commissariat à l'Energie Atomique (CEA), Cadarache, Direction des Sciences du Vivant (DSV), Institut de Biologie Environnementale et Biotechnologie (IBEB), Lab Ecol Microb Rhizosphere and Environ Extrem (LEMiRE), Saint-Paul-lez-Durance, France.
Centre National de la Recherche Scientifique (CNRS), UMR Biol Veget and Microbiol Enviro, Saint-Paul-lez-Durance, France.
Aix-Marseille Université, Saint-Paul-lez-Durance, France.

Anahit Penesyan
School of Biotechnology and Biomolecular Sciences and Centre for Marine Bio-innovation, University of New South Wales, Sydney, Australia.

Paolo Pesaresi
Dipartimento di Produzione Vegetale, University of Milan c/o Fondazione Parco Tecnologico Padano, via Einstein—Località Cascina Codazza, I-26900 Lodi, Italy.

Brian K. Petroff
Department of Internal Medicine, University of Kansas Medical Center, 3901 Rainbow, USA.

Shui Yen Phang
School of Biotechnology and Biomolecular Sciences and Centre for Marine Bio-innovation, University of New South Wales, Sydney, Australia.

Barbara Pinto
Department of Experimental Pathology, Medical Biotechnology, Infectivology and Epidemiology, University of Pisa, Via San Zeno 37, 56127 Pisa, Italy.

Jennifer S. Powers
Department of Ecology, Evolution and Behavior, University of Minnesota, St. Paul, MN, USA.
Department of Plant Biology, University of Minnesota, St. Paul, MN, USA.
Department of Soil, Water and Climate, University of Minnesota, St. Paul, MN, USA.

Bhakti Prinsi
Dipartimento di Produzione Vegetale, University of Milan, via Celoria 2, I-20133 Milano, Italy.

Chris Pröschel
Department of Biomedical Genetics, University of Rochester Medical Center, Rochester, New York, USA.

Daniela Reali
Department of Experimental Pathology, Medical Biotechnology, Infectivology and Epidemiology, University of Pisa, Via San Zeno 37, 56127 Pisa, Italy.

Jerome Rose
Centre Européen de Recherche et d'Enseignement des Géosciences de l'Environnement (CEREGE), UMR 6635 CNRS—Universite Aix Marseille, IFR 112 PMSE Europole de l'Arbois, Aix en Provence, France.

Lesley Rushton
Department of Epidemiology and Public Health, Imperial College London, Faculty of Medicine, Norfolk Place, London, W2 1PG, UK.

Julian P. Sachs
School of Oceanography, University of Washington, Seattle 98195-5351, USA.

Roswitha Saenz
Biology Department, El Paso Community College, P.O. Box 20500, El Paso, TX 79998, USA.

Emma J. Sayer
Department of Plant Sciences, University of Cambridge, Cambridge, UK.
Smithsonian Tropical Research Institute, Balboa, Ancon, Panama, Republic of Panama.

Peter J. Schupp
Marine Laboratory, University of Guam, Mangilao, Guam, USA.

Zhanquan Shi
Department of Internal Medicine, University of Kansas Medical Center, 3901 Rainbow, USA.

Abdullah Solomon
Department of Physical Science, Peninsula Technicon, P.O. Box 1906, Bellville 7535, South Africa.

Peter Steinberg
School of Biological, Earth and Environmental Sciences and Centre for Marine Bio-innovation, University of New South Wales, Sydney, Australia.

Zainulabeuddin Syed
Honorary Maeda–Duffey Laboratory, Department of Entomology, University of California Davis, Davis, CA, USA.

Edmund V. J. Tanner
Department of Plant Sciences, University of Cambridge, Cambridge, UK.

Adrian Valles
Biology Department, El Paso Community College, P.O. Box 20500, El Paso, TX 79998, USA.

Alejandro Vazquez
Biology Department, El Paso Community College, P.O. Box 20500, El Paso, TX 79998, USA.

Jeremy S. Webb
School of Biological Sciences, University of Southampton, Southampton, UK.

Wei Xu
Honorary Maeda–Duffey Laboratory, Department of Entomology, University of California Davis, Davis, CA, USA.

List of Abbreviations

AADT	Annual average daily traffic
ABR	Auditory brainstem response
ACN	Acetonitrile
ADIs	Acceptable daily intakes
Ah	Aryl hydrocarbon
AhR	Aryl hydrocarbon receptor
ARA	Antibiotic resistance analyses
ARC	Agricultural Research Council Laboratory
BCA	Bicinchoninic acid
BFT	Back-Fourier transformation
BIM-1	Bisindolylmaleimide 1
BMP-4	Bone morphogenetic protein-4
BrdU	Bromodeoxyuridine
BRI	Bedoukian Research Incorporated
BTEX	Benzene, toluene, ethylbenzene, m/p-xylene and o-xylene
cCBB	Coomassie Brilliant Blue G-250
CCL	Contaminant Candidate List
CD	Circular dichroism
CDM	Chemically-defined medium
CNS	Central nervous system
CO_2	Carbon dioxide
3D	3-Dimensional
DA	Dissemination area
DDT	Dichlorodiphenyltrichloroethane
DEMs	Digital elevation models
DES	Diethylstilbestrol
DMSO	Dimethylsulphoxide
DMTI	Desktop Mapping Technologies Inc.
DN	Dominant-negative
DO	Dissolved oxygen
E2	17b-Estradiol
ECD	Electron capture detector
EDCs	Endocrine disrupting chemicals
EDX	Energy dispersive X-ray
EEC	Exceeded the European Community

EGF	Epidermal growth factor
EGFR	Epidermal growth factor receptor
EPA	Environmental Protection Agency
ERE	Estrogen-responsive element
ERK1/2	Extracellular signal-regulated kinase 1 and 2
ESEM	Environmental scanning electron microscopy
ESRF	European Synchrotron Radiation Facility
EXAFS	Extended X-ray absorption fine structure
FGF-2	Fibroblast growth factor-2
FID	Flame ionization detector
G6PD	Glucose-6-phosphate dehydrogenase
GB	Great Britain
GC	Gas chromatography
GC-EAD	Gas chromatography-electroantennographic detection
GC-ECD	Gas chromatographer with electron capture detector
GC-MS	Gas chromatography mass spectrometry
GISs	Geographic information systems
GOGAT	Glutamate synthase
GPRF	General Practice Research Framework
GPs	General practitioners
GS	Glutamine synthetase
HAPs	Hazardous air pollutants
HBSS	Hank's buffered saline solution
hERa	Human estrogen receptor
HGF	Hepatocyte growth factor
HPLC	High-performance liquid chromatography
HRP	Horseradish peroxidase
IARC	International Agency for Research on Cancer
IBWC	International Boundary and Water Commission
IC	Inhibition capacity
ICES	International Council for the Exploration of the Seas
ICM	Inner cell mass
IHC	Immunohistochemistry
IMGs	Integrated microbial genomes system
IRGA	Infra-red gas analyzer
JE	Japanese encephalitis
JMPR	Joint WHO/FAO Meeting on Pesticide Residues
LC-ESI/MS	Liquid chromatography-electrospray ionization mass spectrometry
LCS	Laboratory control sample

LOX	Lipoxygenase
LUR	Land use regression
MDA	Malondialdehyde
MDHAR	Monodehydroascorbate reductase
MDR	Multidrug resistance
2-ME	2-Methoxyestradiol
MeHg	Methylmercury
MI	Moran I
MICs	Minimal inhibitory concentrations
MOP	Mosquito oviposition pheromone
MPN	Most probable number
MTCs	Maximal tolerated concentration
MUG	4-Methyl-umbelliferyl
NAC	N-Acetyl-L-cysteine
NADPH	Nicotinamide adenine dinucleotide phosphate
NAPS	National Air Pollution Surveillance
NI	Nitric oxide
NiR	Nitrite reductase
NMR	Nuclear magnetic resonance
NPN	N-Phenyl-1-naphthylamine
NR	Nitrate reductase
NT-3	Neurotrophin-3
O-2A	Oligodendrocyte-type-2 astrocyte
OBP	Odorant-binding protein
OEC	Oxygen-evolving complex
OEE2	Oxygen-evolving enhancer protein 2
ONPG	O-Nitrophenyl β-D-galactopyranoside
OSD	Olfactory signal/defense
PAL	Phenylalanine ammonia-lyase
PCA	Perchloric acid
PCA	Principal components analysis
PCBs	Polychlorinated biphenyls
PDC	Pyruvate decarboxylase
PDGF	Platelet-derived growth factor
PDGFRα	Platelet-derived growth factor receptor α
PENTECH	Peninsula Technikon
PEPCase	Phosphoenolpyruvate carboxylase
PGAM-1	Phosphoglycerate mutase
6PGD	6-Phospho-gluconate dehydrogenase

PI	Propidium iodide
PKC	Protein kinase C
PLS	Partial least square
PM_{10}	Particulate matter (particles with diameter less than 10 μm)
$PM_{2.5}$	Fine particles (particles with diameter less than 2.5 μm)
POM	Particulate organic matter
POPs	Persistent organic pollutants
ppm	Part per million
PVDF	Polyvinyl difluoride
QA	Quality assurance
RDFs	Radial distribution functions
RMSE	Root mean squared error
RNAi	Inhibitory RNA
RTK	Receptor tyrosine kinase
RT-PCR	Reverse transcriptase-polymerase chain reaction
SD	Standard deviation
SF	State Forensic
siRNA	Small interfering RNA
SLS	St. Louis encephalitis
SPME	Solid-phase micro extraction
SRE	Serum response element
TCA	Trichloroacetic acid
TCDD	2,3,7,8-Tetrachlorodibenzo-p-dioxin
TCEQ	Texas Commission on Environmental Quality
TDSs	Total dissolved solids
TE	Trophectoderm
TEM	Transmission electron microscopy
TH	Thyroid hormone
TMA	Trimethylamine
TOC	Total organic carbon
2-DE	2-Dimensional gel electrophoresis
UB	Ubiquitin
USEPA	United States Environmental Protection Agency
VC	Visual conspicuousness
VEE	Venezuelan equine encephalitis
VMT	Vehicle miles traveled
VNSS	Vaatanen nine-salt solution
VOCs	Volatile organic compounds
WEE	Western equine encephalitis

WHO	World Health Organization
WNV	West Nile Virus
WWTPs	Waste Water Treatment Plants
XANES	X-Ray absorption near edge structure
XAS	X-Ray absorption spectroscopy

Contents

1. **Protein Pattern Changes in *Zea mays* Plants from Nitrate** 1
Bhakti Prinsi, Alfredo S. Negri, Paolo Pesaresi, Maurizio Cocucci, and Luca Espen

2. **Pesticide-related Illness Reported to and Diagnosed in Primary Care** 22
Lesley Rushton and Vera Mann

3. **Effects of Increased Litterfall in Tropical Forests** 36
Emma J. Sayer, Jennifer S. Powers, and Edmund V. J. Tanner

4. **Evolution of Warning Coloration** 46
Jostein Gohli and Göran Högstedt

5. **Chemically Diverse Toxicants Effects on Precursor Cell Function** 53
Zaibo Li, Tiefei Dong, Chris Pröschel, and Mark Noble

6. **Estrogen-like Activity of Seafood from Chemical Contaminants** 82
Sonia Garritano, Barbara Pinto, Marco Calderisi, Teresa Cirillo, Renata Amodio-Cocchieri, and Daniela Reali

7. **Microbial Contamination and Chemical Toxicity** 95
Jose Mendoza, James Botsford, Jose Hernandez, Anna Montoya, Roswitha Saenz, Adrian Valles, Alejandro Vazquez, and Maria Alvarez

8. **Heavy Metal Tolerance in *Stenotrophomonas maltophilia*** 111
Delphine Pages, Jerome Rose, Sandrine Conrod, Stephane Cuine, Patrick Carrier, Thierry Heulin, and Wafa Achouak

9. **Chemical Defense in Marine Biofilm Bacteria** 120
Carsten Matz, Jeremy S. Webb, Peter J. Schupp, Shui Yen Phang, Anahit Penesyan, Suhelen Egan, Peter Steinberg, and Staffan Kjelleberg

10. **Reverse and Conventional Chemical Ecology in *Culex* Mosquitoes** 132
Walter S. Leal, Rosângela M. R. Barbosa, Wei Xu, Yuko Ishida, Zainulabeuddin Syed, Nicolas Latte, Angela M. Chen, Tania I. Morgan, Anthony J. Cornel, and André Furtado

11. **Morphogenesis of the Rat Pre-implantation Embryo and Environmental Toxicant** 150
Karla J. Hutt, Zhanquan Shi, David F. Albertini, and Brian K. Petroff

12. **Endosulfan in Farming Areas of the Western Cape, South Africa** 163
Mohamed A. Dalvie, Eugene Cairncross, Abdullah Solomon, and Leslie London

13. **Land Use Regression in Sarnia, "Chemical Valley," Ontario, Canada** 181
Dominic Odwa Atari and Isaac N. Luginaah

14. Chemical and Physical Properties of Saline Lakes in Alberta and Saskatchewan 197
Jeff S. Bowman and Julian P. Sachs

Permissions 216

References 218

Index 256

Chapter 1

Protein Pattern Changes in *Zea mays* Plants from Nitrate

Bhakti Prinsi, Alfredo S. Negri, Paolo Pesaresi, Maurizio Cocucci, and Luca Espen

INTRODUCTION

Nitrogen nutrition is one of the major factors that limit growth and production of crop plants. It affects many processes, such as development, architecture, flowering, senescence, and photosynthesis. Although the improvement in technologies for protein study and the widening of gene sequences have made possible the study of the plant proteomes, only limited information on proteome changes occurring in response to nitrogen amount are available up to now. In this work, two-dimensional gel electrophoresis (2-DE) has been used to investigate the protein changes induced by NO_3 concentration in both roots and leaves of maize (*Zea mays* L.) plants. Moreover, in order to better evaluate the proteomic results, some biochemical and physiological parameters were measured.

Through 2-DE analysis, 20 and 18 spots that significantly changed their amount at least two folds in response to nitrate addition to the growth medium of starved maize plants were found in roots and leaves, respectively. Most of these spots were identified by Liquid Chromatography Electrospray Ionization Tandem Mass Spectrometry (LC-ESI-MS/MS). In roots, many of these changes were referred to enzymes involved in nitrate assimilation and in metabolic pathways implicated in the balance of the energy and redox status of the cell, among which the pentose phosphate pathway. In leaves, most of the characterized proteins were related to regulation of photosynthesis. Moreover, the up-accumulation of lipoxygenase (LOX) 10 indicated that the leaf response to a high availability of nitrate may also involve a modification in lipid metabolism.

Finally, this proteomic approach suggested that the nutritional status of the plant may affect two different post-translational modifications of phosphoenolpyruvate carboxylase (PEPCase) consisting in monoubiquitination and phosphorylation in roots and leaves, respectively.

This work provides a first characterization of the proteome changes that occur in response to nitrate availability in leaves and roots of maize plants. According to previous studies, the work confirms the relationship between nitrogen and carbon metabolisms and it rises some intriguing questions, concerning the possible role of nitric oxide (NO) and LOX 10 in roots and leaves, respectively. Although further studies will be necessary, this proteomic analysis underlines the central role of post-translational events in modulating pivotal enzymes, such as PEPCase.

Under field conditions, nitrogen nutrition is one of the major factors that influence plant growth [1, 2]. The availability of this nutrient affects many processes of the plant, among which development, architecture, flowering, senescence, photosynthesis, and photosynthates allocation [1-7].

The low bio-availability of nitrogen in the pedosphere with respect to the request of the crops has spawned a dramatic increase in fertilization that has detrimental consequences on environment such as water eutrophication and increase in NH_3 and N_2O in the atmosphere [6, 8].

Moreover, this side-effect is severe in the case of cereals, which account for 70% of food production worldwide. Indeed, in these crops the grain yield is strictly correlated with N supply but the use efficiency is not higher than 50% [9].

Because of the economical relevance, the feasibility to combine extensive physiological, agronomic and genetic studies as well as the high metabolic efficiency of C_4 plants, maize (*Zea mays* L.) was proposed as the model species to study N nutrition in cereals [10].

Among nitrogen inorganic molecules, nitrate is the predominant form in agricultural soils, where it can reach concentrations three or more orders of magnitude higher than in natural soils [11, 12].

In root cells, the uptake of this mineral nutrient involves inducible and constitutive transport systems [13]. Both systems mediate the transport of the anion by H^+ symport mechanisms [14-19] sustained by H^+-ATPase [20-22].

The first step of nitrate assimilation, that occurs in both roots and shoots, involves its reduction to ammonia by nitrate reductase (NR) and nitrite reductase (NiR) enzymes, followed by transfer of ammonia to α-chetoglutaric acid by the action of glutamine synthetase (GS) and glutamate synthase (GOGAT) [23-25]. The pathway is induced in the presence of nitrate and shows many connections with other cellular traits, among which carbohydrate and amino acid metabolism, redox status, and pH homeostasis [6, 19, 26, 27]. Hence, nitrate and carbon metabolisms appear strictly linked and co-regulated, both locally and at long distance for the reciprocal root/leaf control, in response to the nutritional status of the plant and environmental stimuli [3, 6, 26-28].

In the last years, some transcriptomic analyses have been conducted to shed light on the molecular basis of these regulatory mechanisms. Wang and co-workers studied the transcriptomic changes occurring after exposure to low and high nitrate concentrations in whole plants of *Arabidopsis thaliana*, by means of microarray and RNA gel blot analysis [29]. Besides the genes already known to be regulated by the presence of nitrate, the authors found new candidate genes encoding for regulatory proteins such as a MYB transcription factor, a calcium antiporter, putative protein kinases, and several metabolic enzymes. Another study conducted by Scheible and co-workers [7] reports a comparative transcriptomic analysis of *Arabidopsis thaliana* seedlings grown in sterile liquid culture under nitrogen-limiting and nitrogen-replete conditions by using Affymetrix ATH1 arrays and reverse transcriptase-polymerase chain reaction (RT)-PCR. The authors observed that the response to nitrogen availability involved a deep reprogramming of primary and secondary metabolisms. These data well describe the complexity of nitrogen pathway as well as the direct and/or indirect consequences that nitrogen availability exerts on the whole metabolism of the plant.

Starting from these results it should be now desirable to deepen the knowledge about the changes at translational and post-translational levels in response to nitrogen availability. In the last decade, the improvement in technologies for protein study and the widening of gene sequences made possible the study of the plant proteomes [30-34].

In this context, the availability of a large EST assembly and the efforts in sequencing maize genome [35] contributed to improve the use of maize, as highlighted by a large number of studies conducted on this species, among which the proteomic characterizations of leaf [36], of chloroplasts in bundle sheath and mesophyll cells [37] and of pericycle cells of primary roots [38].

At the present time, to the best of our knowledge no studies on nitrogen nutrition in maize were conducted by this approach. The only two proteomic works regarding this issue in cereals are based on the use of 2-DE to compare the leaves [39] and the roots [40] of two wheat varieties exposed to different levels of nitrogen. These works pointed out some significant differences, correlated to N availability during the plant growth, in the protein profiles of both organs.

In order to obtain further information, in this work we investigated protein accumulation changes induced by nitrate in both roots and leaves of *Zea mays* plants. The attention was focused on the changes in the pattern of protein soluble fractions caused by the addition of 10 mM nitrate to the hydroponic solution, after a period in which the plants were grown in the absence of nitrogen. First, the changes of some biochemical parameters were measured to describe the physiological response occurring after nitrate addition and were used to define the sampling time for proteomic analysis. These experiments led to compare the proteomes of plants previously grown for 17 days in absence of nitrogen and incubated for further 30 hr without the nutrient or in the presence of 10 mM nitrate. Through 2-DE and LC-ESI-MS/MS analyses a first characterization of the proteome changes occurring in maize plants in response to an increase in nitrate availability was obtained. The results show how many of these changes were related to enzymes of the nitrate assimilation or metabolic pathways strictly linked to it (e.g., pentose phosphate pathway and photosynthesis), but also reveal new proteins that may play a role in the nitrate responses.

RESULTS AND DISCUSSION

Experimental Design and Biochemical Parameters

The aim of this work was to apply a proteomic approach to study the changes in protein patterns of root and leaf organs of maize plants in the first phase of exposure to high availability of nitrate, comparable to agricultural conditions, after a growth period under nitrogen starvation. This is a typical condition in which the addition of nitrate induces an increase in uptake and assimilation of this nutrient [5, 28].

The need for a simultaneous analysis of the root and the leaf organs of starved plants, with completely developed but not stressed leaf apparatus, led to the definition of the experimental design showed in Figure 1. Briefly, seedlings were transferred into a hydroponic system after 3 days of germination and grown for further 14 days in a solution deprived of nitrogen. After that, at the beginning of the light period (T_0), some plants were maintained in the same nutritional condition (control, C) whereas others were transferred in a nutrient solution containing 10 mM NO_3^- (N). In order to define the sampling time for proteomic analysis, the changes of biochemical parameters in response to NO_3^- were firstly evaluated. Roots and leaves were collected at T_0 time and after 6, 30, and 54 hr of nitrate exposure.

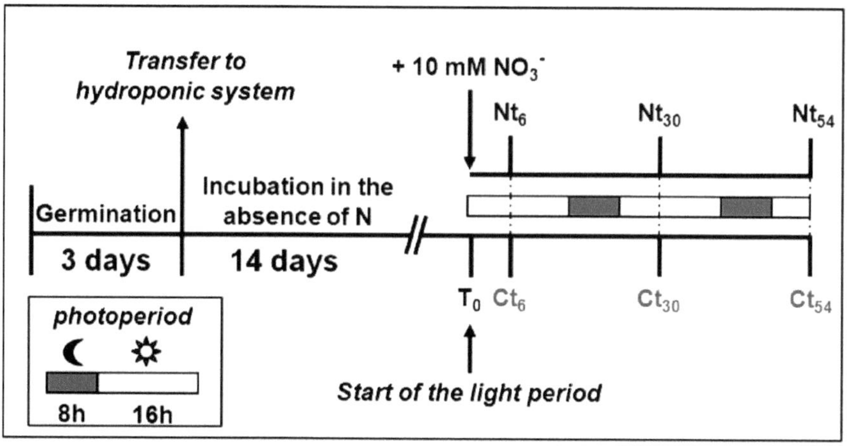

Figure 1. Experimental design. *Zea mays* seeds were germinated in the dark. After 3 days, the seedlings were transferred in a hydroponic system and grown for 14 days in the absence of nitrogen (T_0), afterwards the plants were incubated for further 54 hr in the same condition (Control, C) or in the presence of 10 mM KNO_3 (N). For details see the Materials and Methods section.

At these sampling times, the plants achieved the developmental stage corresponding to the complete expansion of the third leaf. The qualitative comparison between the C and N plants revealed some morphological differences. In particular, while the plants appeared very similar at the T_0 sampling time, after 30 hr the expansion of the fourth leaf was slightly more evident in N plants with respect to the C ones. This trend was more pronounced at 54 hr and, only in C plants, was accompanied by the comparison of faint yellow areas in the leaf blades. In the tested conditions, no significant differences were observed in root system.

In order to characterize the physiological status of the plants, the changes in nitrate content and NR activity (Figure 2) as well as the levels of proteins, amino acids, reducing sugars, sucrose, and chlorophyll were evaluated (Figure 3).

In roots and leaves of starved plants, both nitrate and NR activity were undetectable. After the addition of the nutrient to hydroponic solution the levels of nitrate progressively increased in plant tissues, reaching a level of 32.6 and 10.3 μM of NO_3^- g^{-1} FW after 54 hr in roots and leaves respectively (Figure 2A). A parallel dramatic increase of NR activity was measured until the 30th hr of NO_3^- exposure, while at the longest time considered (54 hr) a decreased activity was observed (Figure 2B).

This trend was more evident in the roots in which a more rapid and large availability of nitrate took place. The total protein levels did not change significantly in all the conditions tested (Figures 3A and B), while a sharp increase in free amino acids was detected in both organs after nitrate addition (Figures 3C and D). Moreover, the levels of amino acids were higher in the leaves than in the roots. Although many factors are involved in the overall amino acid levels, these results may suggest a contribution of translocation of nitrogen compounds between the two organs. Nitrate exposure also induced a decrease in reducing sugars in both organs (Figures 3E and F), while only in the roots of the plants exposed for 54 hr to 10 mM NO_3^- a drop of sucrose took place (Figure 3G).

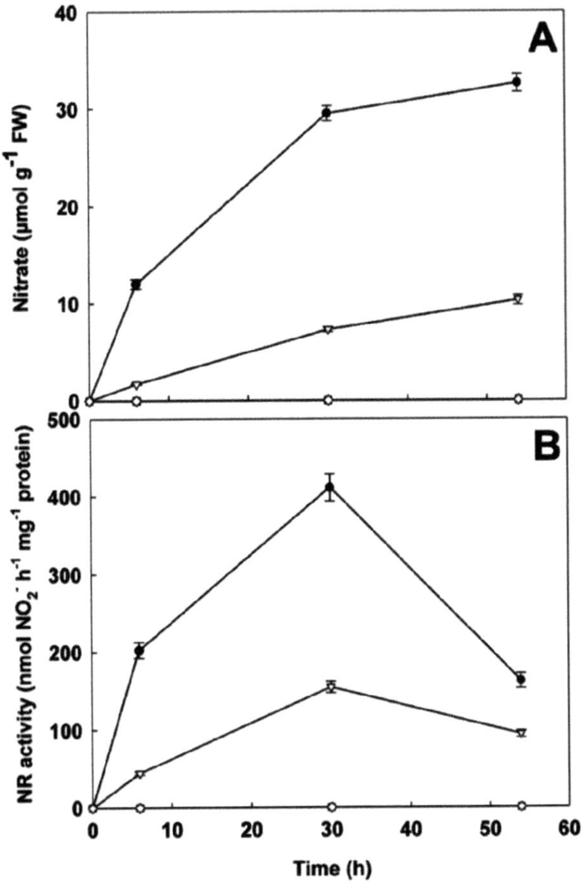

Figure 2. Nitrate content and nitrate reductase (NR) activity. Time course of the changes in nitrate content (A) and NR activity (B) in roots (close circles and closed squares) and leaves (open triangle and open rhombuses) of *Zea mays* plants, previously grown for 17 days under nitrogen starvation (T_0) and incubated for further 6, 30, and 54 hr in the absence (closed squares and open rhombuses) or in the presence (closed circles and open triangles) of 10 mM NO_3^-. In roots and leaves of starved plants, both nitrate and NR activity were undetectable. Values are the mean ± SE of three independent biological samples analyzed in triplicate (n = 9).

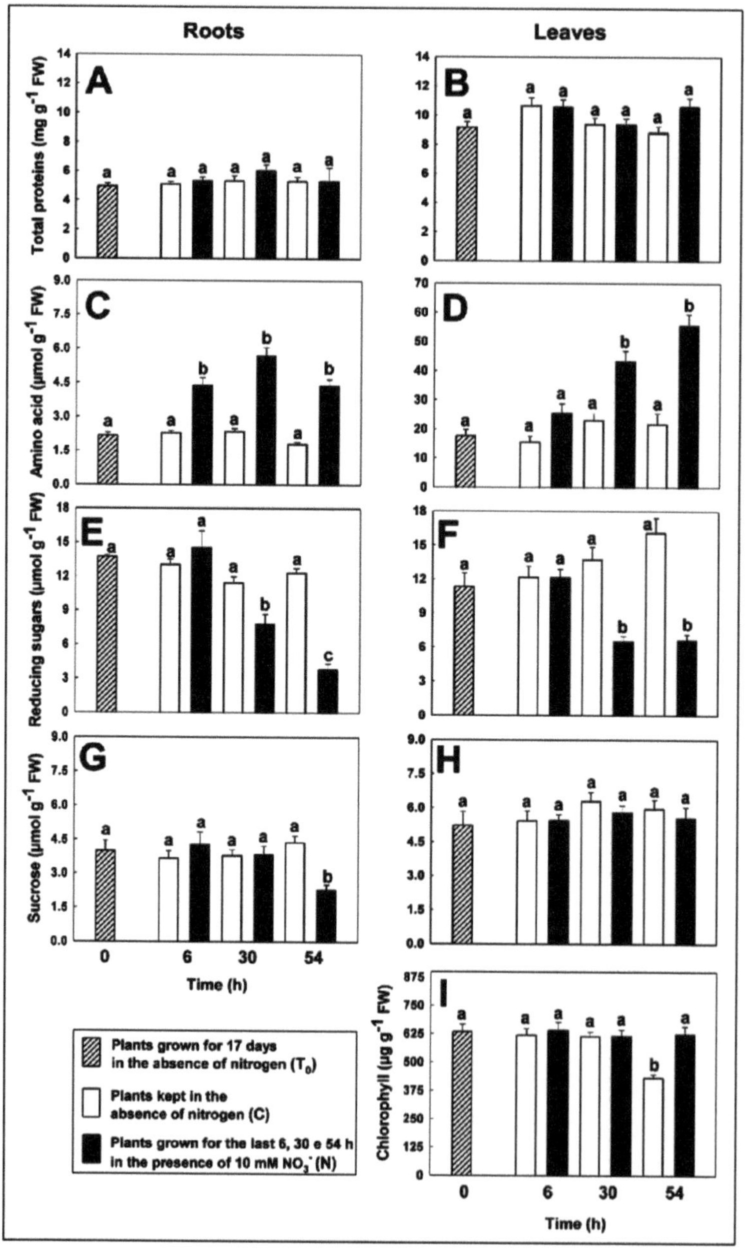

Figure 3. Total proteins, amino acids, reducing sugars, sucrose, and chlorophyll content. Time course of the changes in the content of total proteins, amino acids, reducing sugars and sucrose in roots (A, C, E, and G) and leaves (B, D, F, and H) and chlorophyll content in leaves (I) of *Zea mays* plants, previously grown in the absence of nitrogen for 17 days (T_0) and incubated for further 6, 30, and 54 hr in the absence (C) or presence of 10 mM NO_3^- (N). Values are the mean ± SE of three independent biological samples analyzed in triplicate (n = 9). Samples indicated with the same letters do not differ significantly according to Tukey's test ($p < 0.01$).

Taken together, these results well describe the induction trend of NO_3^- assimilation pathway, as suggested by the increase of NR activity and amino acids accompanied by the consequent decrease of reducing sugars, the main source of carbon skeletons [41]. In roots, where photosynthesis cannot satisfy this request and/or the demand of carbon skeleton is high, sucrose pool was also affected. The changes in carbohydrate availability and the increase of amino acid levels also explain the decrease in NR activity observed in roots at the 54th hr. In fact, these data are in agreement with the inhibitory effect on NR evocated by an increase of some amino acids, mainly asparagine and glutamine [5, 42]. Moreover, it is know that NR activity increases after sucrose addition while the low sugar content, condition that we observed in the roots of N plants, affects the nitrate reduction system [5, 42, 43]. The results suggested that this feedback mechanism was activated in roots of the plants exposed for 54 hr to 10 mM NO_3^-. Finally, only at the 54th hr, a significant decrease in chlorophyll content (Figure 3I) was measured in the leaves of starved plants, thus suggesting that the first symptoms of stress were appearing.

2-DE Analysis and Protein Identification

The biochemical and physiological data showed that the plants incubated for the last 30 hr in the presence of 10 mM NO_3^- were in a condition in which nitrogen metabolism is completely activated in both root and leaf organs and that, at the same time, no stress symptoms were detectable in the control plants. Starting from these results, the proteomic study was conducted by analyzing the soluble protein fractions extracted from roots and leaves of plants incubated for the last 30 hr in the absence or in the presence of 10 mM NO_3^-.

The ratio between dry and fresh weight as well as the total protein content appeared similar both in the roots and in the leaves of C and N samples (Table 1). The adopted protocol permitted to obtain an extraction yield of soluble proteins of about 14% and 20% for roots and leaves, respectively. Moreover, no significant differences were observed between C and N plants.

Table 1. Evaluation of the procedure for the extraction of soluble proteins from roots and leaves of plants grown in the two conditions compared in the proteomic analysis.

Organ	Condition	FW/DW	Total proteins (mg g^{-1} fW)	Extraction yield of soluble proteins (%)
Root	C plants	8.63 ± 0.05	5.32 ± 0.34	13.39 ± 0.72
	N plants	8.43 ± 0.34	6.04 ± 0.41	14.63 ± 0.69
Leaf	C plants	8.45 ± 0.09	9.39 ± 0.45	19. 17 ± 0.99
	N plants	8.64 ± 0.15	9.35 ± 0.43	20.96 ± 0.43

In the table, the fresh/dry weight (FW/DW), the content of total protein (mg g^{-1} FW) and the% yield of the extraction of soluble proteins (% of extracted soluble proteins respect to the total content) for the roots and the leaves of the plants compared by proteomic analysis are reported. The fresh weight of the roots was 0.56 ± 0.03 and 0.60 ± 0.04 g in C and N plants, respectively. The fresh weight of the leaves was 0.79 ± 0.03 and 0.86 ± 0.04 g in C and N plants, respectively. C plants: plants kept in the absence of nitrogen; N plants: plants grown for the last 30 hr in the presence of 10 mM NO_3-. Values are the mean ± SE of three independent biological samples analyzed in triplicate (n = 9).

The 2-DE representative gels of the soluble fractions of root and leaf samples are shown in Figure 4. The electrophoretic analyses detected about 1,100 and 1,300 spots in roots and leaves gels, respectively. To ascertain the quantitative changes in the proteomic maps, the relative spot volumes (%Vol) were evaluated by software-assisted analysis. The Student's t-test ($p < 0.05$), coupled with a threshold of 2-fold change in the amount, revealed that 20 spots in roots and 18 spots in leaves were affected by nitrogen availability.

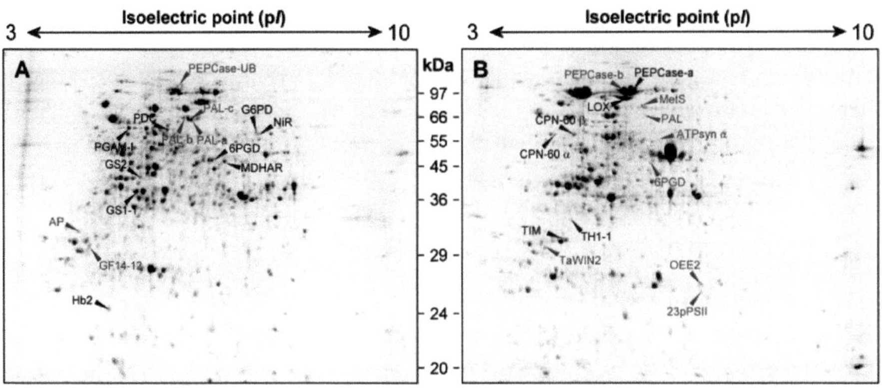

Figure 4. The 2-DE maps. Representative 2-DE maps of soluble protein fractions extracted from roots (A) and leaves (B) of *Zea mays* plants. Proteins (400 μg) were analyzed by IEF at pH 3–10, followed by 12.5% SDS-PAGE and visualized by Coomassie Brilliant Blue G-250 (cCBB)-staining. Name abbreviations, corresponding to those in Tables 2 and 3, indicate the spots, identified by LC-ESI-MS/MS, showing significant changes of at least 2-fold in their relative volumes (t-test, $p < 0.05$) after the exposure to 10 mM nitrate for 30 hr.

The analysis of these spots by LC-ESI-MS/MS allowed to identify 15 and 14 proteins in root and leaf patterns, respectively. These proteins and the changes in their accumulation are shown in Tables 2 and 3.

Table 2. List of the spots identified in the roots and their change in abundance after the exposure to 10 mM nitrate for 30 hr.

Spot ID	Accession number	Protein description	Abbr.[a]	M_r^b/pI^b	M_r^c/pI^c	Change in level [Relative volume (%)]	
						Control	10 mM NO_3
Glycolysis, gluconeogenesis, C-compound and carbohydrate metabolism							
53	BAA28170 P69319	Phosphoenolpyruvate carboxylase Ubiquitin	**PEPCase-UB**	*115.4/5.7*	*109.4/5.7* *8.5/6.6*	**0.223** ± 0.022	**0.084** ± 0.032
216	P30792	2,3-bisphosphoglycerate-independent phosphoglycerate mutase	**PGAM-1**	*63.0/5.1*	*60.6/5.3*	**0.124** ± 0.086	**0.245** ± 0.0 I I
231	AAL99745	Pyruvate decarboxylase	**PDC**	*62.4/5.5*	*65.0/5.7*	**0.080** ± 0.043	**0.167** ± 0.024
392	EAZI8378	6-phosphogluconate dehydrogenase[d]	**6PGD**	*50. 1/6.1*	*50. 1/5.5*	**0.080** ± 0.031	**0.275** ± 0.033

Table 2. *(Continued)*

Spot ID	Accession number	Protein description	Abbr.[a]	M_r^b/pI^b	M_r^c/pI^c	Change in level [Relative volume (%)]	
						Control	10 mM NO_3
1162	NP 196815	Glucose-6-phosphate 1-dehydrogenase	**G6PD**	*60.3/6.7*	*67.2/8.5*	**0.002** ± 0.001	**0.010** ± 0.014
Nitrogen metabolism, amino acid metabolism and protein/peptide degradation							
268	ACG29734	Ferredoxin-nitrite reductase	**NiR**	*59.7/6.7*	*66.2/6.5*	**0.035** ± 0.054	**0.124** ± 0.084
483	P25462	Glutamine synthetase, chloroplastic	**GS2**	*42.2/5.2*	*41.015.4*[e]	**0.066** ± 0.015	**0.137** ± 0.059
538	P38559	Glutamine synthetase root isozyme I	**GSI-1**	*38.7/5.1*	*39.2/5.6*	**0.210** ± 0.010	**0.480** ± 0.039
707	BAA06876	Aspartic protease	**AP**	*31.6/4.6*	*54.1/5.1*	**0.051** ± 0.043	**0.0 15** ± 0.065
Secondary metabolism							
171	AAL40137	Phenylalanine ammonia-lyase	**PAL-a**	*68.6/5.9*	*74.9/6.5*	**0.476** ± 0.034	**0.184** ± 0.012
172	AAL40137	Phenylalanine ammonia-lyase	**PAL-b**	*68.6/5.8*	*74.9/6.5*	**0.904** ± 0.136	**0.277** ± 0.026
1160	AAL40137	Phenylalanine ammonia-lyase	**PAL-c**	*68.0/5.8*	*74.9/6.5*	**0.713** ± 0.103	**0.275** ± 0.034
Cell rescue, defense and virulence							
390	NP 001061002	Putative monodehydroascorbate reductase [d]	**MOHAR**	50.1/6.2	*52.8/6.8*	**0.127** ± 0.016	**0.275** ± 0.033
960	AAZ96.790	hemoglobin 2	**Hb2**	*24.8/4.9*	20.6/5.0	**0.018** ± 0.061	**0.099** ± 0.068
Unknown							
774	001526	14-3-3-like protein GF 14-12	**GFI4-12**	*29.6/4.6*	*29.614.7*	**0.345** ± 0.034	**0.146** ± 0.028

Statistical information about LC-ESI-MS/MS analysis are reported in Additional files 2 and 3. Changes in the relative spot volumes are the mean ± SE of six 2-DE gels derived from three independent biological samples analyzed in duplicate (n = 6). Proteins were classified according to MIPS funcat categories.

[a]: Protein abbreviation

[b]: Experimental molecular weight (kDa) or isoelectric point

[c]: Theoretical molecular weight (kDa) or isoelectric point

[d]: Information obtained by alignment of the sequence through BLAST analysis against NCBI nr database

[e]: Values referred to the mature form of the protein

Table 3. List of the spots identified in the leaves and their change in abundance after the exposure to 10 mM nitrate for 30 hr.

Spot ID	Accession number	Protein description	Abbr.[a]	M_r^b/pI^b	M_r^c/pI^c	Change in level [Relative volume (%)]	
						Control	10 mM NO_3
Nitrogen and amino acid metabolism							
1094	BABII740	TaWIN2	TaWIN2	*29.9/4.7*	*28.7/4.8*	**0.182** ± 0.009	**0.090** ± 0.014
254	AAL73979	Methionine synthase protein	MetS	*83.4/5.9*	*83.8/5.9*	**0.148** ± 0.020	**0.073** ± 0.008
C-compound and carbohydrate metabolism							
650	AAC27703	Putative cytosolic 6-phosphogluconate dehydrogenase	6PGD	*47.4/6.0*	*52.9/6.2*	**0.088** ± 0.005	**0.043** ± 0.007

Table 3. *(Continued)*

Spot ID	Accession number	Protein description	Abbr.[a]	M_r^b/pI^b	M_r^c/pI^c	Change in level [Relative volume (%)]	
						Control	10 mM NO_3
Photosynthesis							
134	P04711	Phosphoenolpyruvate carboxylase I	PEPCase·a	*104.4/5.8*	*109.3/5.8*	**0.990** ± 0.083	**2.770** ± 0.295
138	P04711	Phosphoenolpyruvate carboxylase I	PEPCase-b	*104.4/5.7*	*109.3/5.8*	**2.220** ± 0.278	**1.090** ± 0.205
500	P05022	ATP synthase subunit alpha, chloroplastic	ATPsyn α	*55.9/6.1*	*55.7/5.9*	**0.042** ± 0.007	**0.015** ± 0.003
1065	NP 00 I 063777	Putative triosephosphate isomerase chloroplast precursor [d]	TIM	*31 .0/4.9*	*32.4/7.0*	**0.028** ± 0.009	**0.088** ± 0.014
1244	000434	Oxygen-evolving enhancer protein 2 chloroplast precursor	OEE2	*26.6/6.5*	*27.3/8.8*	**0.201** ± 0.013	**0.090** ± 0.011
1612	BAA08564	23 kDa polypeptide of photosystem II	23pPSII	*26.3/6.5*	*27.0/9.5*	**0.147** ± 0.008	**0.055** ± 0.006
Protein folding and stabilization							
462	NP 001056601	RuBisCO subunit binding-protein beta subunit [d]	CPN-60 β	*58.51/5.1*	*64.1/5.6*	**0.079** ± 0.014	**0.164** ± 0.015
467	AAP44754	Putative rubisco subunit binding-protein alpha subunit precursor	CPN-60 α	*58.2/4.8*	*61.4/5.4*	**0.046** ± 0.004	**0.096** ± 0.004
Metabolism of vitamins, cofactors, and prosthetic groups							
999	041738	Thiazole biosynthetic enzyme 1-1, chloroplast precursor	THI-1	*33.0/5.1*	*32.814.9*[e]	**0.010** ± 0.001	**0.048** ± 0.003
Secondary metabolism							
313	AAL40137	Phenylalanine ammonia-lyase	PAL	*70.2/6.0*	*74.9/6.5*	**0.076** ± 0.008	**0.023** ± 0.002
Lipid metabolism							
219	ABC59693	Lipoxygenase	LOX	*94.6/5.8*	*102.1/6.1*	**0.023** ± 0.011	**0.149** ± 0.011

Statistical information about LC-ESI-MS/MS analysis are reported in Additional files 2 and 3. Changes in the relative spot volumes are the mean ± SE of six 2-DE gels derived from three independent biological samples analyzed in duplicate (n = 6). Proteins were classified according to MIPS funcat categories.

[a]: Protein abbreviation

[b]: Experimental molecular weight (kDa) or isoelectric point

[c]: Theoretical molecular weight (kDa) or isoelectric point

[d]; Information obtained by alignment of the sequence through BLAST analysis against NCBI nr database

[e]: Values referred to the mature form of the protein

Functional Role and Quantitative Change of the Proteins Identified in Roots

Many of the spots identified in roots were enzymes involved in nitrogen and carbon metabolisms (Table 2). According to the induction of the NO_3^- assimilation pathway, in the roots of the plants incubated for the last 30 hr in the presence of the nutrient, we observed an increase in the accumulation of nitrite reductase (spot 268, NiR) and of glutamine synthetase plastidial isoform (spot 483, GS2).

Moreover, in response to the demand of carbon skeletons and Nicotinamide adenine dinucleotide phosphate (NADPH), which is used in non-green tissues for ferredoxin reduction [44], an increase in the levels of phosphoglycerate mutase (spot 216, PGAM-1), glucose-6-phosphate dehydrogenase (spot 1162, G6PD) and 6-phosphogluconate dehydrogenase (spot 392, 6PGD) took place. These results well agree with previous array data that describe the responses to nitrate exposure in *Arabidopsis* and tomato [7, 29, 45].

An increase in accumulation of the cytosolic isoform of glutamine synthetase (spot 538, GS1-1) was also detected in roots of N plants. On the basis of identified peptides by MS analysis it was possible to discriminate among the five GS1 isoforms known in *Zea mays* (SwissProt reviewed database) and to restrict the possible identification to two of them (GS1-1 (Swiss-Prot:P38559) and GS1-5 (Swiss-Prot:P38563) [46]). The fact that Li and co-workers [46], through a Northern blot hybridization analysis, found that the transcript of *GS1-1* gene was the only one expressed in roots, conducted to the specific identification of GS1-1 protein. Moreover, Sakakibara and co-workers [47] showed that GS1-1 transcript was the only induced by NO_3^-. The proteomic approach used in the present work allows to confirm these results at the translational level, demonstrating that in maize roots a cytosolic ammonia assimilation pathway can be activated also in response to nitrate.

Other spots that were found to increase their relative volumes in response to nitrate were a non-symbiotic hemoglobin and a monodehydroascorbate reductase (spot 960, Hb2 and spot 390, MDHAR). In a previous work on *Arabidopsis*, it was found that NO_3^- induced AtHB1 and AtHB2, two genes that encode for non-symbiotic hemoglobins [7, 29]. Scheible and co-workers [7] suggested that these proteins could change their abundance in relation to the redox status, whereas Wang and co-workers [29] speculated on the possibility that the induction of hemoglobin could aim at reducing oxygen concentration during NR synthesis, since molybdenium can be sensitive to oxygen. Besides, hemoglobin and MDHAR are known to be involved in the scavenging of NO that can be produced by cytosolic and/or plasmamembrane NR when nitrite is used as substrate [48, 49]. The NO is a signaling molecule which is involved in many biochemical and physiological processes [50]. It has been reported that in plant roots, NO plays a role in growth, development and in some responses to environmental conditions, such as hypoxia [51]. Recently, a possible involvement of NO in the mediation of nitrate-dependent root growth in maize has been suggested [52]. According to this work, that describes a reduction of endogenous NO at high external NO_3 concentration, the observed concomitant up-accumulation of Hb2 and MDHAR in our experimental condition supports the hypothesis that they might contribute in controlling NO levels in root tissues after exposure to NO_3 [48, 49, 52].

The last protein found to be present in higher amount in N plants was a pyruvate decarboxylase (spot 231, PDC). This enzyme catalyzes the decarboxylation of pyruvic acid into acetaldehyde, the first step of the alcoholic fermentation. In particular, we identified the PDC isoenzyme 3 that has been previously found to be induced in hypoxia condition [53]. Although further studies are required to understand why PDC is induced by NO_3^-, we can observe that fermentation pathways are induced in response to redox status changes and that this condition could be also linked to the activation of the Hb/NO cycle (see above) [49, 54].

Among the spots identified in roots, six showed a down-accumulation in N plants (Table 2). Three of them were identified as phenylalanine ammonia-lyase (spots 171, 172, and 1160, PAL-a, PAL-b, and PAL-c). The MS analysis indicated for all three spots the same protein [GenBank:AAL40137] while the electrophoretic data showed some differences in M_r and p_I, suggesting that post-translational modification events may have occurred. It has been shown as low nitrogen availability induces transcripts encoding enzymes of phenylpropanoid and flavonoid metabolism, such as PAL, chalcone synthase, and 4-coumarate:coenzyme A ligase, while after nitrogen repletion these activities are down-regulated [7, 55]. Our proteomic data appear to be in agreement with these studies.

Previously, it was found that under low nitrogen availability four proteases (e.g., serine, aspartate/metalloproteases and two cysteine proteases) increased their activity to degrade non-essential proteins in order to remobilize this nutrient [56]. In this work, we found an aspartic protease belonging to the A1 family (spot 707, AP) that was down-regulated after NO_3^- exposure. Moreover, the experimental M_r appeared lower with respect to that expected for this protein, thus suggesting that this spot is referable to the active form of the enzyme [57]. These data support a new possible role for A1 protease family [57, 58].

The PEPCase activity is known to increase during nitrate assimilation, having a role in cell pH homeostasis and an anaplerotic function [14, 19, 59-61]. In addition, the monoubiquitination of this enzyme was recently well described in germinating castor oil seeds by Uhrig and co-workers [62]. It was found that this event is non-destructive and that this reversible post-translational modification of the enzyme reduces its affinity for PEP and its sensitivity to allosteric activators and inhibitors. The MS analysis of spot 53 identified eight peptides, seven of which matched with a PEPCase (DDBJ:BAA28170) (theoretical M_r/p_I equal to 109.4/5.7), while the last peptide belonged to an ubiquitin (UB) (Swiss-Prot:P69319) (theoretical M_r/p_I equal to 8.5/6.6). The experimental M_r and p_I of spot 53, that were 115.4 and 5.7 respectively, were in agreement with the monoubiquitination of the PEPCase (PEPCase-UB, theoretical M_r/p_I equal to 117.9/5.8 respectively). Moreover, the domain responsible to bind ubiquitin previously identified in PEPCase of other vascular plants is present in this maize PEPCase [62]. These results suggest that in maize roots the modulation of PEPCase activity in response to nitrogen availability could occur also through reversible monoubiquitination.

The last spot identified in roots that was down-regulated by NO_3^- was the 14-3-3-like protein GF14-12 (spot 774, GF14-12). Previously, it was found that this protein is localized in the nucleus where it binds the DNA at the G-box regions in association with transcription factors and that it is involved in the regulation of gene expression [63, 64]. More recently, it was described an interaction of 14-3-3 proteins with some transcription factors such as VP1, EmBP1, TBP, and TFIIB [65]. Further studies are required to clarify the effective role of GF14-12, for which the functional information are still lacking.

Functional Role and Quantitative Change of the Proteins Identified in Leaves

Many of the spots identified in leaves by LC-ESI-MS/MS analysis were proteins linked to the NO_3^- assimilation as well as to the photosynthetic activity (Table 3).

The activity of NR can be modulated also at post-translational level through a phosphorylation event followed by binding of inhibitory 14-3-3 protein [66, 67]. One of the spots analyzed in the leaves was identified as TaWIN2 (Table 3, spot 1094, TaWIN2), that was previously described to be involved in the NR inactivation [67]. We found that the level of this protein decreased in leaves of N plants, where NR activity was induced (Table 3).

According to the well known relationships existing between nitrogen and carbon metabolism, the changes in accumulation of some spots after NO_3^- addition are consistent with an increase of photosynthesis rate. Two spots that raise after NO_3^- addition were identified as CPN-60α and CPN-60β (spot 467 and 462, CPN-60α and CPN-60β, respectively), that are chaperonin proteins involved in folding of ribulose-1,5-bisphosphate carboxylase [68]. Moreover, a chloroplastic triosephosphate isomerase was up-regulated by NO_3^- (spot 1065, TIM), while a cytosolic 6-phosphogluconate dehydrogenase (spot 650, 6PGD) was down-regulated, as expected when the request of reducing power could be satisfied by the increase in photosynthetic activity [69].

Spot 500 was identified as the α subunit of the chloroplastic ATP synthase (ATPsyn α), but unexpectedly it was more abundant in leaves of C plants. Although only a speculative interpretation of this result can be made, we could hypothesize that in leaves of the N plants ATP synthase should be activated and this process requires the reconstitution of the enzymatic complex in the thylakoid membranes [70]. Hence, to clarify this point, it should be necessary to investigate if the decrease of ATPsyn α observed in the soluble fraction of N plants is effectively accompanied by an increase of this protein in the membrane fraction.

Thiamine (i.e., vitamin B1) is required in many pathways, such as the Calvin cycle, the branched-chain amino acid pathway and pigment biosynthesis [71]. Along with higher request of this vitamin in leaves of N plants, where the activation of these pathways could take place, we identified, among the spots up-regulated by N, the thiazole biosynthetic enzyme (spot 999, TH1-1) that is known to be involved in thiamine biosynthesis [71].

Two spots were identified as PEPCase (spot 134 and 138, PEPCase-a and PEPCase-b respectively). In C_4 plants such as maize, this enzyme plays a central role in photosynthesis, because it catalyzes the primary fixation of atmospheric CO_2 [72]. The catalytic activity and sensitivity of this enzyme are mediated by a reversible phosphorylation [73]. The experimental p_Is of the spots 134 and 138 were 5.8 and 5.7, respectively. Moreover, these two PEPCase forms showed opposite changes in abundance in the leaves of plants grown in the last 30 hr in the presence of NO_3^- with respect to the controls. The results obtained in our work suggest that the two spots of PEPCase are referable to the phosphorylated (spot 138) and to the unphosphorylated (spot 134) form with a predicted p_I of 5.7 and 5.8, respectively, that are known to correspond to the more and less active states of this enzyme [74]. Interestingly, despite the fact that data suggest an increase in the photosynthetic activity, the phosphorylated form was more abundant in the proteomic map of C plants. These results support the immunological observation by Ueno and co-workers [73] that the diurnal regulation of phosphorylation state of PEPCase appears delayed in nitrogen-limited conditions, suggesting that the circadian control of PEPCase is affected by nitrogen starvation.

Two of the spots down-regulated in leaves of N plants were a phenylalanine ammonia-lyase (spot 313, PAL) and a methionine synthase (spot 254, MetS). The decrease of PAL, observed also in root tissue (see above), is a further evidence that phenylpropanoid and flavonoid metabolisms are affected by nitrogen availability [7, 55]. On the other hand, the change in accumulation of MetS is contrasting with a recent proteomic study performed on wheat by Bahrman and co-workers [39]. These authors found that the induction of this enzyme was positively related to nitrogen availability. This discrepancy could be associated to different genetic traits of the two species, as well as it could be linked to different experimental approaches adopted in the two studies. Nevertheless, it should be observed that in both these works a single spot referable to MetS was detected, while further information on total level and/or on activity of this enzyme is necessary to clarify this point.

The spot 219, which considerably increased in N plants, was a LOX. In particular, the analysis of the MS spectra identified the LOX codified by *ZmLOX10* gene, which was found to be a plastidic *type 2* linoleate 13-LOX [75]. The expression analysis of this gene revealed that its transcript was abundant in leaves and was regulated by a circadian rhythm with a trend strictly linked to the photosynthetic activity. Moreover, it has been proposed that ZmLOX10 is involved in the hydroperoxide lyase-mediated production of C_6-aldehydes and alcohols and not in the biosynthesis of JA [75]. Although some evidences suggest a role of ZmLOX10 in the responses to (a)biotic stresses, its involvement in the diurnal lipid metabolism was also proposed [75, 76].

At the same time, we identified two proteins as an oxygen-evolving enhancer protein 2 (spot 1244, OEE2) and a 23 kDa polypeptide of photosystem II (spot 1612, 23pPSII), which were down-accumulated in leaves of N plants (Table 3). Both have been classified as members of PsbP family that is one of the three extrinsic protein families composing the oxygen-evolving complex (OEC) of photosystem II in higher plants [77-79]. In addition, it was recently demonstrated that PsbP proteins are essential for the normal function of PSII and play a crucial role in stabilizing the Mn cluster *in vivo* [80]. Moreover, the stability of this class of protein seems related to the lipid composition of chloroplastic membranes that is also affected by nitrogen availability [81, 82].

In order to elucidate the physiological meaning of these variations and to verify if they could be related to a stress status or to an alteration in photosynthetic performance, changes of both maximum quantum yield of photosystem II (F_V/F_M; dark adapted plants) and effective quantum yield of photosystem II (Φ_{II}; light adapted plants), dry weight and MDA levels of shoot were measured (Figure 5). Although the F_V/F_M parameter, measured on over-night dark adapted plants at time points 0, 24, and 48 hr, resulted in very similar values between C and N plants (about 0.80; see also Figure 5A), the Φ_{II} values showed a very slight decrease in C plants during the second period of illumination (C plants Φ_{II}, 0.71 vs. N plants, 0.73) and the difference became more marked between 48 and 54 hr of nitrogen starvation. Similar data could be obtained by monitoring biomass production at the different time points (Figure 5B), indicating that photosynthetic performances are highly impaired in C plants after 48–54 hrs of treatment. Nevertheless, no changes in MDA were detected in all the conditions tested (Figure 5C).

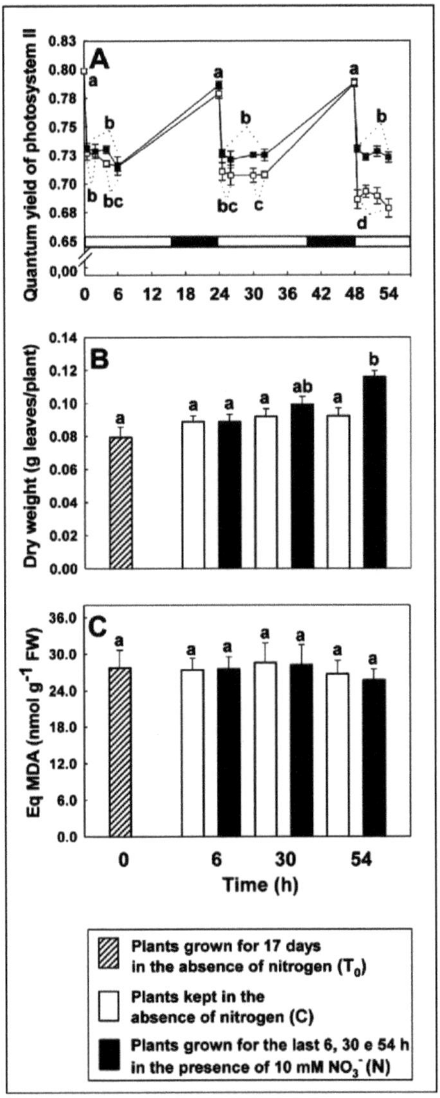

Figure 5. The F_V/F_M, Φ_{II}, dry weight and MDA in leaves. Time course of the changes in F_V/F_M and ΦII (A), dry weight in leaves (B) and MDA levels (C) of *Zea mays* plants, previously grown for 17 days under nitrogen starvation (T_0) and incubated for further 6, 30, and 54 hr in the absence or presence of 10 mM NO_3^-. Symbol in Figure A: open squares, Control; closed squares, 10 mM NO_3^-; horizontal bar: white bars, light periods; black bars, dark periods (for details see Figure 1). Values are the mean ± SE of three independent biological samples analyzed in triplicate (n = 9). Samples indicated with the same letters do not differ significantly according to Tukey's test ($p < 0.01$).

Taken together these results indicate that at the 30th hr, the time point chosen for proteomic analysis, plants start feeling the different nitrogen content in the growth media without developing major stress symptoms and the associated pleiotropic effects.

These data sustain the hypothesis that ZmLOX10 could be involved in lipid metabolism of the chloroplast that is strictly depending on photosynthetic activity [75, 76]. Further analyses are needed to unravel this possible intriguing role of ZmLOX10.

Considering the PsbP proteins, the change in accumulation of OEE2 and 23pPSII could indicate that OEC stability is affected by the N availability. Through time-course experiments, it will be possible to better correlate the relationship among N nutritional status, lipid metabolism, PsbP protein level, and PSII functionality.

MATERIALS AND METHODS

Plant Material and Growth Conditions

Maize (*Zea mays* L.) seeds of T250 inbred line, kindly provided by Prof. Zeno Varanini of Udine University—Italy, were germinated in the dark at 26°C on blotting paper saturated with deionized water. After 72 hr, seedlings were transferred to a hydroponic system placed in a growth chamber with a day/night regime of 16/8 hr and a PPFD of 200 μM m^{-2} s^{-1} at plant level, with a temperature of 22°C in the dark and 26°C in the light and with a relative humidity of 70%. Seedlings were grown using of the following solutions: (i) 4 mM $CaSO_4$ for the first 48 hr; (ii) 0.4 mM $CaSO_4$, 0.2 mM K_2SO_4, 0.175 mM KH_2PO_4, 0.1 mM $MgSO_4$, 5 μM KCl, 20 μM Fe-EDTA, 2.5 μM H_3BO_3, 0.2 μM $MnSO_4$, 0.05 μM $CuSO_4$, 0.2 μM $ZnSO_4$, 0.05 μM Na_2MoO_4 (growing solution) for the following 12 days. After this 17 days-long period of N starvation (T_0), plants were transferred in a fresh growing solution added (N) or not (C) with 10 mM KNO_3. The pH of all the growth solutions was adjusted to 6.1 and the solutions were changed every 3 days. All hydroponic solutions were continuously aerated by an electric pump.

At T_0 stage and after a period of 6, 30, and 54 hr plants were harvested, washed with distilled water and then blotted with paper towels. Finally, roots and leaves were separated and the samples were frozen in liquid N_2 and stored at –80°C. The roots used for determining nitrate content were rinsed twice in ice-cold 0.4 mM $CaSO_4$ solution for 15 min for removing the anion present in the apoplast before sampling.

Levels of Nitrate

Nitrate was extracted from the tissues by homogenizing the samples previously boiled in 4 volumes of distilled water for 15 min. The homogenate was centrifuged at 12,000 g for 20 min to obtain a clarified supernatant. Nitrate content was measured by adding 0.8 ml of 5% (w/v) salicylic acid in concentrated sulfuric acid solution to 0.2 ml of the supernatant. The mixture was stirred vigorously and allowed to react over 20 min, afterwards 19 ml of 2 N NaOH were slowly added and the resulting color was read at 410 nm [83].

Nitrate Reductase Activity

The NR was extracted by using 4 volumes of ice-cold 50 mM pH 7.8 MOPS-KOH buffer containing 5 mM EDTA, 5 mM NaF, 2 mM MSH, 1 mM PMSF, 10 μM FAD, 1 μM leupeptin, and 10 μM chymostatin. The homogenates were centrifuged at 13,000 g for 15 min at 4°C. The NR activity was measured as described by Ferrario-Méry et al. [84] using a reaction mixture consisting of 50 mM pH 7.5 MOPS-KOH buffer, 1 mM NaF, 10 mM KNO_3, 0.17 mM NADH, 10 mM $MgCl_2$, and 5 mM EDTA. The reaction was blocked after 10 or 20 min by adding an equal volume of sulphanilamide (1%, w/v in 3 M HCl) followed by n-naphtylethylethylenediamine dihydrochloride (0.02%, w/v). 30 min later, the concentration of NO_2^- was determined spectrophotometrically at 540 nm. The protein concentration was determined by 2-D Quant Kit (GE Healthcare).

Determination of Reducing Sugars, Sucrose, Amino Acids, Total Proteins, and Chlorophyll

Reducing sugars, sucrose and amino acids were extracted by homogenizing frozen tissues in 4 volumes of ice-cold 0.5 M perchloric acid (PCA). The homogenate was centrifuged for 10 min at 13,000 g at 4°C and the resulting pellet was washed with the same volume of PCA and then centrifuged again in the same conditions. The KOH was added to the collected supernatant (to pH 7.6) to remove excess PCA. Reducing sugars were measured according to the colorimetric method by Nelson [85]. Total soluble sugars were determined by the same method boiling an aliquot of PCA extract for 1 hr before neutralization. Sucrose was estimated from the difference between total soluble and reducing sugars. Total amino acids were measured by the ninhydrin method [86].

Total proteins were extracted as previously described by Martínez and co-workers [87] by homogenizing the samples, previously powdered in liquid nitrogen, in 4 volumes of a 125 mM pH 8.8 Tris HCl buffer containing 1% (w/v) SDS, 10% (w/v) glycerol, 50 mM $Na_2S_2O_5$. The homogenate was centrifuged at 13,000 g for 20 min to obtain a clarified supernatant. The protein content was measured by using 2-D Quant Kit (GE Healthcare).

Chlorophyll was extracted by homogenizing the leaves, previously powdered in liquid nitrogen, in 4 volumes of 80% pre-cooled acetone (v/v). The homogenate was centrifuged at 13,000 g for 20 min at 4°C to obtain a clarified supernatant. Chlorophyll concentration was measured according to Lichtenthaler [88].

Determination of Malondialdehyde and Chlorophyll Fluorescence of the Leaves

Malondialdehyde (MDA) was assayed by the method of Heath and Packer [89]. Frozen samples were homogenized with 4 volumes of ice-cold 0.1% (w/v) trichloroacetic acid (TCA) and centrifuged at 13,000 g for 20 min at 4°C. An equal volume of 20% (w/v) TCA plus 0.5% (w/v) thiobarbituric acid was added to the supernatants, which were subsequently heated at 95°C for 30 min. The extracts were then clarified by centrifugation at 13,000 g for 10 min, and the difference between the absorbance at 532 and 600 nm was measured. The MDA equivalent was calculated from the resulting difference using the extinction coefficient of 155 mM^{-1} cm^{-1}.

In order to determine the photosynthetic performance, the chlorophyll fluorescence was measured by using a portable continuous-excitation type fluorometer (Handy-PEA,

Hansatech Instrument). The maximum quantum efficiency of photosystem II (F_V/F_M) was calculated on over-night dark adapted plants, according to the equation (F_M-F_0)/F_M, where F_0 and F_M are the fluorescence levels when plastoquinone electron acceptor pool (Qa) is fully oxidized and transiently fully reduced, respectively [90]. The photosynthetic performance of light adapted plants was evaluated by monitoring the effective quantum yield of photosystem II (Φ_{II}) defined as (F_M¢-F_0¢)/F_M¢ [91], where F_M'and F_0¢ represent the maximal and minimal fluorescence emission of photosystem II under light conditions.

Statistical Analyses of Biochemical and Physiological Measurements

For all the biochemical and physiological measurements, the experimental design consisted in three independent biological samples each analyzed in triplicate (n = 9).

One-way analysis of variance (ANOVA) followed by the post hoc Tukey's test ($p < 0.01$) was used to verify the significance of the variations measured among all the tested parameters. This statistical analysis was performed using the software STATISTICA 7.

Extraction of Protein Samples for 2-DE Analysis

Three independent biological replicates were extracted for each condition. Frozen samples, each composed by leaves or roots of six plants, were finely powdered in liquid nitrogen using a pestle and mortar, added with PVPP (0.5% and 1% (w/w) for roots and leaves samples, respectively), homogenized in 4 volumes of extraction buffer (0.5 M Tris HCl pH 8, 0.7 M sucrose, 10 mM EDTA, 1 mM PMSF, 1 μM leupeptin, 0.1 mg ml^{-1} Pefabloc (Fluka), 0.2% (v/v) MSH) and centrifuged at 13,000 g at 4°C for 20 min. The resultant supernatant was centrifuged at 100,000 g at 4°C for 38 min to obtain the soluble fraction. Proteins were then purified using the method previously described by Hurkman and Tanaka [92] by adding an equal volume of ice-cold Tris buffered phenol (pH 8) to the supernatant. Samples were shaken for 30 min at 4°C, incubated for 2 hr at 4°C and finally centrifuged at 5,000 g for 20 min at 4°C to separate the phases. Proteins, grouped in the upper phenol phase, were precipitated by the addition of 5 volumes of –20°C pre-cooled 0.1 M ammonium acetate in methanol and the incubation at –20°C overnight. Precipitated proteins were recovered by centrifuging at 13,000 g at 4°C for 30 min and then washed again with cold methanolic ammonium acetate and three times with cold 80% (v/v) acetone. The final pellet was dried under vacuum and dissolved in IEF buffer (7 M urea, 2 M thiourea, 3% (w/v) CHAPS, 1% (v/v) octylphenoxy polyethoxy ethanol (NP-40), 50 mg ml^{-1} DTT and 2% (v/v) IPG Buffer pH 3–10 (GE Healthcare)) by vortexing and incubating for 1 hr at room temperature. Samples were centrifuged at 10,000 g for 10 min and the supernatants stored at –80°C until further use. Protein concentration was determined by 2-D Quant Kit (GE Healthcare).

2-DE Analysis

Protein samples (400 μg) were loaded on pH 3–10, 24 cm IPG strips passively rehydrated overnight in 7 M urea, 2 M thiourea, 3% (w/v) CHAPS, 1% (v/v) NP-40, 10 mg ml^{-1} DTT and 0.5% (v/v) IPG Buffer pH 3–10. The IEF was performed at 20°C with current limit of 50 μA/strip for about 90 kVh in an Ettan IPGphor (GE Healthcare).

After IEF, strips were equilibrated by gentle stirring for 15 min in an equilibration buffer (100 mM Tris HCl pH 6.8, 7 M urea, 2 M thiourea, 30% (w/v) glycerol, 2% (w/v) SDS) added with 0.5% (w/v) DTT for disulfide bridges reduction and for an additional 15 min in the same equilibration buffer to which 0.002% (w/v) bromophenol blue and 4.5% w/v iodoacetamide for cysteine alkylation were added. Second-dimensional SDS-PAGE [93] was run in 12.5% acrylamide gels using the ETTAN DALT six apparatus (GE Healthcare). Running was first conducted at 5 W/gel for 30 min followed by 15 W/gel until the bromophenol blue line ran off. Two replicates were produced for each biological replicate, thus obtaining six gels per condition (n = 6).

Proteins were stained using the colloidal Coomassie Brilliant Blue G-250 (cCBB) procedure, as previously described by Neuhoff and co-workers [94]. The gels were scanned in an Epson Expression 1680 Pro Scanner and analyzed with ImageMaster 2-D Platinum Software (GE Healthcare). Automatic matching was complemented by manual matching. The molecular weights of the spots were deduced on the basis of the migration of SigmaMarkers™ wide range (MW 6.500–205.000), while p_Is were determined according to the strip manufacturer's instructions (GE Healthcare) reporting on the reference gel of the software-assisted analysis the values of p_I predicted for any given length of the strip. Both M_r and p_I of the spots of interest were then determined by using software-automated algorithm.

Relative spot volumes (%Vol) of the six replicate gels per condition were compared and were analyzed according to the Student's t-test to verify whether the changes were statistically significant ($p < 0.05$). This analysis was performed by using SigmaStat software. Only spots showing at least a 2-fold change in their relative volumes were considered for successive analyses.

Protein In-gel Digestion and LC-ESI-MS/MS Analysis

Spots excised from gels stained with cCBB were digested as described by Magni and co-workers [95] with some refinements. In detail, after the destaining procedure, spots were dried under vacuum on a centrifugal evaporator and incubated in 10 mM DTT, 100 mM NH_4HCO_3 for 45 min at 56°C. The solution was replaced with 55 mM iodoacetamide, 100 mM NH_4HCO_3 and the spots were incubated for 30 min in the dark at room temperature. After that, spots were briefly washed with 100 mM NH_4HCO_3 and again incubated for 15 min in 50% (v/v) acetonitrile (ACN), for 3 min in 100% ACN, for 3 min in 100 mM NH_4HCO_3, for 15 min in 50 mM NH_4HCO_3 in 50% (v/v) ACN and finally dried under vacuum. The following phases consisting in the protein digestion with trypsin [Sequencing grade modified Trypsin V5111, Promega, Madison] and in the recovery of peptides were carried out as described in the article above cited.

The LC-ESI-MS/MS experiments were conducted using a Surveyor (MS pump Plus) high-performance liquid chromatography (HPLC) system directly connected to the ESI source of a Finnigan LCQ DECA XP MAX ion trap mass spectrometer (ThermoFisher Scientific Inc., Waltham, USA). Chromatography separations were obtained on a BioBasic C18 column (180 μM I.D × 150 mm length, 5 μM particle size), using a linear gradient from 5% to 80% solvent B [solvent A: 0.1% (v/v) formic acid; solvent B: ACN containing 0.1% (v/v) formic acid] with a flow of 2.5 μl/min. The ESI was performed in positive ionization mode with spray voltage and capillary temperature set at 3 kV and at 220°C, respectively. Data were collected in the full-scan and data dependent MS/MS mode with collision energy of 35% and a dynamic exclusion window of 3 min.

Spectra were searched by TurboSEQUEST® incorporated in BioworksBrowser 3.2 software (ThermoFisher Scientific Inc., Waltham, USA) against the *Zea mays* protein subset, *Zea mays* EST subset and against the protein NCBI-nr database, all downloaded from the National Center for Biotechnology Information [96]. The searches were carried out assuming parent ion and fragment ion mass tolerance of ± 2 Da and ± 1 Da, respectively, two possible missed cleavages per peptide, fixed carboxyamidomethylation of cysteine, and variable methionine oxidation. Positive hits were filtered on the basis of peptides scores [Xcorr ≥ 1.5 (+1 charge state), ≥2.0 (+2 charge state), ≥2.5 (≥3 charge state), ΔCn ≥ 0.1, peptide probability $<1 \times 10^{-3}$ and Sf ≥ 0.70] [97]. If needed, identified peptides were used in protein similarity search performed by alignment analyses against the NCBI-nr database using the FASTS algorithm [98]. Physical properties of the characterized proteins were predicted by *in silico* tools at ExPASy [99].

CONCLUSION

Many of the proteins found to change in accumulation in response to NO_3^- were directly involved in the assimilation of this mineral nutrient. Moreover, the results underline the strict relationship between nitrogen and carbon metabolisms. The experimental design chosen for this proteomic study allows to emphasize some intriguing metabolic activities in both organs. Besides a dramatic increase of NO_3^- assimilation pathway, the exposure to a high NO_3^- concentration after a starvation period seems to induce a modification in NO metabolism in roots, that could depend on the need of responding to the new nutritional status. In leaves, many proteins were found to be (in)directly involved in the photosynthesis reactivation and in the maintenance of the chloroplastic functionality.

In addition, this proteomic analysis confirms the modulation by phosphorylation of the PEPCase in the leaves, suggesting that nitrogen availability could affect the circadian rhythms, as well as it shows that the form of this enzyme operating in roots could be modulated by monoubiquitination. Although further efforts are required to elucidate these results, the present study underlines the central role of post-translational events to modulate pivotal enzymes in plant metabolic response to NO_3^-.

KEYWORDS

- **Glutamine synthetase**
- **Nitrate reductase**
- **Nitrite reductase**
- **Phosphoenolpyruvate carboxylase**
- **Two-dimensional gel electrophoresis**
- ***Zea mays***

AUTHORS' CONTRIBUTIONS

Bhakti Prinsi contributed to the conception of the experimental design, carried out the determination of biochemical and physiological parameters, protein extraction, 2-DE, protein characterization by LC-ESI-MS/MS and analyzed the MS data, participated in writing the Materials and Methods section of the manuscript. Alfredo S. Negri analyzed the gels and performed statistical analyses. Paolo Pesaresi measured fluorescence parameters. Maurizio Cocucci contributed to the interpretation of the results and took part in the critical revision of the manuscript. Luca Espen conceived the study, coordinated the experiments, participated to the determination of biochemical and physiological parameters, wrote and edited the manuscript. All authors read and approved the final manuscript.

ACKNOWLEDGMENTS

This work was supported by grants from the Italian Ministry of Education, University and Research (MIUR-PRIN 2007). The authors wish to thank Dr. Chiara Fedeli for her valuable contribution during the writing of this manuscript.

Chapter 2

Pesticide-related Illness Reported to and Diagnosed in Primary Care

Lesley Rushton and Vera Mann

INTRODUCTION

In Great Britain (GB), data collected on pesticide associated illness focuses on acute episodes such as poisonings caused by misuse or abuse. This study aimed to investigate the extent and nature of pesticide-related illness presented and diagnosed in primary care and the feasibility of establishing a routine monitoring system.

A checklist, completed by general practitioners (GPs) for all patients aged 18+ who attended surgery sessions, identified patients to be interviewed in detail on exposures and events that occurred in the week before their symptoms appeared.

The study covered 59,320 patients in 43 practices across GB and 1,335 detailed interviews. The annual incidence of illness reported to GPs because of concern about pesticide exposure was estimated to be 0.04%, potentially 88,400 consultations annually, approximately 1,700 per week. The annual incidence of consultations where symptoms were diagnosed by GPs as likely to be related to pesticide exposure was 0.003%, an annual estimate of 6,630 consultations that is about 128 per week. Forty-one percent of interviewees reported using at least one pesticide at home in the week before symptoms occurred. The risk of having symptoms possibly related to pesticide exposure compared to unlikely was associated with home use of pesticides after adjusting for age, gender and occupational pesticide exposure (odds ratio (OR) = 1.88, 95% CI 1.51–2.35).

The GP practices were diverse and well distributed throughout GB with similar symptom consulting patterns as in the primary care within the UK. Methods used in this study would not be feasible for a routine surveillance system for pesticide-related illness. Incorporation of environmental health into primary care education and practice is needed.

Pesticides, pesticide products, and related chemicals have been found to have a wide range of health effects. They include: mutagenic substances, carcinogens or probable carcinogens, endocrine disrupters, reproductive toxic substances, and neurotoxic substances [1, 2]. The effect of low-level, long-term exposure has been of recent concern, with the organophosphate pesticides as a group receiving a great deal of medical research interest, particularly with regard to their potential effects on farmers using sheep dips [2-4]. Although pesticides have undoubted acute health effects, these usually occur as a result of accidental or deliberate misuse. In the UK, information on acute events can be obtained, for example, from the NHS Hospital Episode Statistics [5]. Much less is known about the incidence of ill-health due to low-levels of pesticide exposure and there are currently no surveillance schemes in primary care in GB that identify an illness as possibly due to pesticide exposure, whether occupational or environmental.

The aim of this study was to fill this gap in knowledge and in particular to: estimate the annual national incidence and prevalence of pesticide-related illness presented to and diagnosed by GP; investigate associations between symptoms and possible pesticide exposure, evaluate the feasibility and practicalities of setting up permanent arrangements to collect data on pesticide-related illness in primary care. As few patients (92: 7% of interviewees) reported occupational use of pesticides, this chapter focuses on results related to home and environmental pesticide exposure, a fuller technical report is available elsewhere [6].

There are several challenges in surveillance of pesticide-related illness in general practice including the large number of symptoms (often rather vague) that could potentially result from low-level exposure to pesticides and the lack of readily available sensitive, specific, and interpretable biological tests to confirm exposure. Most GPs, presented with a patient reporting such symptoms in the absence of reported exposure to pesticides, do not routinely consider the possibility of pesticide exposure. Any surveillance system must therefore encourage the GP to consider further a possible relationship but without over-prompting. The project thus focused on patients who consulted to report recent pesticide exposure, with or without current symptoms, and on patients who presented with symptoms that were "unusual" for them that is a new occurrence and not a chronic recurring problem and that the GP considered could potentially be related to a (possible) recent pesticide exposure. Although it was thought that acute severe illness related to high exposure was likely to lead to presentation and treatment in secondary care rather than primary care it was also thought important to capture any acute symptoms.

MATERIALS AND METHODS

The study was carried out between 2003 and 2006 in general practices from the UK General Practice Research Framework (GPRF), an organization of almost 1,100 general practices throughout the UK involved in epidemiological and health service research [7]. The GPRF network covers over 9% of UK practices, giving access to 12% of the population, with sufficient number of all types and in all areas to provide representative samples of the UK practices with regard to demographic characteristics and agricultural practices. Of particular relevance to this project is that 14% of the practices are in areas classified by the Office of National Statistics as remote rural and 17% are in mixed urban/rural areas. Information about the study and an invitation to participate was sent to all GPs on the GPRF database. Each participating GP completed a one page screening checklist for all patients aged 18 years or over consulting during a surgery session. Our pilot study indicated that GPs would be unwilling to use the screening checklist at all surgery sessions. They were therefore requested to do this for at least two sessions per week during a year of data collection. The practice research nurse organized a continuous flow of both checklists and interviews throughout the data collection period assuring that the consultation sessions occurred on different days and in both mornings and afternoons to ensure representation of patient consulting patterns (e.g., not always on Monday mornings or Friday afternoons when more acute or urgent consultations might take place).

An information pack was provided for GPs on pesticide-related illnesses with instructions on how to complete the checklist. The research nurses attended training days and, during the project, back-up training or additional support was provided by GPRF regional training nurses who were also responsible for quality control during the study ensuring standardization and checking that the practice was fulfilling all the requirements of research governance. The GPs were asked to carry out their "normal" consulting practice and to complete the checklists at the end of the consultation.

The checklist was used to identify patients who attended, with or without reporting symptoms, because of their concern about exposure to pesticides and those patients consulting with symptoms that were unusual for them. On each checklist GPs were asked to record their opinion of whether the patient's symptoms were likely, possibly, unlikely to be or definitely not related to pesticide exposure. A computer algorithm selected patients who were eligible for an invitation to an in-depth interview with the research nurse if:

- They consulted because of their concern about exposure to pesticides regardless of the opinion of the GP about potential relation to pesticide exposure.
- They had serious acute symptoms such as blurring of vision, vertigo, respiratory compromise that were not definitively attributed by the GP to a cause other than pesticides exposure.
- They had newly occurring flu type, respiratory, gastro-intestinal, skin, eye, or neurological symptoms, which were unusual for the patient and were not definitively attributed by the GP to a cause other than pesticides exposure.
- They had the above symptoms which were not unusual, that is recurring symptoms, for the patient but which the GP thought were likely or possibly related to pesticide exposure.

Initially ethical approval was only given to approach eligible patients about an interview by a single letter without a reminder. As about 40% of the eligible patients in the pilot study did not respond to the invitation to attend for interview, the Chairman of the Ethics Committee was approached and approval was obtained for sending reminder invitation letters to non-responders. Later permission was also obtained to carry out the interview over the telephone although this method was seldom preferred by patients.

The computer-based interview questionnaire consisted of sections on:

1. Occupational exposure, including applying and mixing pesticides during work, formulations, frequency, and duration of potential exposure, substance(s) including chemical and/or brand name and use of personal protective equipment.
2. Amateur use at home and in the garden including the use and mixing of pesticides, occurrences of professional pest control in the home and the storage and disposal of pesticides at home.

3. Hobbies and leisure during which exposure to hazardous materials or pesticides might have occurred.
4. Other suspected exposure to pesticides, including incidents such as accidental exposures from spray drift and incidents occurring in public places and near farmland.
5. Demographic, medical, and miscellaneous information.

All the questions focused on exposures and events that occurred in the week before the symptoms appeared. Patients were asked about the numbers of days' use of pesticides and whether this was more use than usual. Information on use of other substances in the home that potentially might lead to similar symptoms as those expected from pesticides (e.g., detergents, solvents, paints, etc.) and whether they had been used more frequently, in a larger quantity or with a change of brand was also collected.

Before the main study, a pre-pilot phase was carried out in Northern Ireland to test the feasibility of the GP administered checklist followed by a pilot study in nine practices in England and Wales to pilot both the use of the checklist and the interview questionnaire. Some changes were made for the main study to improve accuracy of responses, to ensure questions could not be omitted and to improve clarity.

To ensure that duplicate interviews did not take place, nurses were instructed not to invite a patient more than once if the patient had consulted more than once within a short period of time for the same illness and more than one checklist had been completed. However, a consultation could be considered as a separate episode if there was a gap of at least 2 weeks between consultations and it was clearly for a new problem.

Ethical approval was obtained through the UK Multi-centre Research Ethics Committees (Wales) and research governance approval was obtained from all the relevant Primary Care Organizations.

Statistical analyses were carried out using statistical software Stata version 9.2. Estimations of annual incidence proportion (cumulative incidence) and prevalence (proportion) of pesticide-related illness presented to GPs used the total number of patients screened as denominator, while the estimations of incidence and prevalence of possibly or likely pesticide-related illness diagnosed by GPs, without the patients specifically mentioning exposure, excluded the number of patients who reported exposure from the denominator. Annual incidence estimations included only patients with newly occurring symptoms in the nominators; annual prevalence estimations also included those who presented with recurring symptoms. Ninety-five percent confidence intervals for incidence and prevalence were estimated using the normal approximation or the exact method if the numbers were small.

Statistical methods included descriptive analyses together with univariable and multivariable logistic regressions to assess the association between risk factors and potential pesticide-related illness using robust standard error estimation to take account of clustering of patients within GP practices.

Participating Practices

The 157 GPs and seven nurse practitioners participated from 43 practices between November, 2004 and July, 2006 (not necessarily all GPs in a participating practice).

Six practices withdrew within 4 months or less. Reasons given for withdrawal included illness of the research nurse or GP, changes of GPs resulting in replacements being unwilling to participate and heavy practice workloads. The majority carried out the study for well over 6 months giving an average of 52.6 weeks. The number of check-lists completed per session was generally quite consistent between practices, varying between 6 and 14 giving an average of 11 per session.

The practices were spread throughout GB; industrial areas (5), cities or urban areas including outer London and metropolitan districts (13), mixed urban and rural areas including new towns and coastal resorts (12), rural areas (13) (Figure 1). The number of partners in the practices ranged from 1 to 8 or more, with practice sizes ranging from about 4,000 patients to over 13,000.

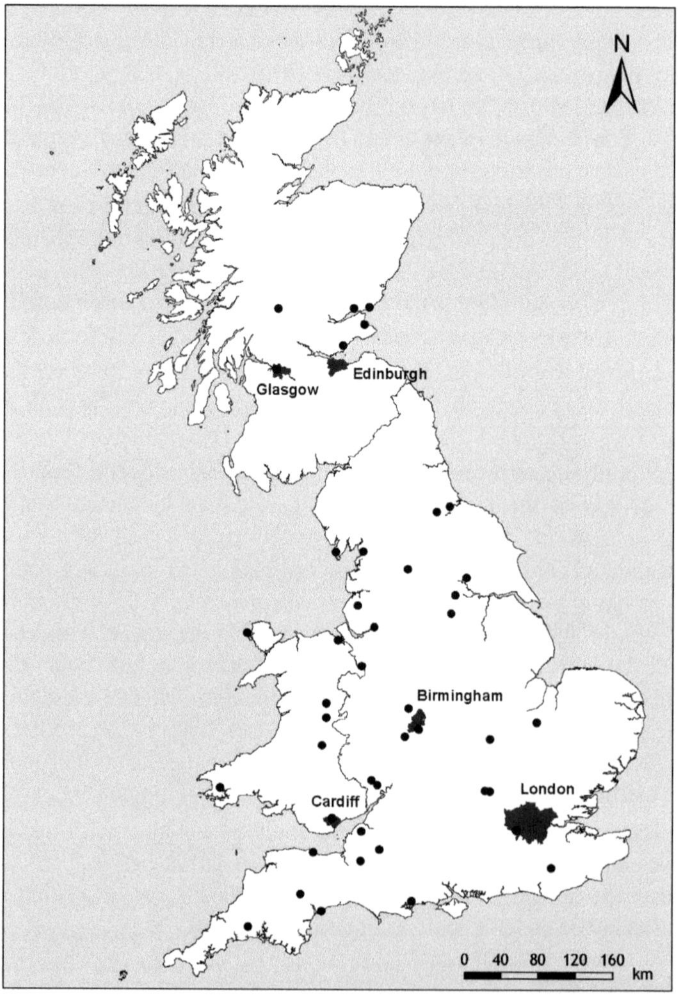

Figure 1. Locations of the participating general practices locations of the participating general practices.

Eligibility for and Response to Interview Invitation

The 59,320 checklists (male patients 40.4%) were completed during 5446 GP surgery sessions. Nearly 25% (14,490) of patients were asymptomatic and consulting the GP for other reasons and one patient, though asymptomatic, consulted the GP because of exposure to pesticides. The 4,741 (8%) patients were identified as eligible for invitation for interview, all of whom were received an invitation. The 1,335 patients (28.2% of those eligible and 49.8% of those who did not refuse) completed the interview, 2,060 (43.5%) refused to be interviewed and 1,346 (28.4%) had not replied by the end of the study. The distributions of symptoms were similar in all three groups regardless whether the patients agreed, refused to be interviewed or did not respond. On average, non-responders and those who refused to be interviewed were slightly younger than those who were not eligible and those who completed the interview (mean ages 46.3, 51.1, 54.2, 56.7 years respectively). Similar proportions of men and women were eligible for an interview (8.7%, 7.5% respectively). Of those who were eligible proportions who refused to be interviewed were similar for men (44%) and women (43%) as were proportions who did not respond (29.1% men, 27.8% women).

Incidence and Prevalence of Symptoms Related to Pesticide Exposure

Forty-two people consulted because of their concern about exposure to pesticides. The GP thought the symptoms were likely to be related to exposure to pesticides for 13 of these patients and also for a further seven patients who did not report exposure (Table 1).

Table 1. General practitioner's opinion on likelihood of symptoms being related to pesticides.

Reason for consultation	New or recurring symptoms	GP's opinion on likelihood of symptoms being pesticide-related				
		Likely	Possible	Unlikely	Definitely not	Total
Because of concern about pesticide exposure	Asymptomatic	0	0	0	0	1
	Recurring	2	4	6	1	13
	Newly occurring	10	11	3	0	24
	Not known	1	3	0	0	4
	Total	13	18	9	1	42
Pesticide exposure not mentioned by patient	Asymptomatic	0	0	0	0	14490
	Recurring	5	528	13451	20207	34191
	Newly occurring	2	972	4165	4020	9159
	Not known	0	81	932	425	1438
	Total	7	1581	18548	24652	59278
Overall total		20	1599	18557	24653	59320

The GPs also thought that 1,599 patients (2.7% overall) had symptoms that were possibly related to pesticide exposure (33.7% of all eligible patients). The 18,557 patients (31.3% overall) were thought by the GP to have symptoms that were unlikely to be related to pesticide exposure, 3,120 of these patients were eligible for in-depth interview (65.8% of all eligible patients).

As the average number of weeks for which the study was carried out was approximately a year, no weighting was carried out for the prevalence and incidence estimation. The denominator for estimating the annual incidence and prevalence for pesticide-related illness presented to GPs was thus the total number of patients screened (59,320) and for pesticide-related illness diagnosed by GPs it was 59,278 that is excluding the 42 patients who reported exposure to pesticides (Table 1).

The annual prevalence of consultations because of concern by the patient about pesticide exposure is estimated to be 0.07% (42/59,320) (95% CI 0.05%, 0.09%) and the estimate of the annual incidence (i.e., new cases) is 0.04% (24/59,320) (95% CI 0.02%, 0.06%) (Table 1). For people who did not consult because of concern about exposure to pesticides, the estimate of the annual prevalence of consultations for which the GP thought the symptoms were likely to be related to pesticide exposure is extremely small, 0.01% (7/59,278) (95% exact CI 0.004%, 0.02%), with an estimate of those with symptoms thought by the GP to be possibly related to pesticide exposure being 2.7% (1,581/59,278) (95% CI 2.5%, 2.8%). The annual incidence of consultations for which the GP though symptoms were likely to be related to pesticide exposure is 0.003% (2/59,278) (95% exact CI 0%, 0.01%) and the annual estimate of those with symptoms thought by GP to be possibly related to pesticide exposure is 1.6% (972/59,278) (95% CI 1.5%, 1.7%) respectively.

In 2001, GPs in the UK carried out about 221 million consultations (patients aged 16 years or more), [8]. Although our study was based on people aged 18 years or more in GB (excluding Northern Ireland) the estimate of an annual incidence of 0.04% for consultations made by patients because of concern about pesticides translates to an annual estimate of 88,400 consultations that is approximately 1,700 per week for people aged 16 years or over. Similarly the annual incidence of 0.003% for those patients not consulting because of concern about pesticide exposure but for whom the GP thought their symptoms were likely to be related to pesticide exposure translates to an annual estimate of 6,630 consultations that is about 128 per week for people aged 16 years or over.

Home Use of Pesticides

The 547 (41%) interviewees reported using at least one pesticide in the home environment in the week before symptoms occurred (273 patients, 20% used two or more). The most frequently used pest control chemicals used at home in the week before symptoms occurred were slug and snail pellets and weed killer (Table 2). For most of the chemicals in Table 2 over half the patients reported that they had used it more than usual during that week. Almost a third of the pesticides were applied with an aerosol or spray, 25% as liquid and 21% as pellets or granules. The 65.4% (358) of home pesticide users had used no personal protective equipment when applying the chemicals, although 284 patients (51.9%) reported that their arms and legs were covered.

Table 2. Pest control chemical use at home.

Pest-control chemical	Number of patients	Percentage of pesticide users	Reported more use than usual number	%
Weed killer	131	23 .9	74	56.5
Kill root/nettles etc.	60	11 .0	32	53.3
Kill aphids/greenfly etc.	89	16.3	47	52.8
Kill wasp/fly	78	14.3	43	55.1
Kill ant/cockroach etc.	80	14.6	57	71.3
Fungicidal paint	7	1.3	5	71.4
Mould/mildew treatment	48	8.8	30	62.5
Tick/flea control	98	17.9	44	44.9
Head lice treatment	17	3.1	11	64.7
Insect repellent	35	6.4	20	57.1
Other animal repellent	12	2.2	9	75.0
Rat/mouse poison	30	5.5	16	53.3
Slug/snail pellets	149	27.2	81	54.4
Creosol/cuprinol	57	10.4	42	73.7
Dry rot treatment	2	0.4	2	100.0
Kill algae/lichen/moss	15	2.7	7	46.7
Intestinal worm treatment	55	10.1	22	40.0
Other	22	4.0	15	68.2

In deciding on the quantity of pesticide to use 103 patients (44.6%) reported that they followed the label exactly, 103 patients (18.8%) used it as guidance, and 210 patients (38%) used their previous experience. The majority of pesticides stored at home were kept in the kitchen (15.7%) and/or in the garage or shed (63.3%).

The 61.5% of interviewees reported that they never disposed of pesticides and 25.5% disposed of them in the household rubbish bin. Relatively few reported that they used a chemical waste disposal site (2.2%) or other waste disposal site (7.1%).

The proportions of patients presenting with eye, skin, gastrointestinal, or respiratory symptoms were almost identical for patients who used or did not use pesticides. A slightly higher proportion of the users had neurological and a slightly lower proportion of them had multiple symptoms (data not shown). However, almost twice as many of non users visited their GP because of flu symptoms compared with users.

The number of patients who reported some sort of change in use (more frequent use, change of brand, and/or larger quantity) of hazardous materials in the home other than pesticides was small. However, there was a slight tendency based on small numbers for an increased proportion of respiratory symptoms among patients who also used pesticides compared with those who did not use pesticides.

A follow-up questionnaire was sent to GPs after the study to investigate the criteria they used to categorize the symptoms of each presenting patient as possibly or likely to be related to pesticide exposure. It appeared that the decision by the GPs to use the category “possibly related to pesticide exposure” was more often based on a discussion of both symptoms and activities (48% vs. 41% of GPs), but less often based on pesticide alone (1% vs. 19% of GPs) or included a discussion on specific pesticide exposure (10% vs. 27% of GPs) compared to the decision to use the category "likely". Eligibility for invitation for an interview could thus have been biased by consideration of exposures, since those patients categorized by the GP as likely were automatically invited for an interview. For this reason logistic regression analysis was carried out excluding the likely group and those who consulted because of concern about pesticide exposure, thus comparing the risk of patients being categorized by their GP as having symptoms possibly related to pesticide exposure with those classified as unlikely to have symptoms related to pesticide exposure (1,316 patients). Home use of pesticides and the change in use of several commonly used substances at home in the week before symptoms occurred showed statistically significant increased risk (Table 3 univariable results, based on 1,316 patients categorized by the GP as having symptoms possibly related to pesticides).

Table 3. Univariable logistic regression models for the likelihood of being categorized by the GP as having symptoms possibly related to pesticide exposure.

Factors affecting pesticide related illness		Proportion of participants (%) (n=l316)	Odds ratio (95% Confidence Interval)
Occupational pesticide use versus no use		6	1.17 (0.62, 2.20)
Home pesticide use versus no use		41	1.83 (1.49, 2.26)
Age (one year increase)		-	0.99 (0.985, 0.997)
Male versus female		42	1.0 (0.81, 1.24)
Proximity of farmland	100 m-1 km vs < 100m	29	0.65 (0.45, 0.93)
	>I kmvs< lOOm	40	I. 79 (0.55, 5.83)
	Don't know vs < I 00 m	2	0.59 (0.26, 1.36)
Proximity of chemical plant	100 m-1 km vs < 100m	2	0.91 (0.19, 4.32)
	>I kmvs< lOOm	62	1.23 (0.25, 6.05)
	Don't know vs < I 00 m	35	0.69 (0. 19, 2.49)
Proximity of landfill site	100 m-1 km vs < 100m	3	2.92 (0.81, I 0.49)
	>I kmvs< lOOm	80	1.84 (0.39, 8.68)
	Don't know vs < I 00 m	16	1.38 (0.30, 6.35)
Proximity of heavy traffic	100 m-1 km vs < 100m	37	0.93 (0.66, 1.31)
	>I kmvs< lOOm	39	0.80 (0.46, 1.37)
	Don't know vs < I 00 m	1	0.53 (0.12, 2.23)
Proximity of railway	100 m-1 km vs < 100m	30	0.96 (0.65, 1.43)
	>I kmvs< lOOm	58	1.41 (0.70, 2.84)
	Don't know vs < I 00 m	4	0.84 (0.27, 2.63)
Area of living	Surburban versus urban	37	0.29 (0.08, 1.02)
	Rural versus urban	39	0.29 (0.07, 1.1)
Change in use of Laundry detergent		9	1.60 (0.93, 2.76)
Change in use of Disinfect-ant/bleach		8	1.26 (0.81, 1.95)
Change in use of Cleaning agent		8	1.43 (0.88, 2.34)

Table 3. *(Continued)*

Factors affecting pesticide related illness	Proportion of participants (%) (n=1316)	Odds ratio (95% Confidence Interval)
Change in use of White spirit	9	1.60 (1.06, 2.41)
Change in use of Polish/ varnish	6	1.83 (0.92, 3.66)
Change in use of Air freshener	10	1.25 (0.89, 1.77)
Change in use of Paint	11	1.74 (1.19, 2.54)
Change in use of Toiletries	7	1.85 (1.22, 2.80)
Change in use of Stain remover	4	1.36 (0.63, 2.91)
Change in use of Furniture renovator	4	1.86 (1. 19, 2.93)
Change in use of Oil/grease	3	1.12 (0.49, 2.57)
Change in use of Insulation material	5	1.06 (0.64, 1.74)

In multivariable models that included occupational and home use of pesticides, age and gender, and change in use of each of the 12 commonly used substances in Table 3 in turn, none of the 12 substances altered the ORs for occupational or home use of pesticides substantially (ORs for occupational pesticide use ranged from 0.97 to 1.01 and ORs for home use of pesticides ranged from 1.83 to 1.88). These variables do not appear therefore to be confounding the effect of occupational and home pesticide use. The risk associated with changed use of paint, toiletries, and furniture renovator remained significantly raised in these analyses, although based on small numbers.

The results of a multivariable model including occupational and home use of pesticides, age as a continuous variable and gender are given in Table 4 and show a statistically significant increased risk for being categorized as having possible pesticide related symptoms if the pesticide was used at home (OR = 1.88, 95% CI 1.51–2.35) (Table 4).

Table 4. Multivariable logistic regression model for the likelihood of being categorized by the GP as having symptoms possibly related to pesticide exposure.

Factors affecting pesticide related illness	Multivariable model OR (95% CI)
Occupational pesticide use versus no use	0.99 (0.53, 1.88)
Home pesticide use versus no use	1.88 (1.51, 2.35)
Age (I year increase)	0.99 (0.99, 1.00)
Male versus female	1.02 (0.81, 1.28)

DISCUSSION

The results from this study suggest that the incidence and prevalence of ill health presenting to and diagnosed in primary care as related to pesticides in GB are small but that this could result in considerable numbers of consultations. There are no directly comparable studies in the primary care sector in GB. In 2005/2006 there were 169 hospital episodes of accidental poisoning by exposure to pesticides in the UK, 93% of which were emergency admissions and 70% occurred to children under the age of 15 years [5].

Other countries also monitor acute pesticide poisoning and there has been some attempt by the World Health Organization to provide a standardized case definition and classification scheme with regard to exposure, health effects, and causality [9]. Underreporting of acute events can occur in monitoring systems, as illustrated in a cross-sectional survey Nicaragua [10]. Pesticide-related illness as an important cause of acute morbidity among migrant workers has also been found in multi-state standardized occupational surveillance programs such as those led by the US National Institute for Occupational Safety and Health which include primary care physicians [11].

In designing our study we found few UK-based statistics on which to base sample size calculations but anticipated that the incidence of possible pesticide-related illness could potentially be as low as 1% (as indeed was the case, 1.6%). An early study on deaths from pesticide poisoning in England and Wales indicated that such deaths represented 1.1% of all deaths from poisoning over 1945–1989 [12] and at least 73% of these pesticide fatalities were due to suicide. We originally aimed to recruit 70 practices from the GPRF giving access to approximately 420,000 patients, about 0.75% of the UK population but, although the GPRF network consists of practices interested in participating in research, it proved difficult to recruit them. We did, however, manage to recruit 43 practices with over 160 GPs and nurse practitioners participating in the study.

These practices were well spread geographically throughout GB between urban, suburban, and rural areas and between different areas of deprivation, with the average list size per GP in the practices being 2,210, varying from under 1,000 in a rural area to over 2,500 in two city practices. The average list size of GPs during our study period in England and Wales was 1,666 [13]. The average number of surgeries held per week per practice, used for the pesticide project in the participating practices, was 2.4 over the study period. In the UK generally the average number of surgery sessions held weekly by GPs is about eight. The study thus included about 30% of the consulting workload of each participating GP. Overall, of all those patients screened, 15.5% of those not asymptomatic (6,930 of 44,829 patients with symptoms) had a respiratory problem (including flu-like symptoms). The same proportion of patients was estimated to consult their GP for respiratory condition problems in the UK in 2002 [13]. The corresponding figures from our study and for the UK respectively are: skin symptoms 12.2%, 10.9%; eye problems 2.2%, 4.5%; gastrointestinal (our study)/digestive system (UK figures) 8.7%, 7.2%; neurological (our study)/nervous system (UK figures) 1.8%, 3.4%. Although in our study all participants were over 18 years old and the UK estimates include all ages the similarity of these figures suggests that our study closely mirrors the general pattern of symptoms within the UK consultations.

Over 40% of those interviewed in our study had use a household pesticide in the week before symptoms developed. Other studies have tended to ask about past use of household pesticides over a longer period. For example, a case-control study of childhood haematopoietic malignancies in France found that about 50% of mothers of cases had used a household pesticide during pregnancy; use varied by type of residence and area of residence for example for fathers use during pregnancy was 72%, 61%, and 44% for rural, mixed, and urban areas respectively and 86%, 67%, and 26% for living in a farm, house, and flat respectively [14]. A similar study of childhood leukaemia found that more than half of households had used insecticides or indoor pesticides during the first year after the birth of the child [15]. A UK survey of a sample of parents from the Avon Longitudinal Study of Parents and Children found that 93% had used at least one pesticide product in the last year with a range of frequencies [16]. For example, 33% of parents used pesticides to treat pets with a median use of four times per year; other use included insecticides (42% of parents, 1–90 applications per year, median 2.5), slug pellets (44% of parents, 1–60 applications per year, median 4), weed killer (27% of parents, 1–18 applications per year, median 2). A UK observational study found that few participants read the label, that they often found it hard to understand and that compliance with instruction was low [17].

Our study is limited in some aspects. It was felt that it would be too impractical and costly to try and interview in depth a random sample of patients consulting their GPs throughout a year at a large number of practices. We also wanted to gain some knowledge of GP diagnosing of pesticide-related illness. The screening checklist was thus designed to include this and to screen out patients who were asymptomatic, consulting for on-going/chronic health problems or whose symptoms, in the opinion of the GP, were definitely not related to pesticide exposure. This reduced the proportion of patients eligible for an interview to 8%. Some of those not invited for interview could have had a pesticide-related problem not attributed by either the patient or GP to pesticides, leading to potential underestimation of incidence and prevalence.

At the completion of the study a follow-up questionnaire was sent to GPs (86 GPs from 32 of the 43 practices responded) to investigate how they decided on their classification of the likelihood of pesticide-related illness. The 71 GP (83%) of the respondents classified patients to have possibly pesticide-related illness on the basis of symptoms and/or activities before the symptoms occurred. Only one GP reported using only a mention of pesticides to categorize a patient as "possible case". These results confirm that very few GP specifically discussed pesticides with the patient when categorizing them as "possible related to pesticide exposure".

A high proportion (44%) of those invited for interview refused to participate and this did not improve after we had received ethical approval to carry out telephone interviews. However, overall, 50% of those who did not refuse were interviewed. The interview questionnaire was fairly lengthy. It was felt important to consider total exposure to pesticides from all sources. The interview thus attempted to capture these data. The information on actual chemicals and active ingredients of pesticides is, however, limited as it was thought that patients would either not know this or be unable to recall it accurately. The study was thus limited in its ability to define a definitive pesticide-related case of ill-health. However, other systems such as the SENSOR system in the US and the Pesticides Incident Reporting Scheme in the UK also use some element of self-reporting and expert judgment, particularly in defining a possible case [18, 19]. In addition we did not, in this study, follow-up patients categorized as having likely or possible pesticide-related illness with regard to treatments or the results of this. The establishment of a definite causal relationship from these systems, as in our study, would thus require careful consideration.

A clear outcome from the study is that it would not be feasible to implement the methods used in this study for a wider surveillance system in GB. Screening of patients at nearly 60,000 consultations yielded relatively few likely or probable cases of pesticide-related symptoms. The project also required constant monitoring, and motivation and encouragement of the practices. This would be infeasible as part of a routine monitoring system.

Throughout the world primary care surveillance networks have been developed that monitor voluntarily one or more specific illness problems on a regular or continuing basis [20, 21]. The main objective of these is on disease surveillance, many focusing on infectious diseases nationally or as part of international networks such as GeoSentinel and TropNetEurope [22, 23]. Primary care physicians may also detect sentinel cases of occupationally or environmentally caused diseases. For example, a cluster of Guillain–Barre syndrome cases was observed in relation to aerial organophosphate insecticides [24].

In the UK, it has been shown that none of the existing GP morbidity recording schemes routinely recorded occupation although it would be feasible to add procedures to obtain this information [25], as it would for environmental factors. The importance of incorporating environmental health into primary care education and practice has been recognized in other countries. In the US, the National Strategies for Health Care Providers: Pesticides Initiative and the National Environmental Public Health Tracking Program have been launched [26, 27]. The former aims to raise awareness among GPs and nurses of potential exposures to pesticides. Cities like New York are investigating how to develop their capacities to track and link environmental public health indicators such as pesticide sales and applications, housing and building information, and medical data. However, data collected for surveillance need to be relevant and action-driven [28]. In many studies worldwide there has been recognition of the under-reporting of pesticide-related poisoning and the need to improve surveillance [29]. Although the prevalence and incidence results from our study are small these potentially give fairly large number of patients presenting to GPs each week with possible pesticide-related illnesses, albeit mild ones. Our study has thus shown that there is a need to extend surveillance to include primary care.

CONCLUSION

The use of pesticides was widespread among interviewees in the home environment and there was unsatisfactory use of product labels and precautionary measures; storage and disposal of pesticides was also poor. Although the estimated annual prevalence and incidence of pesticide-related illness reported to and diagnosed in primary care was small, this implies large numbers of consultations potentially concerning pesticide exposure. The study methods used here are not feasible for routine pesticide illness monitoring system. However, there is a need to incorporate environmental health into primary care education and practice.

KEYWORDS

- **General practice research framework**
- **General practitioners**
- **Great Britain**
- **Monitoring system**
- **Primary care**

AUTHORS' CONTRIBUTIONS

Lesley Rushton was one of the initiators and had overall management of the study and participated in its design and coordination. Vera Mann participated in the design and conduct of the study and performed the statistical analysis. Both authors contributed to the drafting of the manuscript and approved the final version.

ACKNOWLEDGMENTS

This study was funded by the UK Health and Safety Executive. We thank Nicky Fasey, Eric Bissoon, Madge Vickers, Ken Whyte, Morag McKinnon (GPRF), Justin Fenty, Len Levy, and Alex Capleton (MRC Institute of Environment and Health) and Helen Pedersen (Imperial College London) for their involvement throughout the study and all the practices, GPs, and nurses who participated.

Ethical approval was obtained from the UK Multi-centre Research Ethics Committee (MREC for Wales; MREC03/9/056) and research governance approvals were obtained from each of the Primary Care Organizations the practices belong to.

COMPETING INTERESTS

The authors declare that they have no competing interests.

Chapter 3

Effects of Increased Litterfall in Tropical Forests

Emma J. Sayer, Jennifer S. Powers, and Edmund V. J. Tanner

INTRODUCTION

Aboveground litter production in forests is likely to increase as a consequence of elevated atmospheric carbon dioxide (CO_2) concentrations, rising temperatures, and shifting rainfall patterns. As litterfall represents a major flux of carbon from vegetation to soil, changes in litter inputs are likely to have wide-reaching consequences for soil carbon dynamics. Such disturbances to the carbon balance may be particularly important in the tropics because tropical forests store almost 30% of the global soil carbon, making them a critical component of the global carbon cycle; nevertheless, the effects of increasing aboveground litter production on belowground carbon dynamics are poorly understood. We used long-term, large-scale monthly litter removal and addition treatments in a lowland tropical forest to assess the consequences of increased litterfall on belowground CO_2 production. Over the second to the 5th year of treatments, litter addition increased soil respiration more than litter removal decreased it; soil respiration was on average 20% lower in the litter removal and 43% higher in the litter addition treatment compared to the controls but litter addition did not change microbial biomass. We predicted a 9% increase in soil respiration in the litter addition plots, based on the 20% decrease in the litter removal plots and an 11% reduction due to lower fine root biomass in the litter addition plots. The 43% measured increase in soil respiration was therefore 34% higher than predicted and it is possible that this "extra" CO_2 was a result of priming effects, that is stimulation of the decomposition of older soil organic matter by the addition of fresh organic matter. Our results show that increases in aboveground litter production as a result of global change have the potential to cause considerable losses of soil carbon to the atmosphere in tropical forests.

Changes in litter quantity as a consequence of global climate change are becoming increasingly likely; recent FACE experiments have shown that litterfall increases with elevated atmospheric CO_2 concentrations [1-5] and predicted changes in rainfall distribution patterns [6] and temperature [7] may also affect litterfall by altering leafing phenology. As litterfall represents a major pathway for carbon and nutrients between vegetation and soil it seems likely that changes in aboveground litter production will have consequences for belowground processes. However, despite the increasing recognition that research on terrestrial ecosystem dynamics needs a combined aboveground-belowground approach [8], the potential impact of changes in litterfall on belowground carbon dynamics has been largely ignored [9].

Tropical forests are a critical component of the global carbon cycle as they store 20–25% of the global terrestrial carbon [10, 11]. Ongoing debates about whether tropical forests are a source or sink for atmospheric carbon have led to increased interest in the belowground components of their carbon cycle [12] because they also contribute almost 30% to global soil carbon storage [13]. Soil respiration from root and heterotrophic respiration alone releases approximately 80 Pg of carbon into the atmosphere per year to which tropical and subtropical forests contribute more than any other biome [14]. Recent studies have investigated the direct effects of elevated CO_2 [15], rising temperature [16, 17], and fertilizer [18-20] on soil carbon cycling but we know very little about how soil respiration will be affected by the predicted changes in aboveground production caused by global climate change. We believe that an increase in aboveground litterfall may have a large impact on belowground carbon and nutrient cycling, as annual litterfall is closely correlated with soil respiration on a global scale [21, 22], and the amount of litter on the forest floor also affects soil nutrient status, soil water content, soil temperature, and pH [9], all of which can influence soil respiration rates. To investigate this, we conducted an experiment consisting of large-scale monthly litter removal (L–) and litter addition (L+) treatments in a lowland tropical forest. Our results show that an increase in aboveground litterfall caused a disproportionate increase in soil respiration, reduced the amount of carbon allocated to fine root biomass and thus has the potential to cause substantial losses of carbon belowground.

Soil Respiration

Soil respiration showed a seasonal pattern with low rates (c. 130 mg C $m^{-2}h^{-1}$) in the dry season and much higher rates (c. 260 mg C $m^{-2}h^{-1}$) in the wet season. There was no effect of litter manipulation during 2003, the first year of treatments, but from May, 2004 (17 months after litter manipulation commenced), respiration from the mineral soil in the litter removal (L–) plots was on average 20% lower than in the control (CT) plots (Figure 1).

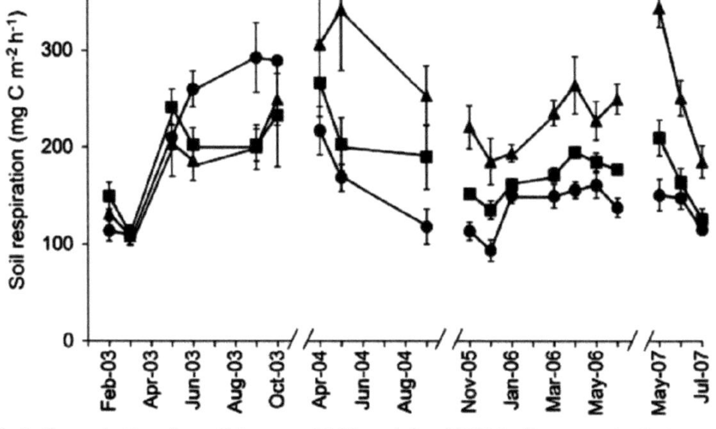

Figure 1. Soil respiration from February, 2003 to July, 2007 in litter manipulation treatments in tropical rainforest. Soil respiration rates were measured over bare soil in all treatments; squares are controls, triangles are litter addition treatments, and circles are litter removal treatments; error bars show standard errors of means for n = 5.

From the second year of treatments, litter addition increased respiration from the mineral soil more than litter removal decreased it. On average, soil respiration in the litter addition (L+) plots was 43% higher than in the controls and the increase was significant or marginally significant in 11 out of 12 months while respiration in the L– plots was significantly lower than the CT plots in only 2 out of 12 months (Figure 2). The smallest increase in the L+ plots relative to the CT plots was during the dry season (20% in January, 2006; Figure 2), while the greatest increases were observed during the dry-to-rainy season transition (69% in May, 2004 and 64% in May, 2007; Figure 2). The strong increase in soil respiration in the L+ plots was sustained until the end of the study in July, 2007 (Figure 1).

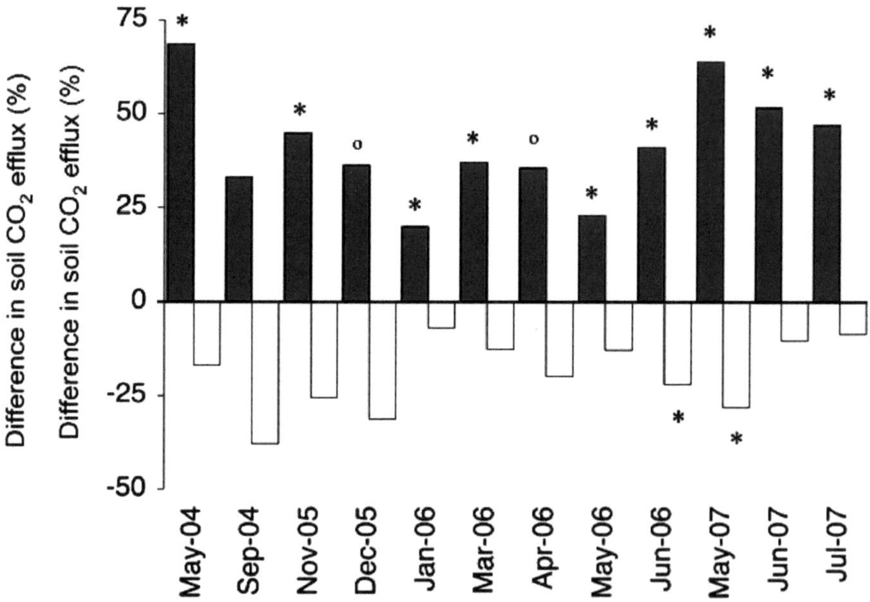

Figure 2. Differences in soil respiration between litter manipulation treatments and controls in tropical rainforest. The differences are calculated as a percentage of the average respiration measured in the control plots for each month; gray bars are litter addition plots and white bars are litter removal plots; a star above a bar denotes a significant treatment effect ($P < 0.05$) compared to the controls, a circle above a bar denotes a marginally significant treatment effect ($P < 0.065$) compared to the controls.

Soil Temperature and Soil Water Content

Soil temperature (0–100 mm depth) varied very little throughout the year; litter removal decreased soil temperature by c. 0.5°C relative to the CT and L+ plots during the rainy season only in 2003 and 2004 ($P = 0.002$); soil temperature did not differ between the L+ and CT plots except in June and July, 2007, when it was

0.3°C (P = 0.019) and 0.4°C higher (P = 0.002), respectively, in the L+ plots. Soil water content from 0 to 60 mm depth was not affected by litter manipulation in any season or year.

Fine Root and Microbial Biomass

Fine root biomass in the mineral soil (0–100 mm depth) was 37% lower in the L+ plots than in the CT plots in June and July, 2004, after 19 months of litter addition and removal treatments (P < 0.01; Figure 3) and 28% lower in August, 2006, after 41 months (P = 0.05; Figure 3). There was no significant difference in fine root biomass between CT and L– plots in either year.

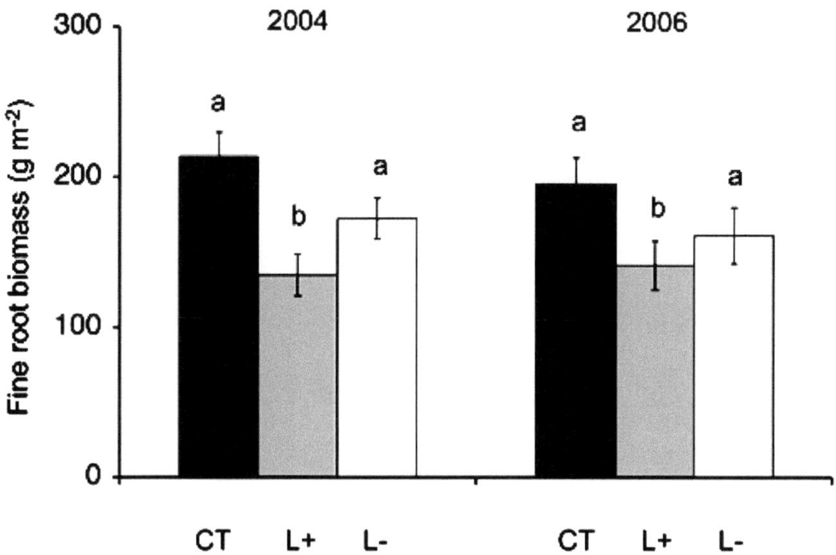

Figure 3. Fine root biomass in the mineral soil (0–100 mm) in litter manipulation treatments in tropical rainforest. The CT is control, L+ is litter addition, L– is litter removal; error bars show standard errors of means for n = 5; different letters above bars indicate a significant difference between treatments at P < 0.05. Data for 2004 has been previously published in a different form [25].

Total microbial C and N (0–100 mm depth) had decreased by 23% in the L– plots relative to the control treatment in August, 2004 (P = 0.011 and P = 0.003 for C and N, respectively; Figure 4a) and microbial N in the L– plots was 18% lower than in the CT plots in June, 2006 (P = 0.006; Figure 4b).

Litter addition had no significant effect on microbial biomass C and N in either year.

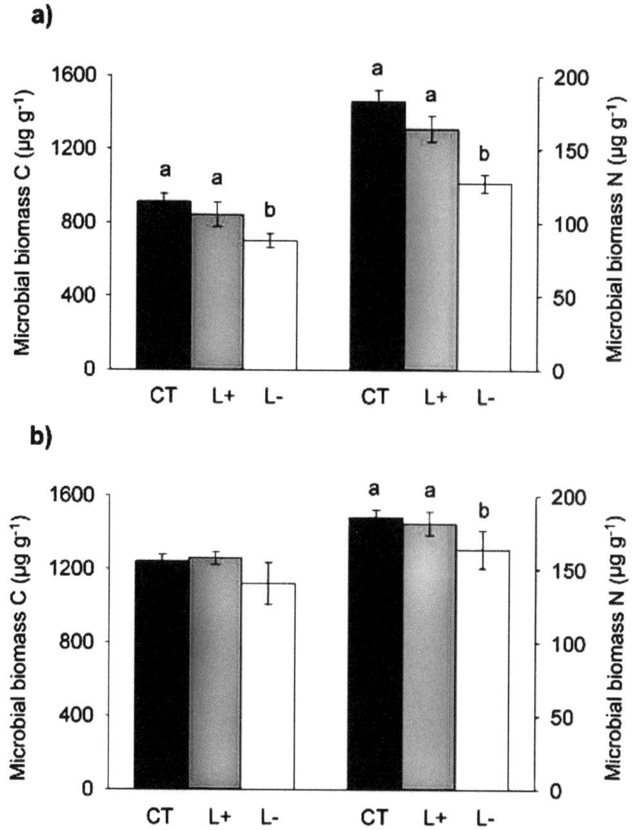

Figure 4. Microbial biomass in the mineral soil (0–100 mm) in litter manipulation treatments in tropical rainforest. Data are given as microbial C and N in a) August, 2004, and b) June, 2006; CT is control, L+ is litter addition, L– is litter removal; error bars show standard errors of means for n = 5; different letters above bars indicate a significant difference between treatments at $P < 0.05$.

DISCUSSION

We attribute the lack of significant responses in soil respiration to the experimental treatments during the first year to a combination of 2 factors: first we started treatments during the dry season when decomposition is limited by the lack of moisture [23] and second, we did not include the litter layer in our measurements of soil respiration. Thus, we would not expect CO_2 efflux from the mineral soil to be affected by our treatments until decomposition processes were sufficiently advanced to affect the input of carbon and nutrients to the mineral soil.

The 20% reduction in soil respiration observed in the litter removal treatment from the second year of treatments until the end of the study is similar to the 28% decrease reported in plots in young regrowth forest in Brazil after 1 year of litter removal, where controls included the litter layer in CO_2 efflux measurements [24], but lower than the 51% decrease after 7 years of litter removal in lower montane forest in Puerto Rico

[25]. We can attribute the decrease in our study principally to a reduction in heterotrophic respiration due to the withdrawal of fresh substrate, as there were no differences in fine root biomass in the upper 100 mm of the mineral soil between CT and L– plots in 2004 or in 2006 (Figure 3). Furthermore, we found no significant differences in soil water content between treatments and the small ($\leq$0.5°C) and inconsistent differences in soil temperature were unlikely to affect soil respiration.

We expected an average increase of 20% in soil respiration in the L+ plots during the period from May, 2004 to July, 2007, as CO_2 efflux from the mineral soil decreased by this percentage in the L– plots. However, soil respiration in the L+ plots was on average 43% higher than the controls and therefore 23% higher than expected by the addition of litter alone. Furthermore, this increase was sustained from May, 2004 until the end of the study in July, 2007 (Figure 2). The increase in soil respiration in the L+ plots is considerable and greater than the effects of fertilization with 150 kg ha^{-1} yr^{-1} of phosphorus in a study in Costa Rica [20]. While fertilization treatments are thought to boost soil respiration by removing the nutrient limitation of decomposition processes [20, 26], and increasing microbial biomass [26], we found no increase in leaf litter decomposition rates in our L+ treatments [27], and no changes in microbial biomass C or N (Figure 4a and b). Furthermore, litter addition decreased fine root biomass in the mineral soil by 37% in 2004 [28] and 30% in 2006 (Figure 3). Fine roots contribute the bulk of root respiration [29, 30], root respiration is proportional to root biomass [31], and it typically makes up at least 30% of total soil respiration in the tropics [31-34]; consequently the lower fine root biomass in the L+ plots would effectively reduce soil respiration by c. 11%. The expected increase in soil respiration in the L+ plots due to litter addition and reduced fine root biomass would therefore only be c. 9% relative to the controls. Thus, measured soil respiration was 34% higher than expected from the extra litter and reduced root biomass (Figure 5). This suggests that litter addition, besides increasing the amount of readily degradable carbon, may also cause substantial losses of CO_2 from the soil. It is likely that this extra CO_2 production is attributable to priming effects, the enhanced microbial decomposition of older, more recalcitrant soil organic matter by the addition of fresh organic matter [35, 36]. Strong pulses of soil respiration have previously been observed in lowland tropical rainforest during the dry-to-rainy season transition and were interpreted as a "natural priming effect" caused by large amounts of water-soluble carbon leaching from the litter that had accumulated during the dry season [20]. Our study shows that additional leaf litter has the potential to sustain priming effects throughout the year.

Estimated annual soil respiration rates were 15.3 t C ha^{-1} yr^{-1}, 19.0 t C ha^{-1} yr^{-1}, and 13.7 t C ha^{-1} yr^{-1} for the CT, L+, and L– treatments, respectively. Thus, the soil carbon lost to the atmosphere in our litter addition treatment is at least 4.4 t C ha^{-1} yr^{-1} and may be as high as 6.5 t C ha^{-1} yr^{-1} (23% and 34%, respectively, of expected soil respiration in the L+ plots). Laboratory incubations have demonstrated that repeated additions of fresh organic matter to soil induce greater priming effects than a single addition [37, 38] and that increased decomposition of soil organic matter continued even when the added fresh organic matter had been completely depleted [39]. We therefore expect that chronic increases in litterfall will induce a substantial release of soil carbon in the medium term.

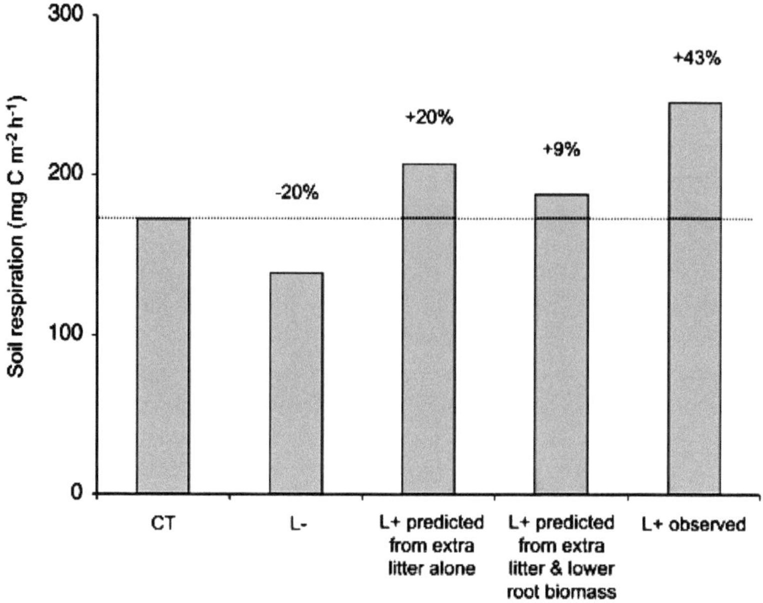

Figure 5. Comparison of predicted and observed soil respiration in litter manipulation plots in tropical rainforest. Differences are expressed as percentages of the mean rate measured in controls from May, 2004 to July, 2007; CT is control, L+ is litter addition, L– is litter removal.

Thus, we show for the first time that increased aboveground litter production in response to global climate change may trigger priming effects and convert considerable amounts of soil carbon to atmospheric CO_2.

MATERIALS AND METHODS

Site Description

The study was carried out as part of an ongoing long-term litter manipulation experiment to investigate the importance of litterfall in the carbon dynamics and nutrient cycling of tropical forests. The forest under study is an old-growth moist lowland tropical forest, located on the Gigante Peninsula (9°06′N, 79°54′W) of the Barro Colorado Nature Monument in Panama, Central America. The soil is an oxisol with a pH of 4.5–5.0, with low "available" phosphorus concentration, but high base saturation and cation exchange capacity [28, 40, 41]. Nearby Barro Colorado Island (c. 5 km from the study site) receives a mean annual rainfall of 2600 mm and has an average temperature of 27°C [42]. There is a strong dry season from January to April with a median rainfall of less than 100 mm per month [43]; almost 90% of the annual precipitation occurs during the rainy season. Fifteen 45 m×45 m plots were established within a 40-ha area (500 m×800 m) of old-growth forest in 2000. In 2001 all 15 plots were trenched to a depth of 0.5 m in order to minimize lateral nutrient—and water movement via the root/mycorrhizal network; the trenches were double-lined with plastic and backfilled. Starting in January, 2003, the litter (including branches ≤100 mm in diameter) in five plots was raked up once a month, resulting in low, but not entirely absent, litter standing crop (L– plots). The removed litter was immediately spread on five further plots, approximately doubling the monthly litterfall (L+ plots); five plots were left as controls (CT plots). The assignment of treatments was made on a stratified random basis, stratified by total litterfall per plot in 2001, that is the three plots with highest litterfall were randomly assigned to treatments, then the next three and so on.

Soil Respiration

In October, 2002 four measurement sites were established in each of the 15 plots; collars made of PVC pipe of 108 mm inner diameter and 44 mm depth were placed 12.5 m into the plot, measured from the centers of each of the fours sides of the plot. The collars were sunk into the soil to 10 mm depth and anchored using small plastic tent pegs, which were attached to the collars by cable binders and sunk diagonally into the ground to avoid channeling water into the soil under the collars. The collars were left undisturbed throughout the experiment. Soil respiration from the mineral soil was measured in the collars in February, March, May, June, September, and October of 2003, and April, May, and September of 2004 using an infra-red gas analyzer (IRGA) Li-6400 with an Li-6400-9 soil chamber attachment (LI-COR, Lincoln, USA). The ambient CO_2 level was determined for each site individually and measurements started at 5 ppm below the ambient CO_2 level. Three measurements were taken over each collar at each time and the values were averaged to give one value per collar per time.

In September, 2005 new measurement sites were set up adjacent to those established in 2002. The PVC collars of 200 mm diameter and 120 mm depth were sunk into the ground to 20 mm and anchored with tent pegs as described above. The collars were set up 2 months before starting measurements and were left undisturbed throughout the experiment. Soil respiration was measured over the new collars in November and December, 2005, January, March, April, May, and June 2006, and May, June, and July, 2007 using the Li-8100 soil CO_2 flux system (LI-COR, Lincoln, USA). The ambient CO_2 level was determined for each site automatically and one measurement of 2 min duration was taken over each collar. As our main research aim was to determine whether the amount of litter on the forest floor affects respiration from the mineral soil, leaf litter was removed from the collars prior to all measurements; great care was taken not to disturb the underlying mineral soil, and the litter was replaced once the measurements had been completed.

All measurements were made during 1–3 days each month between 8.00 hr and 14.00 hr. If measurements could not be completed in one day, they were made over consecutive days whenever possible, but no measurements were taken during or immediately following heavy rainfall. When measurements were taken over several days, an equal number of plots per treatment was measured each day; the plot means (4 collars per plot) were used for statistical analysis of treatment effects.

Annual respiration rates were estimated by obtaining a daily mean each for the rainy and dry seasons from the data collected, multiplying the daily mean by the average number of days in each season (135 days for the dry season, 230 days for the rainy season), and then summing the obtained values.

Soil Temperature and Soil Water Content

Soil temperature was recorded during respiration measurements in 2003, 2004, 2006, and 2007 within 0.5 m of the collars using the IRGA's integrated soil temperature probe inserted to a depth of 100 mm. Volumetric soil water content was measured from 0 to 60 mm depth using a thetaProbe (Delta-T Devices, Cambridge, UK), which was calibrated to the soil type in the plots following the procedure described by Delta-T. Due to technical problems, volumetric soil water measurements were not made in 2004 or 2005 and only in April in 2006. Gravimetric soil water content was determined in May, June, and August, 2004, and in January and June, 2006 from four 20 mm diameter soil cores per plot, taken from 0 to 100 mm depth; volumetric soil water content was then calculated from the gravimetric measurements.

Fine Root and Microbial Biomass

The biomass of fine roots (≤2 mm diameter) from 0 to 100 mm depth in the mineral soil was determined in June and July, 2004 from 10 randomly located 51 mm diameter soil cores per plot [25], and in June, 2006 from 7 randomly located soil cores per plot; live and recently dead fine roots were carefully separated from the soil by washing in a 0.5 mm mesh sieve and then dried to constant weight in the oven at 70°C.

Total microbial biomass of the mineral soil was measured in August, 2004 and June, 2006 (during the rainy season). Four soil cores were taken from 0 to 100 mm depth at the 4 corners of the inner 20 m×20 m in each plot using a 20 mm diameter punch-corer; the cores were bulked to give one sample per plot. Subsamples were taken to determine soil gravimetric water content and total microbial biomass was determined by the fumigation extraction method [44, 45]. Briefly, pairs of unfumigated and chloroform-fumigated (exposed to chloroform for 3 days in the dark) soil samples were shaken in 0.5 M K_2SO_4 for 1 hr, filtered through Whatman No. 1 filter paper, and frozen until analyzed. Total organic carbon (TOC) and total nitrogen in the extracts were measured simultaneously on a TOC VCPH/CPN TOC and Nitrogen Analyzer (Schimadzu, Kyoto, Japan). Total microbial C and N were estimated as the difference between fumigated and unfumigated samples (expressed on an oven-dry mass basis), divided by appropriate conversion factors [44, 45].

Statistical Analyses

Using mean values per plot (i.e., n = 5 for each treatment and control) differences among treatments in soil respiration rates, soil temperature, and soil moisture were investigated by separate repeated measures ANOVAs for each year. Fine root biomass and microbial biomass C and N were analyzed with one-way ANOVAs for each year separately. Where treatment effects were found to be significant or marginally significant ($P < 0.07$), post-hoc comparisons were made using Fisher's LSD test. All analyses were carried out in Genstat 7.2 (VSN International Ltd., Hemel Hempstead, UK).

KEYWORDS

- **Control plots**
- **Infra-red gas analyzer**
- **Microbial biomass**
- **Soil carbon**
- **Soil respiration rates**
- **Soil temperature**

AUTHORS' CONTRIBUTIONS

Conceived and designed the experiments: Emma J. Sayer and Edmund V. J. Tanner. Performed the experiments: Emma J. Sayer. Analyzed the data: Emma J. Sayer. Contributed reagents/materials/analysis tools: Jennifer S. Powers. Wrote the chapter: Emma J. Sayer, Jennifer S. Powers, and Edmund V. J. Tanner.

ACKNOWLEDGMENTS

We thank Dr. S. Joseph Wright for the use of the Li 6400, Ruscena Wiederholt, Fabienne Zeugin, Jesus Antonio Valdez, Geraldino Perez, Francisco Valdez, Didimo Ureña, and Arturo Worrell Jr. for help in the field, and Milton Garcia for technical advice. All fieldwork was carried out at the Smithsonian Tropical Research Institute, to whom we are very grateful. We also thank B. L. Turner and W. H. Schlesinger for comments on this.

Chapter 4

Evolution of Warning Coloration

Jostein Gohli and Guran Hugstedt

INTRODUCTION

Several pathways have been postulated to explain the evolution of warning coloration, which is a perplexing phenomenon. Many of these attempts to circumvent the problem of naïve predators by inferring kin selection or neophobia. Through a stochastic model, we show that a secreted secondary defense chemical can provide selective pressure, on the individual level, towards developing warning coloration. Our fundamental assumption is that increased conspicuousness will result in longer assessment periods and divergence from the predators' searching image, thus reducing the probability of a predator making mistakes. We conclude that strong olfactory signaling by means of chemical secretions can lead to the evolution of warning coloration.

The evolution of warning coloration [1] has continued to be a persistent problem for evolutionary biologists. Signals used by aposematic prey increase conspicuousness and/or distinctiveness [2] and will increase the initial probability of attack from predators [3, 4]. If predators are inexperienced, they must sample the aposematic prey to learn the association between the signal and the level of profitability. When aposematism first evolved, all predators were inexperienced and the population of aposematic prey would have been very small. Sampling (killing) would likely have led to an early extinction of this fragile population. A way to circumvent this fundamental problem is to postulate the use of reliable signals, thus removing the need for sampling and learning. It would therefore serve an aposome well to mediate its unpalatability via odorous secretions which can function in such a manner, thus avoiding close contact with the predator [5]. By causing irritation and/or pain when inhaled, such chemicals can give a reliable signal relating to the level of defense. It is difficult to imagine a predator who chooses to attack prey which makes its eyes burn, and causes pain in its respiratory system. In such a case, the chemical secretion is both a signal and a secondary defense component.

Olfactory aposematism [6] has not gone unnoticed by biologists. Both Cott [7] and Rothschild [8] discussed the pungent odors emitted by several aposomes. Cott suggested that odors emitted by aposomes may serve as a noxious defense, in addition to being a warning signal. Rothschild also gave examples of odors which themselves are clearly noxious. Prudic [9] and Eisner, Eisner, and Seigler [10] provide a more recent discussion of smelly defensive secretions. However, none of these discusses the potential effects of such secretions on the evolution of warning coloration.

We explore the possibility that chemical secondary defense could have set the stage for the evolution of warning coloration. By showing that a reliable chemical signal would select for increased visual conspicuousness (VC), we provide a novel explanation to the evolution of visual aposematism. Speed and Ruxton [11] discussed the role of physical secondary defenses in the evolution of aposematism. We modify their simulation model to analyze our hypothesis using a stochastic model.

MATERIALS AND METHODS

The Model

The following is a model of optimal prey defense and signaling based on the factors VC and olfactory signal/defense (OSD). We define OSD as a released chemical toxin that acts both as a secondary defense agent and as an olfactory signal. This secondary defense can cause pain/irritation in the eyes and/or respiratory system, and may even irritate/damage the nervous system. For simplicity, we assume a linear relationship between signal strength and defense strength (a strong defense cannot produce a weak signal and *vice versa*).

Our model assumes no initial aversion towards aposematic traits or conspicuousness, that is, neophobia and/or dietary conservatism are not operating. Our model would work well with neophobia and/or dietary conservatism present, but it is not dependent on it. It is not possible to identify whether neophobia was present before aposematism or if it is an evolutionary response to aposematism, therefore a model explaining the evolution of aposematism cannot build on the assumption of a neophobic response or similar aversions. We discuss the outcome of single interactions between predator and prey, altering only the variables VC and OSD. The VC is given the interval (1.1–1.85) and OSD is given the interval (0.1–0.85). These intervals could be standardized and modified by constants. However, we feel that this would only act to conceal the mechanics of our model. Both variables are dimensionless and are based around the population mean values. In the eventual empirical testing, both variables can be expressed in distance. Values of VC correspond to the distance at which predators locate the prey through sight. Similarly, the ODS value describes the distance at which the predator discover the prey by olfaction. Increased VC and odor intensity would of course result in detection at greater distances. Thus, visual and olfactory conspicuousness are directly correlated to our values. Importantly, the value for ODS also describes the strength of the deterrent effect of the signal. When an olfactory signal is "stronger", more toxins reaches the recipient which results in a stronger deterrence. In nature numerous variables other than the signal strength affect the distance at which the signal is functional, wind affects olfactory signals and vegetation density affects visual signals for instance. Such complicating factors have not been included in our model. The model describes interactions between totally naïve predators and totally egocentric prey (no kin selection).

We include four probabilities in our model:

(1) Pd = Probability of being detected (where k is a constant)

$$Pd = 1 - \exp\left(-\left[\frac{OSD + VCk}{2} + \frac{|OSD - VCk|}{2}\right]\right)$$

(2) Pa = Probability of being attacked once detected

$$Pa = 1 - ([OSD * VC] * [1 - \exp(-OSD)])$$

(3) Pk = Probability of being killed once attacked

$$Pk = 1 - OSD$$

(4) Φ = Total probability of being killed

$$\Phi = Pd * Pa * Pk$$

The *Pd* is based on the one of the two variables (*OSD* and *VC*) exhibiting the largest value (the interval for *VC* is modified by *k*). If, for instance, prey is highly visually conspicuous, a weak olfactory signal will have no effect on *Pd*. On the other hand, should the prey be visually cryptic, a strong olfactory signal will be the governing variable. The intervals are modified in a way that grants *VC* the most power over *Pd*. Although this is not always the case (based on different predators' perceptive abilities and different habitats), we conclude that this is the most realistic scenario.

We treat the probability of being attacked after detection (*Pa*) as solely dependent on the variables *OSD* and *VC*. The probability of attack will be reduced by increasing *OSD* values, because of OSD's chemical defense component. In the model, general conspicuousness (*VC* combined with the olfactory signal component of *OSD*) functions to enhance the effects of the defense, which is expressed through (*OSD***VC*), and is dependent on the intervals given for *OSD* and *VC*. A higher level of conspicuousness with no defense (*OSD*) will result in a higher Φ (see intercept values for different *VC* values in Figure 1). However, an individual with a high *OSD* value will benefit from the longer assessment period provided by higher general conspicuousness (Figure 1). We explain this fact by the following assumptions: the general conspicuousness ties into the length of the assessment period, because predators will detect prey items from longer distances when they are highly conspicuous. Since, a common assumption is that predators may make mistakes, we correlate the assessment period/general conspicuousness to the probability of making a mistake. As the predator will be focused on the prey while moving down a gradient of noxious chemical defense, the prey's low profitability will be highlighted, and mistakes will be less probable. The length of this gradient is tied to general conspicuousness. Prey with a low *VC* value may be detected through the signal component of *OSD* or through visual cues (although the prey is visually cryptic), resulting in a shorter detection distance and assessment period. We base this on the assumption that visual signals work over greater distances than olfactory signals. In spontaneous attacks with short assessment periods, predators may not register the level of secondary defense, fatally injuring or killing the defended prey. Our assumption regarding the effect of the assessment period is supported by Gamberale-Stille [12], who showed that decision time is important in determining attack probability in both naïve and experienced predators.

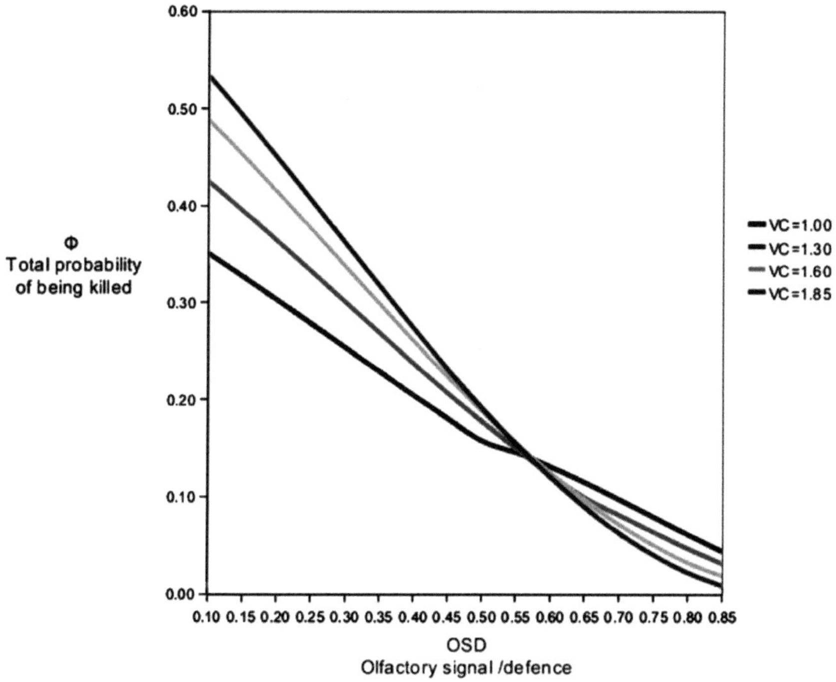

Figure 1. Visual and olfactory components and total probability of being killed. Different fixed values for visual conspicuousness (VC) are plotted against olfactory signal and defense (OSD) values, showing variation in the total probability of being killed (Φ). Selective forces acting on conspicuousness undergo a shift when defense levels reach a critical value (point of intersection). Our model predicts that maximum conspicuousness is the best strategy when the individuals are maximally defended through OSD.

There is a second immediate positive effect of developing increased VC together with chemical secondary defense. Increasing conspicuousness is a sure way of becoming visually distinct from other cryptic prey [13], and no longer coinciding with the predators' searching image. When a prey animal is visually identical to a predators' searching image, a more intense chemical OSD should be required to deter the predator. We are not describing neophobia, simply a divergence from the searching image of predators, which naturally goes hand in hand with divergence from the maximum-crypsis strategy. Many predators will react to prey coinciding with its searching image with immediate attack, which, as previously discussed, will increase the probability of making mistakes.

We assume that OSD is correlated to all forms of chemical defense, thus also affecting the probability of being killed after attack (Pk). This is based on the taste component, where a defended individual has a higher probability of being rejected [14]. The chemical secondary defense component should also reduce the intensity of the attack, further lowering Pk.

The classical problem of the evolution of warning coloration is described by OSD and high VC values. An increase in conspicuousness with OSD at a fixed value of 0.1 will increase the total probability of being killed when predators are naïve (Figure 1; different VC-values, OSD = 0).

As one would expect, an increase in chemical secondary defense decreases the total probability of being killed (Figure 1). Because of the secondary defenses' odor, it also increases the probability of being detected. However, considering the reliable nature of the signal component of OSD, this increased "olfactory conspicuousness" is profitable. Further, increasing the general conspicuousness of the prey through a higher VC will be profitable once OSD reaches a certain level. Figure 1 shows different fixed values for VC combined with increasing values for OSD. The model shows that, given a critical value of OSD, it is also profitable to display warning coloration. This critical value differs slightly for different VC values. The curves also show how VC acts to enhance the effect of OSD. The fact that all values of OSD reduce the total probability of being killed (Φ) (illustrated by the VC = 1 curve) before warning coloration becomes profitable clearly shows an evolutionary path for the development of warning coloration.

Our results and predictions should be possible to test empirically, since the factors VC and OSD can be easily manipulated experimentally.

DISCUSSION

Our model indicates that warning coloration may be profitable when it is coupled to a secreted defense chemical. This statement holds true when predators are totally naïve and exhibit no neophobia. The benefits to defended prey are applicable at the individual level. These facts set our results apart from many of the other attempts to explain the evolution of warning coloration.

In calculating Pa, we make the assumption that general conspicuousness increases the assessment period and that this gives increased protection to prey with certain levels of defense (Figure 1). Higher conspicuousness makes the prey easier to spot, and therefore increases the possibility that the predator is further away from the prey when it is detected. The predator will be focused on the prey for a longer time period when it has to approach it from a distance. Moving towards the prey, the predator will move down a gradient of noxious chemical defense. During this time, the effects of defense will become gradually more apparent. Once the predator reaches the prey, defense will be at its maximum effect. This could be considered a form of "intensive learning"; while keeping its focus on the prey and moving towards it, the unpleasantness caused by defense chemical increases, a fact resulting in the predator learning the association between defense and prey. While, it is unlikely that a predator will choose to attack prey which is defended in this way regardless of VC, the increased assessment period and "intensive learning" will result in fewer mistakes. We base this on the assumption that spontaneous attacks are more prone to result in mistakes than attacks after assessment.

Since the prey may not be aware that a predator is approaching, it might pay to secrete these chemicals continuously. This proposal of continuous secretion might be controversial since insects seen in nature today more often control their defensive secretions and do not release them unless they are first disturbed. However, this apparent problem is created on false grounds because the selective regimes which are prevalent in nature today are not at all similar to those that ruled when aposematism first evolved. If constantly secreted defense compounds result in increased protection against naïve predators (pre-aposematism), it also follows that this would have been the adopted strategy for prey animals. When warning coloration had been fixated in the prey population and learned by the predators/imprinted in the predator psyche, there would no longer be sufficient reason to continuously secrete the defense compounds. In fact, when the selective pressure created by the naïve predators was reduced, the cost of continuous secretion would likely have resulted in a selective pressure towards increased control of secretion. This could explain why we do not often observe prey constantly secreting defensive compounds today.

When prey is conspicuous and chemically defended, predators may learn the association between defense and signals without sampling. When learning has taken place, it may be profitable for the prey to reduce the amount of defense chemical released. Even though naïve predators are not a problem only for the initial evolution of aposematism, the cost of secreting excessive amounts of defense chemicals may outweigh the cost of the odd naïve predator, as mentioned. A simple cost-benefit argument illustrates this point. Initially when all predators are naïve, prey has to be more or less constantly defended, but as predators associate defense with prey traits (for instance visual signals), it no longer pays prey individuals to invest maximally in defense compound secretion. Instead, there will now be a selective benefit of maximizing distant, that is, visual, recognition, paving the way for the evolution of visual/acoustic signals. However, there will always be inexperienced, young predators, a fact explaining the preservation of the reliable olfactory signal.

Conspicuousness gives increased protection for certain values of OSD in our model, but visual signals can give additional advantages beyond those described by our model, once predators have learned the association between signals and defense. Visual and olfactory signals are very different in nature and are affected differently by environmental factors. Once olfactory aposematism is established, based on a reliable signal (OSD), it will probably pay to advertise profitability with more than one signal, taking into consideration the effects of multimodality [15, 16] and the fact that different signals work differently and on different scales. Visual signals may, in certain habitats, be more far-reaching than olfactory signals. Such an increase in signaling distance should decrease the number of close encounters with predators (if the visual aposematic trait is familiar to the predator), which will be stressful no matter what the outcome. Warning coloration could also have evolved based on these advantages once olfactory aposematism had been established. This would not be dependent on our assumptions in calculating Pa (fewer mistakes through a higher assessment period).

As most visual aposomes are insects and their main predators often are birds, a quick word on the olfactory capabilities of birds is in order. Albeit an old and common misconception, the belief that birds are "poor smellers" or that they do not rely heavily on olfaction, is a misconception nontheless. Experimental works on aposematism have shown that odor is an important cue for chickens (*Gallus gallus domesticus*) when foraging [15, 16]. Other works have discussed and provided evidence showing that smell is much more important in birds than previously thought [17, 18].

Through our model we have shown, that given a reliable olfactory signal, VC is a profitable strategy. The element of reliability removes the problem of sampling/killing by naïve predators, making it possible for visual signals to accompany the olfactory element. We conclude that olfactory signals/secreted toxins provide a solution for the evolution of visual aposematic traits.

KEYWORDS

- **Aposematism**
- **Conspicuousness**
- **Naïve predators**
- **Neophobia**

AUTHORS' CONTRIBUTIONS

Conceived and designed the experiments: Jostein Gohli and Göran Högstedt. Performed the experiments: Jostein Gohli. Analyzed the data: Jostein Gohli. Wrote the chapter: Jostein Gohli.

ACKNOWLEDGMENTS

We thank Arne Skorping, Øyvind Fiksen, Espen Strand, and Knut Helge Jensen for helpful comments. A special thanks to Sigrunn Eliassen and Einar Heegaard.

Chapter 5

Chemically Diverse Toxicants Effects on Precursor Cell Function

Zaibo Li, Tiefei Dong, Chris Pruschel, and Mark Noble

INTRODUCTION

Identification of common mechanistic principles that shed light on the action of the many chemically diverse toxicants to which we are exposed is of central importance in understanding how toxicants disrupt normal cellular function and in developing more effective means of protecting against such effects. Of particular importance is identifying mechanisms operative at environmentally relevant toxicant exposure levels. Chemically diverse toxicants exhibit striking convergence, at environmentally relevant exposure levels, on pathway-specific disruption of receptor tyrosine kinase (RTK) signaling required for cell division in central nervous system (CNS) progenitor cells. Relatively small toxicant-induced increases in oxidative status are associated with Fyn kinase activation, leading to secondary activation of the c-Cbl ubiquitin ligase. Fyn/c-Cbl pathway activation by these pro-oxidative changes causes specific reductions, *in vitro* and *in vivo*, in levels of the c-Cbl target platelet-derived growth factor receptor-α and other c-Cbl targets, but not of the TrkC RTK (which is not a c-Cbl target). Sequential Fyn and c-Cbl activation, with consequent pathway-specific suppression of RTK signaling, is induced by levels of methylmercury and lead that affect large segments of the population, as well as by paraquat, an organic herbicide. Our results identify a novel regulatory pathway of oxidant-mediated Fyn/c-Cbl activation as a shared mechanism of action of chemically diverse toxicants at environmentally relevant levels, and as a means by which increased oxidative status may disrupt mitogenic signaling. These results provide one of a small number of general mechanistic principles in toxicology, and the only such principle integrating toxicology, precursor cell biology, redox biology, and signaling pathway analysis in a predictive framework of broad potential relevance to the understanding of pro-oxidant–mediated disruption of normal development.

Determining whether chemically diverse substances induce similar adverse effects at the cellular and molecular level is one of the central challenges of toxicological research. If the structural diversity of different toxicants, and of potential toxicants, means that each works through distinctive mechanisms then this creates a potentially unsolvable challenge in developing means of screening the many tens of thousands of different chemicals for which little or no toxicological information exists. In contrast, the identification of general principles that transcend the specific chemistries of individual substances has the potential of providing broadly relevant insights into the means by which toxicants disrupt normal development. If such principles were found to apply to the analysis of toxicant levels frequently encountered in the environment, this would be of even greater potential importance in providing efficient means of analyzing this diverse array of chemicals.

Of all of the effects associated with toxicant exposure, one of the few that appears to be common to multiple chemically diverse substances is the ability of these agents to cause cells to become more oxidized. The range of toxicants reported to alter oxidative status is very broad, and includes metal toxicants such as methylmercury (MeHg; e.g., [1-6], lead [Pb] [6-9], and organotin compounds [1, 2, 5, 10, 11]), cadmium [12, 13], and arsenic [12, 14]. Ethanol exposure also is associated with oxidative stress [15], as is exposure to a diverse assortment of agricultural chemicals [16], including herbicides (e.g., paraquat [17, 18]), pyrethroids [19-21], and organophosphate and carbamate inhibitors of cholinesterase [22-26]. Thus, the ability to cause cells to become more oxidized is shared by many toxicants, regardless of their chemical structure.

The observations that chemically diverse toxicants share the property of making cells more oxidized is of particular interest in light of the increasing evidence that oxidative regulation is a central modulator of normal physiological function. Although increases in oxidative status in a cell have been most extensively studied in the context of their adverse effects (in particular, the induction of cell death or of cell senescence), multiple studies have demonstrated that changes in redox state as small as 15–20% may be critical in regulating such normal cellular processes as signal transduction, division, differentiation, and transcription (reviewed in, e.g., [27-31]). Although the mechanistic basis for such regulation is frequently unclear, the importance of redox status in modulating cell function makes convergence of different toxicants on this physiological parameter a matter of considerable potential interest.

Despite the observations that many toxicants share the property of making cells more oxidized, multiple questions exist regarding the relevance of such observations for the understanding of toxicant function.

First, there is considerable uncertainty about the relative importance of effects on redox state in the analysis of individual toxicants, and it is generally believed that the major effects of toxicants on cellular function are distinct from any effects on oxidative status. For example, in the context of agents analyzed in the present studies, MeHg-mediated effects on cellular function generally are thought to be mediated through binding to cysteine residues, thus disrupting function of microtubules and other proteins, but may also involve disruption of Ca^{2+} homeostasis (e.g., [32-34]). In contrast, Pb does not bind to cysteine residues and instead is thought to exert its functions through altering normal calcium metabolism by mimicking calcium action and/or by disrupting calcium homeostasis (e.g., [35, 36]). This would lead to alterations in function of multiple proteins, of which the most extensively studied have been members of the protein kinase C (PKC) family of enzymes (e.g., [37, 38]).

A further concern is the general lack of knowledge about whether, or where, oxidation induced by different means would mechanistically converge. For example, MeHg has been suggested to cause oxidative stress by a variety of mechanisms, including by binding to thiols, by causing a depletion in glutathione levels, or by impairing mitochondrial function [39, 40], whereas Pb is thought to disrupt mitochondrial function through its effects on calcium metabolism (e.g., [35, 36, 41-45]) The organic herbicide paraquat (the third agent examined in the present studies) is another example of a toxicant with pro-oxidant activities, but in this case, resulting from initiation of a cyclic oxidation/reduction process in which paraquat first undergoes one electron reduction by NADPH to form free radicals that donate their electron to O_2, producing a superoxide radical; upon exhaustion of NADPH, superoxide reacts with itself and produces hydroxyl free radicals (e.g., [17, 18]). Whether these different means of altering oxidative state would have different mechanistic consequences is unknown.

A further concern regarding the hypothesis that changes in redox state represent an important convergence point of toxicant action is whether oxidative changes are even associated with toxicant exposure at levels frequently encountered in the environment. For example, although several studies have documented the ability of MeHg to cause cells to become more oxidized, effective exposure levels employed in these studies have generally ranged from 1 to 20 μM [2-5], which is 30–600 times the upper range of average mercury concentrations found in the bloodstream of as many as 600,000 newborn infants in the US alone [46]. Similar concerns apply to the analysis of multiple toxicants, for which pro-oxidant effects have largely been studied at exposure levels much higher than those with broad environmental relevance.

In addition, a more general concern regarding the search for general principles of toxicant action is whether such convergence, if it exists, would occur only at exposure levels that induce cell death or whether common mechanisms might be relevant to the understanding of more subtle effects of toxicant exposure, particularly during critical developmental periods. Because development is a cumulative process, the effects of small changes in, for example, progenitor cell division and/or differentiation that are maintained over multiple cellular generations could have substantial effects on the organism. Such changes are poorly understood, however, at both cellular and molecular levels.

Our present studies have led to the discovery of a previously unrecognized regulatory pathway on which environmentally relevant levels of chemically diverse toxicants converge to compromise division of a progenitor cell isolated from the developing CNS. We found that exposure of cells to low levels of MeHg, Pb, or paraquat is sufficient to make cells more oxidized and to activate Fyn kinase, a Src family member known to be activated by increased oxidative status. This first step activates a pathway wherein Fyn activates c-Cbl, an ubiquitin ligase that plays a critical role in modulating degradation of a specific subset of RTKs. The c-Cbl activation in turn leads to reductions in levels of target RTKs, thus suppressing division of glial progenitor cells. The effects of all three toxicants are blocked by co-exposure to N-acetyl-L-cysteine, which is widely used to protect against oxidative stress. We also provide evidence that our *in vitro* analyses successfully predict previously unrecognized effects of developmental MeHg exposure at levels 90% below those previously considered to represent low-dose exposure levels.

Exposure to Environmentally Relevant Levels of MeHg Causes Glial Progenitor Cells to Become More Oxidized and Suppresses Their Division

The progenitor cells that give rise to the myelin-forming oligodendrocytes of the CNS offer multiple unique advantages for the study of toxicant action, particularly in the context of analysis of toxicant effects mediated by changes in intracellular redox state. These progenitors (which are referred to as both oligodendrocyte-type-2 astrocyte (O-2A) progenitor cells ([47] and oligodendrocyte precursor cells, here abbreviated as O-2A/OPCs) are one of the most extensively studied of progenitor cell populations (reviewed in, e.g., [48-52]). They also are among a small number of primary cell types that can be analyzed as purified populations, and at the clonal level, and for which there is both extensive information on the regulation of their development and also evidence of their importance as targets of multiple toxicants (including such chemically diverse substances as Pb [38, 53]), ethanol (e.g., [54-57], and triethyltin [10, 58]).

Another important feature of O-2A/OPCs, in regard to the present studies, is that their responsiveness to small (~15–20%) changes in the intracellular redox state provides a central integrating mechanism for the control of their division and differentiation [59]. The O-2A/OPCs purified from developing animals on the basis of the cell's intracellular redox state exhibit strikingly different propensities to divide or differentiate. Cells that are more reduced at the time of their isolation undergo extended division when grown in the presence of platelet-derived growth factor (PDGF, the major mitogen for O-2A/OPCs [60-62]), whereas those that are more oxidized are more prone to undergo differentiation [59]. Pharmacological agents that make cells slightly more reduced enhance self-renewal of dividing progenitors, whereas pharmacological agents that make cells more oxidized, by as little as 15–20%, suppress division and induce oligodendrocyte generation. Moreover, cell-extrinsic signaling molecules (e.g., neurotrophin-3 (NT-3) and fibroblast growth factor-2 (FGF-2)) that enhance the self-renewal of progenitors dividing in response to PDGF cause cells to become more reduced. In contrast, signaling molecules that induce differentiation to oligodendrocytes (i.e., thyroid hormone (TH) [63, 64]) or astrocytes (i.e., bone morphogenetic protein-4 (BMP-4) [65, 66]) cause cells to become more oxidized [59]. The ability of these signaling molecules to alter redox state is essential to their mechanisms of action, because pharmacological inhibition of the redox changes they induce blocks their effects on either division or differentiation of O-2A/OPCs. Thus, multiple lines of evidence have demonstrated that responsiveness to small changes in redox status represents a central physiological control point in these progenitor cells (as summarized in Figure 1).

We initiated our studies of toxicant effects on O-2A/OPCs with an examination of MeHg, which has been previously studied for its effects on neuronal migration, differentiation, and survival, and on astrocyte function (e.g., [67-74]). Little is known about the effects of MeHg on the oligodendrocyte lineage, despite the fact that there are several reports over the past 2 decades documenting decreases in conduction velocity in the auditory brainstem response (ABR) of MeHg-exposed children [75-78] and rats [79]. Such a physiological alteration has long been considered to be indicative of myelination abnormalities in children whose development has been compromised by iron deficiency (see, e.g., [80, 81]).

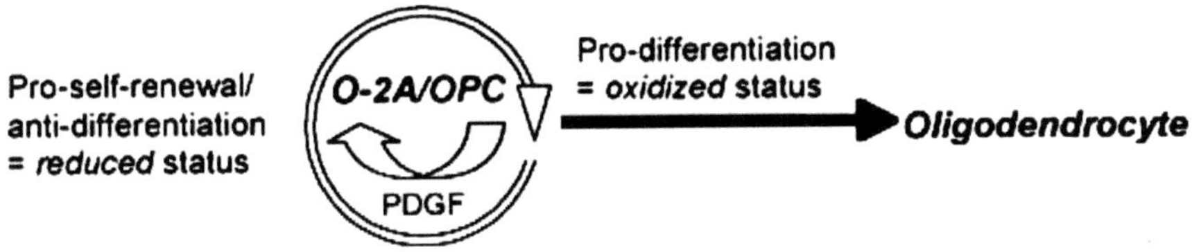

Figure 1. Diagrammatic summary of the role of redox regulation in modulating division and differentiation of O-2A/OPCs progenitor cells are induced to divide by exposure to PDGF. Induced to divide by PDGF alone, progenitors will undergo a limited number of divisions while asymmetrically generating oligodendrocytes. The balance between division and differentiation is modulated, however, by the intracellular redox state [59]. Cells that are more oxidized tend to differentiate, whereas those that are more reduced undergo more self-renewal. Pharmacological agents that make cells more oxidized induce differentiation of dividing O-2A/OPCs into oligodendrocytes. Similarly, signaling molecules that induce differentiation (e.g., TH) make cells more oxidized as a necessary part of their mechanism of action. In contrast, pharmacological agents that make cells more reduced promote self-renewal, and signaling molecules that enhance self-renewal (e.g., NT-3) make cells more reduced as a necessary part of their mechanism of action.

We found that exposure of O-2A/OPCs (growing in chemically defined medium supplemented with PDGF) to environmentally relevant levels of MeHg makes these cells approximately 20% more oxidized (Figure 2A), a degree of change similar to that previously associated with reductions in progenitor cell division [59]. Exposure to MeHg inhibited progenitor cell division as determined both by analysis of bromodeoxyuridine (BrdU) incorporation (Figure 2B) and by analysis of cell division in individual clones of O-2A/OPCs (Figure 2C–E). These oxidizing effects of MeHg were seen at exposure levels as low as 20 nM, less than the 5.8 μg/l or more (i.e., parts per billion (ppb)) of MeHg found in cord blood specimens of as many as 600,000 infants in the US each year [46] and 0.3% or less of the exposure levels previously found to induce oxidative changes in astrocytes [4]. Exposure to 20 nM MeHg was sufficient to cause an approximately 25% drop in the percentage of O-2A/OPCs incorporating BrdU in response to stimulation with PDGF. When examined at the clonal level, MeHg exposure was associated with a reduction in the number of large clones and an increase in the number of small clones, as seen for other pro-oxidant stimuli [59]. Increasing MeHg exposure levels above 50 nM was associated with significant lethality, but little or no cell death was observed at the lower concentrations used in the present studies (unpublished data). Thus, division of O-2A/OPCs exhibits a striking sensitivity to low concentrations of MeHg.

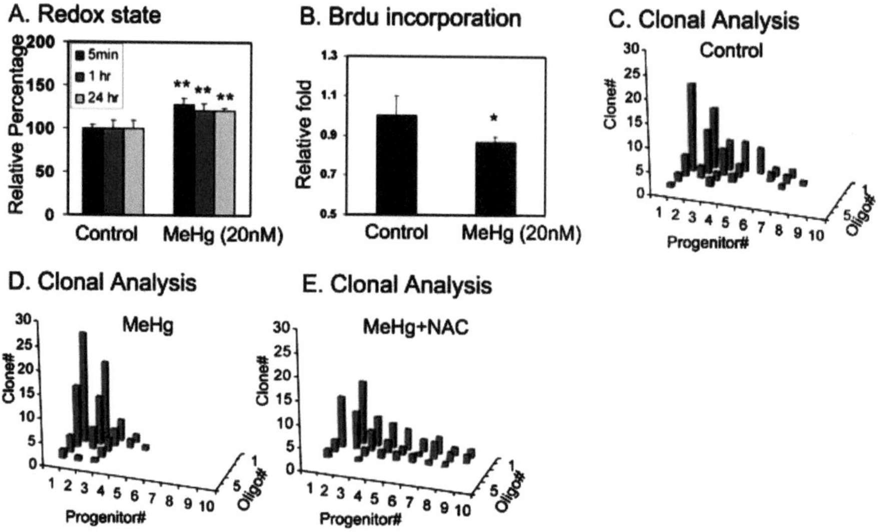

Figure 2. The MeHg exposure makes O-2A/OPCs more oxidized, suppresses BrdU incorporation, and decreases cell division in clonal assays. (A) Purified O-2A/OPCs were grown in the presence of 10 ng/ml PDGF overnight. Effects of MeHg on intracellular redox state were determined by analysis of 2′,7′-dichlorodihydrofluorescein diacetate fluorescence emission in O-2A progenitors exposed to 20 nM MeHg for various lengths of time, as indicated. (B) Cells were plated in medium containing PDGF and then exposed to 20 nM MeHg for an additional 72 hr. During the last 4 hr of exposure, cultures were also exposed to BrdU. Cultures were then stained with A2B5 and anti-BrdU antibodies (to recognize all progenitors and those synthesizing DNA during the BrdU pulse, respectively). Results are presented as comparison with control cultures. (C–E) Suppression of cell division by exposure to MeHg was studied in more detail at the clonal level. Cells were treated as for (B), except for being plated at clonal density (as in [59, 199]). Cultures were maintained for 6 days, and then 100 randomly chosen clones were analyzed for their composition (as in [59, 64, 199]). Data are presented, for all clones analyzed, in three dimensions such that the x-axis equals the number of progenitors per clone, the z-axis equals the number of oligodendrocytes (Oligo) per clone, and the y-axis equals the number of clones with any given composition. In cultures exposed to MeHg, there was a decrease in the representation of large clones and a proportionate increase in the number of small clones and clones containing oligodendrocytes. The effects of MeHg were prevented by co-exposure of cells to 1 mM N-acetyl-L-cysteine (NAC). All experiments were repeated at least three times, and all numerical values represent means ± SD for triplicate data points.

MeHg Exposure Reduces the Effects of PDGF from the Nucleus Back to the Receptor

One possible explanation for the reduced division associated with MeHg exposure would be disruption of PDGF-mediated signaling, and molecular analysis revealed that exposure of O-2A/OPCs to 30 nM MeHg for 24 hr suppressed PDGF-induced signaling pathway activation at multiple points from the nucleus back to the receptor. One pathway stimulated by PDGF binding to the PDGF receptor-α (PDGFRα) leads to sequential activation of Raf-1, Raf-kinase, and extracellular signal-regulated kinase 1 and 2 (ERK1/2), which further leads to activation of the Elk-1 transcription factor and up-regulation of immediate early-response gene expression, at least in part through activation of the serum response element (SRE) promoter sequence [82, 83].

The MeHg exposure was associated with reduced expression of an SRE-luciferase reporter gene (Figure 3A), and reduced ERK1/2 phosphorylation (Figure 3B). The PDGFRα activation also stimulates activity of PI-3 kinase, leading to activation of Akt and induction of NF-κB–mediated transcription (e.g., [82, 84, 85]), both of which also were inhibited by MeHg exposure. Expression of an NF-κB-luciferase reporter gene was decreased (Figure 3C), as was phosphorylation of Akt (Figure 3D). Phosphorylation of PDGFRα, indicating receptor activation, was also reduced in cells exposed to MeHg (Figure 3E). Because O-2A/OPCs growing in these cultures are absolutely dependent upon PDGF for continued division (e.g., [60, 61, 86]), the suppression of PDGF signaling would necessarily cause a reduction in cell division.

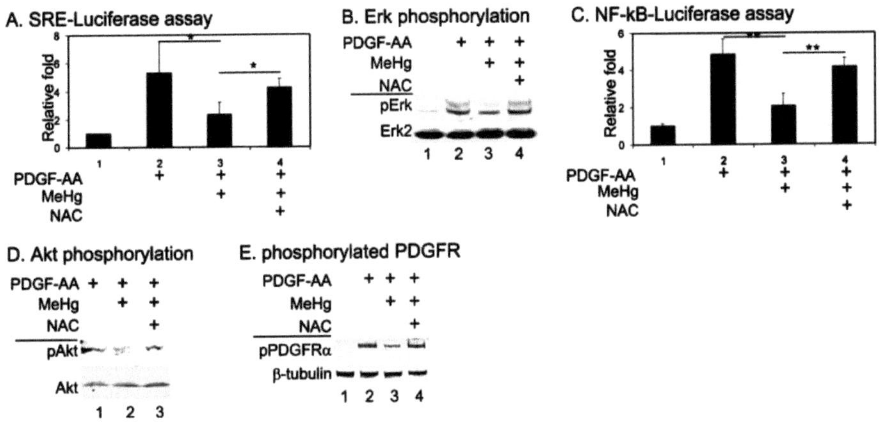

Figure 3. The MeHg suppresses PDGF-mediated signaling from the nucleus back to the receptor. (A) Progenitors transfected with an SRE-luciferase reporter construct and exposed to 30 nM MeHg (24 hr) showed significantly lower levels of reporter activity. *, $p < 0.05$. (B) Cells grown as in (A) and analyzed for phosphorylation of ERK1/2 showed reduced ERK1/2 phosphorylation. (C) Cells transfected with an Nf-κB-luciferase reporter construct and treated as in (A) showed reduced Nf-κB transcriptional activity. **, $p < 0.01$. (D) Cells grown as in (A) showed lower levels of Akt (Thr 308) phosphorylation. (E) Cells grown as in (A) also showed decreased phosphorylation of PDGFRα (as detected with anti-PDGFRα (pY742) antibody). All effects of MeHg were prevented by growth of cells in the additional presence of 1 mM NAC. All experiments were repeated at least three times, and all numerical values represent means ± SD for triplicate data points. The plus symbol indicates exposure of the cells to the indicated substance.

Pathway-specific Disruption of PDGF-mediated Signaling, and Reductions in Levels of PDGFRα, Induced by MeHg

We next found that the effects of MeHg were pathway specific and were associated with reductions in total levels of PDGFRα. The O-2A/OPCs exposed to 30 nM MeHg exhibited no reduction in ERK1/2 phosphorylation induced by exposure to NT-3 (Figure 4A), and no reduction in NT-3–induced expression from an SRE-luciferase reporter construct (unpublished data). This result suggested that the site of action of MeHg was upstream of ERK1/2 regulation, prompting us to look directly at the PDGFRα. We found that the reduction in phosphorylated PDGFRα (Figure 3E) was paralleled by a reduction in levels of the PDGFRα itself (Figure 4B). In contrast, no reduction in levels of TrkC (the receptor for NT-3 [87]) was caused by exposure to MeHg (Figure 4C).

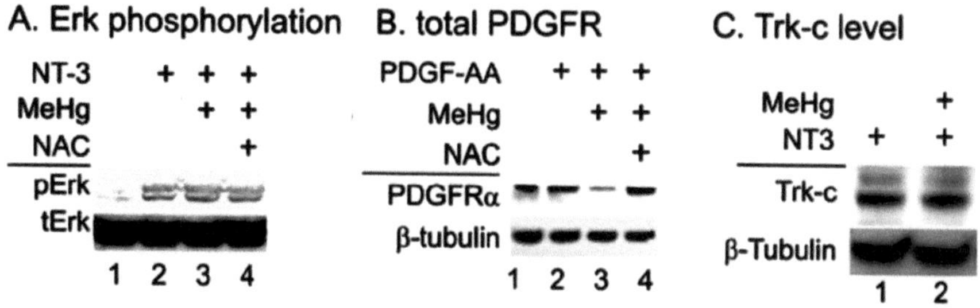

Figure 4. The MeHg effects are pathway specific and are associated with reduced levels of PDGFRα. (A) The MeHg does not inhibit ERK1/2 phosphorylation induced by exposure to NT-3. Cells were grown as in Figure 2A, but exposed to NT-3 instead of PDGF. As shown, MeHg exposure did not reduce the extent of ERK1/2 phosphorylation induced by exposure to NT-3, thus indicating that the site of action of MeHg is not on the level of these kinases. (B) Consistent with the indication from these results that effects of MeHg were mediated further upstream in the PDGF-signaling pathway, analysis of cells treated in the same manner showed reductions in total levels of PDGFRα. The effects of MeHg on total levels of PDGFRα were prevented by co-exposure with NAC. (C) Exposure to MeHg was not associated with reductions in levels of TrkC, indicating that receptor loss was mediated by a mechanism that distinguishes between PDGFRα and TrkC. All experiments were repeated at least three times. The plus symbol indicates exposure of the cells to the indicated substance.

Fyn and c-Cbl Activation, and Enhanced Degradation of PDGFRα Induced by MeHg

One possible explanation for the ability of MeHg to cause a reduction in PDGF-mediated signaling and in total levels of PDGFRα, without affecting NT-3–mediated signaling or TrkC levels, would be that exposure to this toxicant leads to activation of c-Cbl, an E3 ubiquitin ligase that ubiquitylates the activated PDGFRα [88, 89], thus leading to its internalization and potential lysosomal degradation [90-92]. Such a possibility is particularly intriguing in light of multiple reports that c-Cbl can be activated by Fyn kinase (e.g., [93-96]), a Src family kinase that can be activated by oxidative stress [97-100]. The O-2A/OPCs are known to express Fyn, which has been studied in these cells for its effects on regulation of RhoA activity and control of cytoskeletal organization [101, 102]. Because TrkC does not appear to be regulated by c-Cbl, redox-modulated activation of Fyn, leading to c-Cbl activation and enhanced PDGFRα degradation, would provide a potential mechanistic explanation integrating the observations reported thus far.

A variety of data support the hypothesis that MeHg exposure activates Fyn, leading to activation of c-Cbl, followed by degradation-mediated reductions in levels of activated PDGFRα. Exposure of O-2A/OPCs to 30 nM MeHg stimulated Fyn activation and c-Cbl phosphorylation (Figure 5A and B). Activation of Fyn and c-Cbl was blocked by the Src family kinase inhibitors PP1 (Figure 5A and B) and PP2 (unpublished data). We next found that exposure to MeHg enhanced ubiquitylation of PDGFRα (a predicted consequence of c-Cbl activation), an increase readily observed even in the presence of markedly reduced levels of the receptor itself (Figure 5C). Co-exposure to ammonium chloride (NH_4Cl, a lysosomotropic weak base that increases lysosomal pH and disrupts lysosomal protein degradation [103-105]) prevented receptor degradation, and was associated with increased levels of ubiquitylated receptor in treated O-2A/OPCs. The increase in levels of ubiquitylated receptor was as predicted by the lack of effect of NH_4Cl on either Fyn activation or c-Cbl phosphorylation (Figure 5A and B). Treatment with PP1, which inhibits Fyn activity (Figure 5A), was also associated with a marked reduction in the amount of ubiquitylated PDGFRα, particularly in comparison with levels of total receptor (compare upper and lower lanes in Figure 5C). As further confirmation that reductions in levels of PDGFRα were due to protein degradation, exposure to MeHg did not have any significant effects on levels of PDGFRα mRNA, as determined by quantitative PCR analysis. In the presence of cycloheximide, an inhibitor of protein synthesis, MeHg further accelerated receptor loss as compared with that occurring solely due to failure to synthesize new protein. Collectively, these results indicate that MeHg enhances active degradation of PDGFRα, as contrasted with reducing receptor levels as an indirect consequence of altering transcriptional or translational regulation of receptor levels.

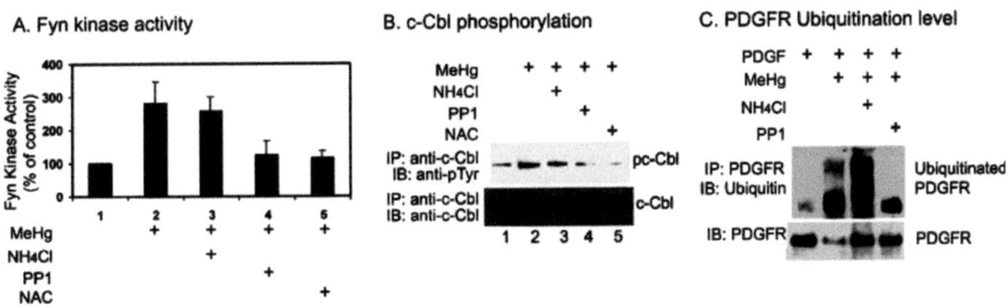

Figure 5. The MeHg exposure causes activation of Fyn, phosphorylation of c-Cbl, and ubiquitylation of PDGFRα leading to reductions in receptor level. (A) The O-2A/OPCs exposed to 30 nM MeHg exhibited higher levels of Fyn kinase activity, as detected by analysis of immunoprecipitated Fyn from these cells using the Universal Tyrosine Kinase Assay Kit (Takara), as described in Materials and Methods. Values (mean ± SD) are expressed as the percent of controls, which were defined from basal Fyn kinase activity without any stimulation. All bars with increased levels of Fyn activity differ from control values at $p < 0.001$. Increased Fyn activity was blocked by pre-treatment of cells with NAC or with the src-family kinase inhibitor PP1, but not by pre-treatment with NH_4Cl (an inhibitor of lysosomal function). (B) As for (A), Increased c-Cbl phosphorylation (see Materials and Methods for immunoprecipitation assay) associated with MeHg exposure was blocked by co-exposure of cells to NAC or PP1, but was not blocked when cells were exposed to NH_4Cl. (C) Exposure to MeHg was associated with a marked increase in the levels of ubiquitylation of PDGFRα, with this increase being apparent even though receptor levels were themselves reduced by toxicant exposure. The NH_4Cl treatment was associated with rescue of levels of total PDGFRα, and therefore was associated with a still more marked increase in the amount of ubqcuitylated receptor detected. As predicted by the hypothesis that both receptor ubiquitylation and receptor loss are due to activation of Fyn, co-exposure of cells to PP1 rescued receptor levels and greatly reduced the extent of receptor ubiquitylation.

The plus symbol (+) indicates exposure of the cells to the indicated substance. The IB = immunoblot; IP = immunoprecipitation.

Molecular confirmation of the role of Fyn and c-Cbl in the effects of MeHg on levels of PDGFRα was obtained by expression of dominant negative c-Cbl, or small inhibitory RNA (RNAi) for Fyn or Cbl, in MeHg-exposed O-2A/OPCs. Expression of the dominant-negative (DN) 70z mutant of c-Cbl [106-108] in O-2A/OPCs prevented

MeHg-induced reductions in levels of PDGFRα (Figure 6A). Reduction in levels of Fyn protein by introduction of Fyn-specific small interfering RNA (siRNA) constructs (Figure 6B) also protected against MeHg-induced reductions in levels of PDGFRα (Figure 6C), as predicted by the hypothesis that MeHg-induced activation of Fyn mechanistically precedes reductions in receptor levels. Similar results were obtained using RNAi constructs for c-Cbl, but are presented later in the chapter, in the context of analysis of other toxicants.

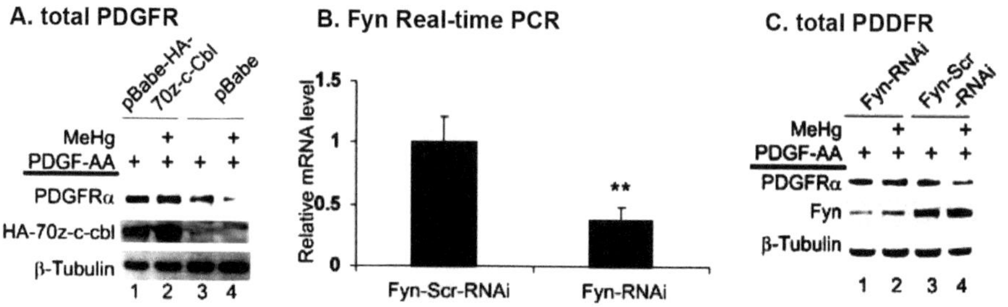

Figure 6. The MeHg-induced reductions in PDGFRα levels were prevented by expression of DN (70Z) c-Cbl and by expression of Fyn-specific RNAi constructs. (A) Expression (by transfection, as described in Materials and Methods) of DN (70Z) c-Cbl prevented MeHg-induced reductions in levels of PDGFRα, an effect not obtained with vector alone (pBabe). (B) Expression of Fyn-specific RNAi (as described in Materials and Methods) caused a reduction in levels of Fyn protein, whereas scrambled (Scr) controls had no effect on levels of this protein. Data are presented as comparisons with levels of Fyn protein in non-manipulated cells. (C) Expression of Fyn-RNAi constructs, but not of Fyn-Scr-RNAi protected O-2A/OPCs from MeHg-induced reductions in levels of total PDGFRα. Fyn RNAi constructs had no effects on levels of tubulin or on levels of c-Cbl (unpublished data).

All experiments were repeated at least three times. The plus symbol indicates exposure of the cells to the indicated substance.

Suppression of Fyn or c-Cbl activity, or overexpression of PDGFRα itself, also protected against the functional effects of MeHg exposure (Figure 7). Pharmacological inhibition of Fyn activity with PP1 enabled analysis of O-2A/OPC division at the clonal level, and demonstrated that PP1 blocked MeHg-induced suppression of cell division (Figure 7A). The O-2A/OPCs expressing DN-70Z-c-Cbl and exposed to MeHg were also protected from effects of MeHg on cell division, as analyzed by BrdU incorporation (Figure 7B). Co-treatment of MeHg-exposed O-2A/OPCs with PP1 or NH_4Cl also blocked MeHg-associated suppression of ERK1/2 phosphorylation (and MeHg-induced reductions in levels of PDGFRα, indicating that ERK1/2 suppression was a secondary consequence of the effects of Fyn and c-Cbl activation (Figure 7C). Overexpression of PDGFRα in MeHg-exposed O-2A/OPCs also protected cells from MeHg-associated reductions in ERK1/2 phosphorylation (Figure 7D).

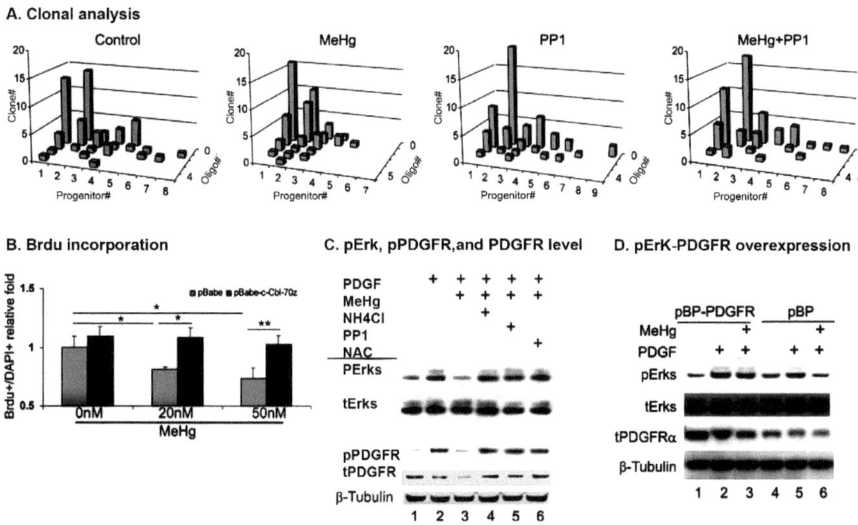

Figure 7. Inhibition of Fyn activity, c-Cbl activity, or of lysosomal function rescues cell division and/or ERK1/2 phosphorylation in MeHg-treated O-2A/OPCs. (A) Purified O-2A/OPCs were plated at clonal density and analyzed as for Figure 2. After 24 hr, MeHg was added to cultures in the presence or absence of PP1. The MeHg by itself was associated with a reduction in the contribution of large clones dominated by progenitors and an increased representation of smaller clones and of oligodendrocytes. Co-exposure to PP1 rescued cells from the effects of MeHg. (B) Expression of DN (70Z) c-Cbl rescued purified progenitors from MeHg-associated suppression of BrdU incorporation. It was notable that expression of DN(70Z) c-Cbl also rescued cells from the effects of 50 nM MeHg, raising the possibility that further exploration of this pathway will reveal additional roles in cell survival (a topic to be explored in future research). *, $p < 0.05$; **, $p < 0.01$. (C) Co-exposure of cells to NH_4Cl or PP1 together with MeHg protected cells from MeHg-induced suppression of ERK1/2 and PDGFRα phosphorylation and reductions in total levels of PDGFRα. (D) Overexpression of PDGFRα also rescued cells from MeHg-induced suppression of ERK1/2 phosphorylation, whereas expression of a control construct (pBP) did not rescue ERK1/2 phosphorylation.

All experiments were repeated at least three times, and all numerical values represent means ± SD for triplicate data points. The plus symbol (+) indicates exposure of the cells to the indicated substance.

Convergence of Chemically Diverse Toxicants on Activation of Fyn and c-Cbl, and Reductions in Levels of PDGFRα

To determine whether effects of MeHg revealed a general mechanism by which chemically diverse toxicants with pro-oxidant activity could alter cellular function in similar ways, we next examined the effects of exposure of dividing O-2A/OPCs to Pb (a heavy metal toxicant) and paraquat (an organic herbicide). As discussed in the Introduction, these toxicants both make cells more oxidized, but through mechanisms that differ between them and also from effects of MeHg.

Despite their chemical differences from MeHg, and from each other, Pb and paraquat had apparently identical effects as MeHg on ERK1/2 phosphorylation, activation of Fyn and c-Cbl, and reductions in levels of phosphorylated PDGFRα and on total levels of PDGFRα (Figure 8). The O-2A/OPCs were exposed to 1 μM Pb (equivalent to the level of 20 μg/dl that is known to be associated with cognitive impairment, and a level of Pb previously found to inhibit O2A/OPC division without causing cell death [38, 53, 109]) or to 5 μM paraquat (an exposure level selected as being in the lowest 0.1% of the range of paraquat concentrations studied by others *in vitro*, which range from 8 μM to 300 mM (e.g., [110-114]). The Pb and paraquat exposure at these levels did not cause cell death, but did make O-2A/OPCs approximately 20% more oxidized, as determined by analysis of cells with the redox-indicator dyes dihydro-chloromethyl-rosamine or dihydro-calcein-AM (unpublished data). Both Pb and paraquat exposure were associated with activation of Fyn (Figure 8A), increased phosphorylation of c-Cbl (Figure 8B), reduced levels of ERK1/2 phosphorylation, and reduced levels of phosphorylated and total PDGFRα (Figure 8C). As for MeHg, the effects of Pb and paraquat on PDGFRα levels were prevented by expression of RNAi for c-Cbl (Figure 8D), DN (70Z) c-Cbl, or RNAi for Fyn (unpublished data).

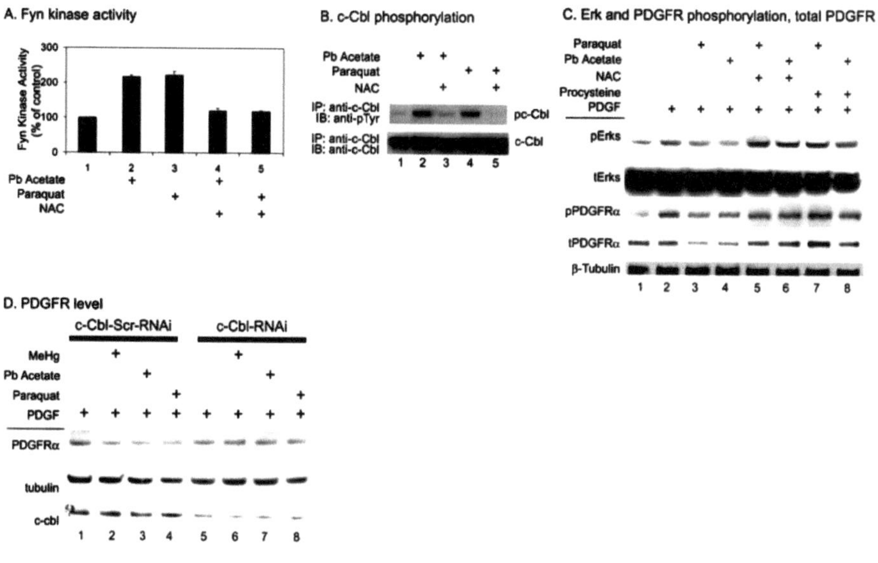

Figure 8. The Pb and paraquat exposure caused activation of Fyn and c-Cbl, suppression of ERK1/2 phosphorylation, and reduction in levels of PDGFRα. (A) Purified O-2A/OPCs were treated as for analysis of MeHg, except that cells were exposed to 1 μM Pb or 5 μM paraquat. Both toxicants caused activation of Fyn, analyzed as in Figure 5. (B) The Pb and paraquat exposure also caused phosphorylation of c-Cbl, as detected by immunoprecipitation of total c-Cbl followed by analysis with anti-phosphotyrosine antibody. (C) The Pb and paraquat exposure cause suppression of ERK1/2 phosphorylation and reductions in total levels of PDGFRα. (D) Expression of c-Cbl RNAi caused a reduction in levels of c-Cbl protein and protected PDGFRα levels from effects of MeHg, Pb, and paraquat, whereas scrambled (Scr) RNAi constructs had no levels of PDGFRα. The NAC (or procysteine, a cysteine pro-drug with no intrinsic anti-oxidant activity) protected against all effects of toxicant exposure. All experiments were repeated at least three times. The plus symbol indicates exposure of the cells to the indicated substance.

It has previously been suggested that the effects of Pb on O-2A/OPCs are mediated through activation of PKC [38], a pathway that has not been implicated in the activity of MeHg or paraquat. To determine whether PKC inhibition could distinguish between effects of Pb versus MeHg or paraquat, and to determine if PKC activation was relevant to the effects of toxicants on Fyn or c-Cbl activation or reductions in PDGFRα levels, we next examined the effects of co-exposure of O-2A/OPCs to bisindolylmaleimide I (BIM-1, a broad-spectrum PKC inhibitor previously used in the analysis of the role of PKC activation in the effects of Pb on O-2A/OPCs [38]). We found that co-exposure of O-2A/OPCs to BIM-1 with Pb, MeHg, or paraquat did not prevent toxicant-mediated activation of Fyn or c-Cbl. The BIM-1 co-exposure also did not protect against MeHg-, Pb-, or paraquat-induced reductions in levels of PDGFRα.

Protection by Cysteine Pro-drugs

If it is correct that Fyn activation, with its consequences, is regulated by the ability of toxicants to make cells more oxidized, then antagonizing such redox changes should prevent Fyn activation. Previous studies have shown that an effective means of preventing the increase in oxidative status and the suppression of cell division caused by exposure of O-2A/OPCs to TH is to treat cells with N-acetyl-L-cysteine (NAC), a cysteine pro-drug that is readily taken up by cells and converted to cysteine [59]. Cysteine is the rate-limiting precursor for synthesis of glutathione, one of the major regulators of intracellular redox status (e.g., [115, 116]). The NAC also possesses anti-oxidant activity, has long been used as a protector against many types of oxidative stress (e.g., [9, 117, 118]), and has been shown to confer protection against a wide range of toxicants, including MeHg (e.g., [119-121]), Pb (e.g., [9, 122, 123]), and paraquat (e.g., [17, 124]), as well as such other substances as aluminum [125], cadmium [126], arsenic [127], and cocaine [128].

As predicted by the hypothesis that the pro-oxidant activities of chemically diverse toxicants are causal in Fyn activation, NAC was equally effective at preventing Fyn activation—and its consequences—induced by exposure to MeHg, Pb, or paraquat (Figures 2–5, 7, and 8). For cells grown at the clonal level, NAC blocked the suppressive effects of MeHg on cell division (Figure 2). The NAC also blocked all effects of MeHg on PDGF-mediated signaling, and rescued normal level of activity of SRE and NF-kB promoter-reporter constructs and levels of phosphorylation of ERK1/2, Akt, and PDGFRα (Figure 3). Consistent with the hypothesis that Fyn is activated when cells become more oxidized [97-100], NAC also blocked MeHg-induced activation of Fyn and phosphorylation of c-Cbl (Figure 5), and prevented MeHg-induced reductions in levels of PDGFRα (Figure 4). Critically, for the hypothesis that Pb and paraquat effects also were mediated by changes in redox state, NAC also blocked the effects of Pb and paraquat on Fyn activation and c-Cbl phosphorylation, and protected against effects of these toxicants on ERK1/2 phosphorylation and levels of PDGFR (Figure 8). Levels of PDGFRα were also protected by exposure of O-2A/OPCs to procysteine (Figure 8), a thiazolidine-derivative cysteine pro-drug that differs from NAC in having no intrinsic anti-oxidant activity [129]. Although it is conceivable that the ability of cysteine pro-drugs to protect against the effects of MeHg, Pb, and paraquat is due to enhanced toxicant clearance associated with elevated levels of glutathione, analysis of

Pb uptake with Leadmium Green AM (a fluorescent indicator of Pb levels) showed no significant difference in Pb levels between cells exposed to Pb as compared with cells exposed to Pb and NAC.

The ability of NAC to block toxicant-induced activation of Fyn raises the question of whether this is due to a true prevention of the effects of toxicant exposure on activation of this kinase or, alternatively, is due to an ability of NAC to independently suppress Fyn activity to such an extent that the apparent block of toxicant effects instead represents the summation of two opposing influences of equivalent magnitude. To evaluate these two possibilities, O-2A/OPCs were exposed to 1 mM NAC in the absence of toxicants, and Fyn and c-Cbl activation were evaluated as in Figure 5. We found that NAC exposure had only a slight, and nonsignificant, effect on the levels of basal Fyn activity in O-2A/OPCs (Figure 9A). In agreement with this outcome, NAC exposure did not have any marked effect on levels of c-Cbl phosphorylation (Figure 9B). Thus, it appears that NAC-mediated counteraction of the effects of toxicants on Fyn activation is far greater in its magnitude than its direct effects on basal levels of Fyn activity.

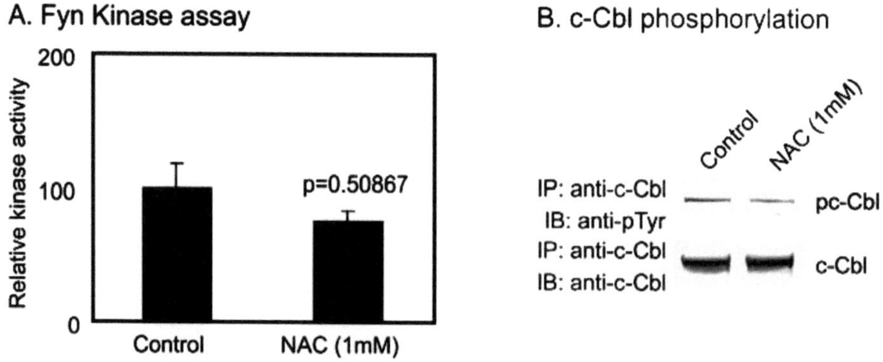

Figure 9. Exposure of O-2A/OPCs to 1 mM NAC has only minimal effects on basal activity of Fyn or phosphorylation of c-Cbl. Assays of Fyn activity and c-Cbl phosphorylation were carried out as in Figure 5, except that cells were exposed only to NAC and not to MeHg. (A) Basal Fyn activity is slightly, but not significantly, lower in cells exposed to NAC. (B) The extent of c-Cbl phosphorylation in O-2A/OPCs exposed to NAC is similar to that seen in cells grown in the presence of PDGF only. IB = immunoblot; IP = immunoprecipitation.

Toxicants Cause Reductions in Levels of Other c-Cbl Targets

If the hypothesis is correct that exposure of O-2A/OPCs to toxicants causes activation of the Fyn/c-Cbl pathway, then other c-Cbl targets should be affected similarly to the PDGFRα. One member of the c-Cbl interactome [92] known to be expressed by O-2A/OPCs is c-Met [130], the receptor for hepatocyte growth factor (HGF; [131, 132]). Oligodendrocytes also have recently been reported to be responsive to epidermal growth factor (EGF) application with morphological changes [133], and microarray analysis confirms that the EGF receptor (EGFR) is expressed by O-2A/OPCs (C. Pröschel and M. Noble, unpublished results). The EGFR is perhaps the most extensively studied

RTK target of c-Cbl [90, 96, 107, 134-137], but c-Met regulation by c-Cbl appears to follow similar principles [106, 138].

As shown in Figure 10, exposure of O-2A/OPCs to MeHg was associated with reductions in levels of c-Met (Figure 10A) and EGFR (Figure 10B). As predicted by the hypothesis that Pb and paraquat converge with MeHg on activation of the Fyn/c-Cbl pathway, levels of c-Met and EGFR were also reduced in O-2A/OPCs exposed to these additional toxicants. Consistent with the hypothesis that such changes were associated with the ability of toxicants to make cells more oxidized, NAC protected both c-Met and EGFR levels from reductions associated with exposure to MeHg, Pb, or paraquat.

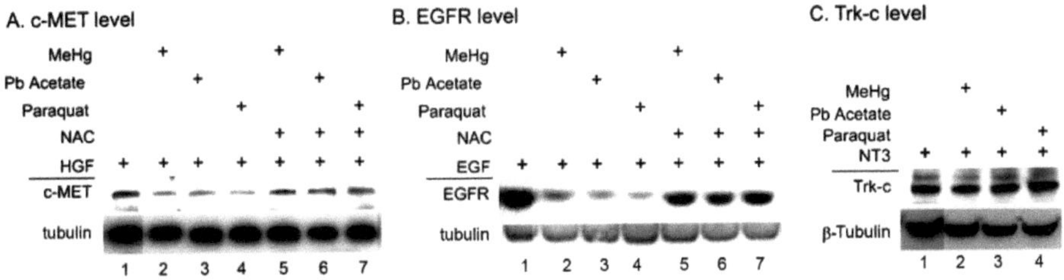

Figure 10. Exposure to MeHg, Pb, and paraquat caused reductions in levels of c-Cbl targets c-Me and EGFR, but not of TrkC. Cells were analyzed as for Figure 5, but with antibodies against c-Met and EGFR. The results for the c-Cbl Targets c-Met (A), EGFR (B), and TrkC (C) are shown. All experiments were repeated at least three times.

Further support for the Fyn/c-Cbl hypothesis of toxicant convergence was provided by observations that neither Pb nor paraquat caused a reduction in levels of TrkC (Figure 10C), just as observed for MeHg (Figure 4C).

Developmental Exposure to Low Levels of MeHg *in Vivo* Causes Reductions in Levels of PDGFRα and EGFR, but Not TrkC, and Causes Reduced Division of O-2A/OPCs

Although the central goal of the present studies was the identification of mechanistic pathways on which chemically diverse toxicants converge, it is important to also consider whether any aspects of our *in vitro* findings are predictive of *in vivo* outcomes. Although detailed *in vivo* investigations will be the subject of future studies, we have tested three of the key findings of our present work for which previous studies are not predictive of likely experimental outcomes.

The three questions we examined *in vivo* were whether toxicant exposure is associated with specific reductions in RTKs that are c-Cbl targets, whether this occurs at levels of toxicant exposure approximating the effects of environmental exposure, and whether such exposure can be shown to cause subtle changes in O-2A/OPC function. These experiments were conducted entirely with MeHg for several reasons. First, there is already extensive evidence that Pb exposure *in vivo* has adverse effects on myelination and on O-2A/OPCs (e.g., [38, 43, 53, 139-142]). In contrast, evidence that MeHg exposure may have any effects on myelination thus far comes only from observations of increased latencies in ABRs [75-79], with no studies examining effects of this toxicant on the function of cells important for myelination (i.e., oligodendrocytes or their ancestral O-2A/OPCs). Third, previous studies on mice have not been conducted using levels of exposure of broad environmental relevance. Instead, such studies have defined a low exposure range as being exposure of animals to MeHg in their drinking water at a concentration of one or more parts per million (e.g., [143-147]), an exposure level considerably higher than what our studies would predict as being necessary to affect progenitor cells of the developing CNS. Thus, the question of whether MeHg exposure levels of broader environmental relevance would have any effects at all *in vivo* appears to be largely unaddressed.

To test the hypothesis that environmentally relevant levels of MeHg exposure can perturb the developing CNS in subtle ways, we exposed SJL mice to 100 or 250 ppb MeHg in their drinking water throughout gestation, and maintained this exposure until sacrifice of pups at 7 and 21 days after birth. As discussed in Materials and Methods, these exposure levels enabled us to approximate the predicted mercury levels in the CNS of 300,000–600,000 infants in the US. The exposure levels examined in our studies are 75–90% below what has otherwise been considered to be low-dose exposure in mice.

We found that developmental exposure of mice to MeHg at either 100 ppb or 250 ppb in the maternal drinking water was associated with clear and significant reductions in levels of PDGFRα and EGFR, but not of TrkC (Figure 11). Treatment of SJL mice with 100 or 250 ppb MeHg in the drinking water during gestation and suckling was associated with reductions in levels of PDGFRα and EGFR in the cerebellum, hippocampus, and corpus callosum when brain tissue was sampled at 7 and 21 days after birth. In contrast, levels of the NT-3 receptor TrkC were not reduced in these animals, as predicted by our *in vitro* analyses. It was particularly striking that exposure even to 100 ppb MeHg in the drinking water was enough to have significant effects on levels of PDGFRα and EGFR. These changes, and the lack of effect of MeHg exposure on TrkC levels, are as predicted from our *in vitro* analyses.

Quantitative data are presented as mean percentage normalized to control animals (n = 3 for each group). Error bars represent ± standard error of the mean. The plus symbol indicates exposure of the cells to the indicated substance.

Analysis of BrdU incorporation revealed that these low levels of MeHg exposure also were associated with statistically significant reductions in the division of O-2A/OPCs *in vivo*. In these experiments, postnatal day 14 (P14) animals were treated as for analysis of receptor levels except that BrdU was administered 2 hr before sacrifice. Sections then were analyzed with anti-BrdU antibodies to identify cells engaged in DNA synthesis and with antibodies to olig2 to identify O-2A/OPCs (as in [148]). The Olig2 is a transcriptional regulator expressed in oligodendrocytes and their ancestral precursor cells (e.g., [50, 149-152]. In white matter tracts of the CNS, $BrdU^+$ cells that express Olig2 are considered to be O-2A/OPCs [153, 154]). In our studies, greater than 90% of all $BrdU^+$ cells in the corpus callosum were also $Olig2^+$. When we analyzed the number of $Olig2^+/BrdU^+$ cells found in the corpus callosum of control and experimental animals (see Materials and Methods for details of analysis), we found a 20% reduction in the number both of total BrdU$^+$ cells and of Olig2$^+$/BrdU$^+$ cells (Figure 11), an outcome in agreement with the results of our *in vitro* studies (Figure 2B).

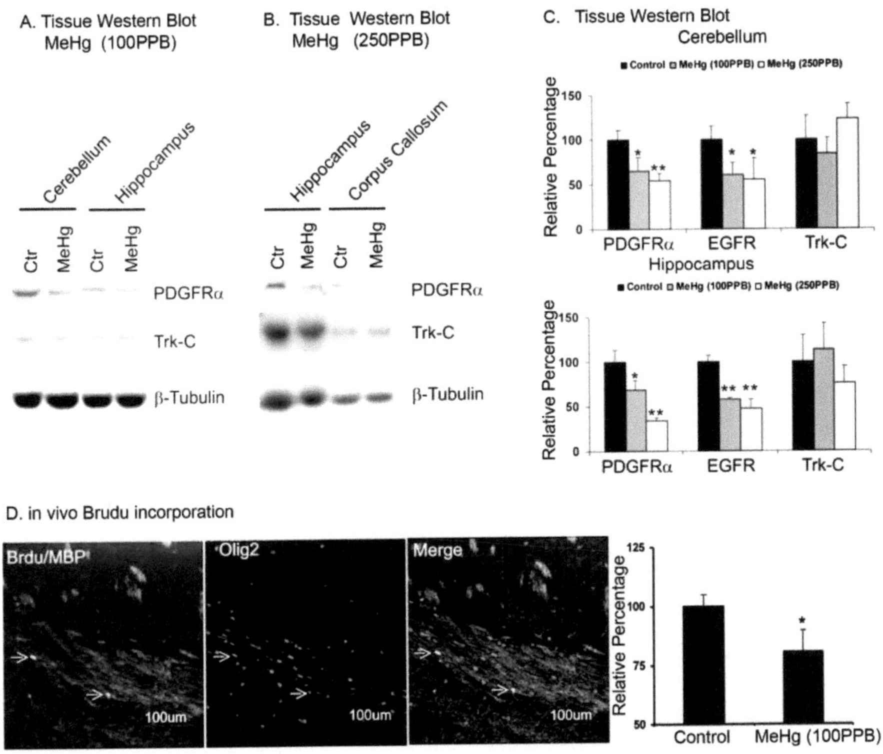

Figure 11. *In Vivo* analysis confirms that developmental exposure to low levels of MeHg is associated with specific reductions in levels of PDGFRα and EGFR, but not of TrkC, and with reductions in O-2A/OPC division. Treatment of SJL mice with 100 or 250 ppb MeHg in the maternal drinking water during gestation and suckling was associated with reductions in levels of PDGFRα in the cerebellum and hippocampus at postnatal day 7 (P7), and in hippocampus and corpus callosum at P21. In contrast, levels of the NT-3 receptor TrkC (which does not appear to be a c-Cbl target) were not reduced in these animals. (A) Animals were treated with 100 ppb MeHg in the drinking water during pregnancy. Pups were sacrificed at 7 days after birth. Analysis of cerebellum and hippocampus showed clear reductions in levels of PDGFRα, but not in TrkC. (B) At P21, enough tissue could also be isolated from corpus callosum for analysis, and was found to have marked reductions in levels of PDGFRα, but not TrkC. (C) Quantitative analysis of receptor levels in P7 mice showed reductions in levels of PDGFRα and EGFR, but not TrkC. Quantitative analysis of changes in receptor expression in tissue from P7 mice. *, $p < 0.05$; **, $p < 0.01$). (D) Analysis of BrdU incorporation in Olig2$^+$ cells reveals a reduction of approximately 20% in the number of double-positive cells in P14 animals born to mothers receiving 100 ppb MeHg in their drinking water beginning 30 days prior to conception and continuing through weaning. The top left figure shows combined labeling with anti-BrdU and anti-MBP antibodies, and the top right figure shows labeling with anti-olig2 antibodies. These images are merged in the bottom left to identify BrdU$^+$/Olig2$^+$ cells. Quantitative analysis of total numbers of double-positive cells reveals that developmental exposure to 100 ppb MeHg via the maternal drinking water is associated with a subtle but significant reduction in the number of O-2A/OPCs engaged in DNA synthesis, consistent with the effects of low-level MeHg exposure *in vitro*.

DISCUSSION

Our studies demonstrate that chemically diverse toxicants converge on activation of a previously unrecognized pathway of cellular regulation that leads from increases in oxidative status to reductions in levels of specific RTKs. Analysis of effects of MeHg on O-2A/OPCs dividing in response to PDGF first demonstrated suppression of PDGF-induced signaling, but no reduction in NT-3–induced phosphorylation of ERK1/2. Further analysis demonstrated that MeHg exposure enhanced degradation of PDGFRα as a consequence of the sequential activation of Fyn and c-Cbl. As predicted by the hypothesis that MeHg exposure activates the redox/Fyn/c-Cbl pathway, exposure to this toxicant was also associated with reductions in levels of EGFR and c-Met (which are c-Cbl targets), but not in levels of TrkC (which is not a c-Cbl target). The redox/Fyn/c-Cbl pathway was also activated by Pb and paraquat, leading to negative modulation of RTK-mediated signaling by regulating receptor degradation and causing reductions in levels of PDGFRα, EGFR, and c-Met, but not of TrkC. Developmental exposure to MeHg was also associated with reduced levels of PDGFRα and EGFR, but not of TrkC, consistent with the hypothesis that this same regulatory pathway is activated in association with *in vivo* toxicant exposure.

The results of our studies are novel in a number of ways, beginning with the identification of a previously unrecognized regulatory pathway activated by chemically diverse toxicants. Although the importance of identifying general principles that apply to chemically diverse toxicants is a widely recognized goal of toxicology research, relatively few such principles have been identified. For example, although toxicants may be classified as hormonal mimetics, mutagens, carcinogens, neurotoxins, and so forth, relatively few mechanistic pathways have been identified on which chemically diverse substances converge.

Our present studies have identified Fyn activation as a common cellular target for the action of chemically diverse toxicants with pro-oxidant activity. Whether oxidative changes are by themselves sufficient to induce sequential activation of Fyn and c-Cbl will be a subject of continued analysis, but existing data make it difficult to imagine a compelling alternative hypothesis to explain our results. Fyn is well established as being activated when cells become more oxidized [97-100], and there is no evidence for any other unifying feature of MeHg, Pb, and paraquat that would cause Fyn activation. Activation of Fyn, and the effects of activation of the Fyn/c-Cbl pathway, were blocked by NAC (which antagonizes oxidative changes in O-2A/OPCs [59]) as effectively as by expression of Fyn-specific RNAi constructs or by pharmacological inhibition of Fyn activity. The NAC protects against physiological stress in two ways, both as an anti-oxidant itself and by providing increased levels of cysteine, the rate-limiting precursor in glutathione biosynthesis (e.g., [115, 116]). The ability of ProCys (which has no intrinsic anti-oxidant properties [129]) to confer similar protection as NAC suggests that it is through their enhancement of glutathione production that these two cysteine pro-drugs exert their protective effects. The relatively small effect of NAC exposure by itself on basal Fyn activity in the experimental conditions used indicates that, at least in these experiments, NAC's protective effect was more likely to be due to protection against increases in oxidative status than due to a direct suppression of

Fyn activity to an extent that would neutralize the activating effects of toxicant exposure. Although increased glutathione levels theoretically could also protect against the effects of toxicants by enabling enhanced cellular export of physiological stressors (reviewed in, e.g., [155, 156]), analysis with Leadmium Green AM (which can detect intracellular Pb in the nM range) revealed no apparent effect of NAC treatment on cellular levels of Pb. Further support for the hypothesis that transport of xenobiotics is not a likely explanation for the protective effects of NAC is also provided by ongoing studies demonstrating that TH and BMP-4 (both of which cause O-2A/OPCs to become more oxidized [59]) also cause activation of Fyn and c-Cbl, with associated reductions in PDGFRα levels (Z. Li and M. Noble, unpublished data). The NAC blocks the effects of TH and BMP on differentiation, and also prevents TH- and BMP-mediated activation of Fyn and c-Cbl ([59]; Z. Li and M. Noble, unpublished data). Changes in intracellular redox state, and the predicted ability to protect with NAC, are the common features linking the activation of Fyn with MeHg, Pb, paraquat, TH, and BMP.

Although Fyn has multiple targets, it seems most likely that activation of c-Cbl provides the explanation for the effects of MeHg on PDGF-mediated signaling. Suppression of c-Cbl activity by expression of DN (70Z) c-Cbl or RNAi protected against the effects of MeHg on cell division and reductions in levels of PDGFRα. Moreover, the induction of PDGFRα ubiquitylation by MeHg, the lack of effects of MeHg on PDGFRα mRNA levels, the rescue of receptor levels by disrupting lysosomal function, and other observations all strongly indicate the importance of c-Cbl regulation in understanding the effects of toxicant exposure. The importance of Fyn in activation of c-Cbl is supported by the ability of expression of Fyn-specific RNAi, or pharmacological inhibition of Fyn activity, to protect against the effects of toxicant exposure. Because Fyn activation in O-2A/OPCs also leads to activation of Rho-GTPase, leading to inhibition of Rho kinase activity [102, 157], we also examined the effects of treatment of cells with the Rho kinase inhibitor Y23762. Although this agent inhibited Rho kinase activity in O-2A/OPCs, it neither protected against nor exacerbated the effects of MeHg on progenitor cell division (as determined by BrdU incorporation). Thus, although it will be of interest to examine the effects of toxicant exposure on other Fyn targets, it currently seems that Fyn-mediated activation of c-Cbl is central to understanding the effects of toxicants on O-2A/OPCs.

The discovery of sequential activation of Fyn and c-Cbl by pro-oxidants provides a new means of integrating the effects of changes in intracellular redox state with the control of the cell cycle. Although the ability of Fyn to be activated by increases in oxidative status [97-100], the functional interaction of Fyn with c-Cbl (e.g., [93-96]), and the regulation of degradation of specific RTKs by c-Cbl (e.g., see [88-90, 96, 106, 107, 134-138]) have all been subjects of study by multiple laboratories, our studies appear to provide the first integration of all of these components into a regulatory pathway of obvious relevance to the regulation of cell function by redox status. This regulatory pathway, summarized in Figure 12, offers a number of clear predictions, some of which have been tested in our present studies. Several studies on different cell types have confirmed our own finding [59] that making dividing cells more oxidized can suppress division and induce differentiation [158-160], and it will be of interest to determine the contribution of the redox/Fyn/c-Cbl pathway in these other cell systems, as well as in modulating other changes in cellular function that have been attributed to increased oxidative status (e.g., [73, 161-166]).

Figure 12. Diagrammatic summary of the Fyn/c-Cbl hypothesis of toxicant convergence. The results of our studies demonstrate a regulatory network in which oxidation causes activation of Fyn. Fyn then phosphorylates c-Cbl. Activation of c-Cbl leads to ubiquitylation of agonist-activated RTKs that are c-Cbl targets, with PDGFRα used here as an example of such a receptor. Reductions in levels of receptor lead to reduced activation of downstream signaling cascades.

It is particularly striking that the changes we observed were seen at environmentally relevant exposure levels for both MeHg and Pb. As many as 600,000 newborn infants in the US each year have cord blood mercury levels greater than 5.8 ppb [46] (i.e., ~30 nM). It is reported that the blood:brain ratio for humans may be as high as 1:5–1:6.7 [167, 168], therefore, *in vivo* levels in brain may be still higher than those we have studied. It is also noteworthy that levels of MeHg exposure at which selective reduction in PDGFRα expression was readily observed *in vivo* were 90% or more lower than exposure levels generally considered to constitute low-to-moderate exposure (e.g., [143-147]). Blood Pb levels may be of concern at levels as low as 10 μg/dl (e.g., [169-176]), which is equivalent to 0.48 μM, but which may be increased to micromolar in the brain by mechanisms relevant to Ca^{2+} transport [171]). Even given equivalence in blood:brain Pb levels, a concentration of 1 μM is equivalent to the approximately 20 μg/dl blood Pb levels known to be associated with cognitive impairment (e.g., [169-177], an exposure level of particular concern in countries where leaded gasoline is still used and in which mean blood lead levels in schoolchildren may be as high as 15 μg/dl [178].

The study of environmentally relevant levels of toxicant exposure is a great challenge, both *in vitro* and *in vivo*, and it may be that analysis of stem and progenitor cell populations will be critical in furthering such analysis. *In vitro*, O-2A/OPCs appear to offer a particularly useful target cell for such studies, in part due to their sensitivity to environmentally relevant exposure levels of toxicants, but also due to the ability to use clonal analysis in quantitative studies on the cumulative effects of small changes in the balance between division and differentiation [179-181]. Such studies have shown that even such potent physiological regulators as TH may only increase the probability of oligodendrocyte differentiation at each progenitor cell cycle from approximately 0.5 to 0.65 [179]. Thus, although their cumulative effects over time may be readily observable, analysis of subtle effects in acute assays may fail to identify important alterations in progenitor cell function. In addition, it will be important to extend analysis on differentiation to other precursor cell populations, as indicated by recent observations that neuronal differentiation of neuroepithelial stem cells may be compromised by MeHg exposure levels as low as 2.5 nM [182]. *In vivo*, the 20% reduction in number of dividing O-2A/OPCs observed in animals exposed to 100 ppb MeHg during development was of particular interest, as such relatively subtle changes might be predicted to reduce myelination in ways that require equally subtle analysis to detect functional outcomes. Analysis of conduction velocity in the auditory system may offer one such analytical tool, and the sensitivity of O-2A/OPCs to toxicant exposure may provide an explanation for the consistency with which increases in ABR latency suggestive of myelination abnormalities are associated with exposure to a variety of toxicants and physiological stressors, including MeHg [75-79], Pb [183, 184]), cocaine [185, 186], and carbamazepine [187].

The general importance of the signaling pathways regulated by Fyn and c-Cbl suggests that the ability of chemically diverse toxicants to converge on this pathway may be of broad relevance to the understanding of toxicant action. Such c-Cbl targets as PDGFRα, EGFR, and c-Met play critical roles in processes as diverse as cell proliferation, survival, and differentiation, cortical neurogenesis, maintenance of the subventricular zone, astrocyte development, development of cortical pyramidal dendrites, motoneuron survival and pathfinding, sympathetic neuroblast survival, and hippocampal neuron neurite outgrowth, as well as having extensive effects on development of kidney, lung, breast, and other tissues (e.g., [60, 61, 130, 188-195]). Indeed, the range of targets of c-Cbl [92, 135] offers a rich fabric of potentially critical regulatory molecules that would be affected by changes in activity of this protein, with the importance of particular proteins being dependent on the cell type and developmental stage under consideration. In addition, Fyn regulation of the Rho/ROCK signaling pathway could be of relevance in understanding toxicant-mediated alterations on such cytoskeletal functions as cell migration, neurite outgrowth, and development of dendritic morphology (e.g., [196-198]). Our studies predict that any toxicant that makes cells and/or tissues more oxidized would activate Fyn, a list that includes substances as chemically diverse as MeHg (e.g., [1-6], Pb [6-9], and organotin compounds [1, 2, 5, 10, 11]), cadmium [12, 13], arsenic [12, 14], ethanol [15, 16], and various herbicides (e.g., paraquat [17, 18], pyrethroids [19-21], and organophosphate and carbamate inhibitors of cholinesterase [22-26]).

In summary, our studies provide a new general principle and evidence of a new regulatory pathway that may be relevant to the understanding of the action of a large number of chemically diverse toxicants and other modulators of oxidative status. Because the outcomes we have identified occur at quite low toxicant exposure levels, they may provide a particularly useful unifying principle for the analysis of toxicant effects. Our present studies, combined with our previous analysis of the central importance of intracellular redox state in modulating progenitor cell function [59], lead to the prediction that any toxicant with pro-oxidant activity will exhibit these effects. Although toxicants of differing chemical structures will also have additional activities, the convergence of small increases in oxidative status on regulation of the redox/Fyn/c-Cbl pathway provides a specific means by which exposure to low levels of a wide range of chemically diverse toxicants might have similar classes of effects on development. Our findings also provide a strategy for rapid identification of such effects by any of the estimated 80,000–150,000 chemicals for which toxicological information is limited or nonexistent, thus enabling a preliminary identification of compounds that would need to be examined *in vivo*. The sensitivity of O-2A/OPCs to environmentally relevant levels of MeHg and Pb provides a great advantage over established cell lines and other such neural cells as astrocytes, for which these low exposure levels may have little effect, and the importance of understanding the effects of toxicants on progenitor cell function provides a direct link between our studies and the broad field of developmental toxicology. In addition, the ability of NAC to protect progenitor cells against the adverse effects of chemically diverse toxicants raises the possibility that this benign therapeutic agent may be of benefit in protecting children known to be at increased risk from the effects of toxicant exposure during critical developmental periods. Finally, the principles indicated by our findings appear likely to have broad applicability in understanding the regulation of cell function by alterations in redox balance, regardless of how they might be generated.

MATERIALS AND METHODS

Cell Isolation, Culture, and Treatment

The O-2A/OPCs were purified from corpus callosum of P7 CD rats as described previously to remove type 1 astrocytes, leptomeningeal cells, and oligodendrocytes [59, 64, 199]. Cells were then grown in DMEM/F12 supplemented with 1-μg/ml bovine pancreas insulin (Sigma, St. Louis, Missouri, US), 100-μg/ml human transferrin (Sigma), 2 mM glutamine, 25-μg/ml gentamicin, 0.0286% (v/v) BSA pathocyte (ICN Biochemicals, Costa Mesa, California, US), 0.2 μM progesterone (Sigma), 0.10 μM putrescine (Sigma), bFGF-2 (10 ng/ml; PEPRO Technologies, London, UK), and PDGF-AA (10 ng/ml; PEPRO) onto poly-l-lysine (Sigma) coated flasks or dishes. Under these conditions, O-2A/OPCs derived from the corpus callosum of P7 rats are predominantly in cell division and do not generate large numbers of oligodendrocytes during the time periods utilized in this analysis.

To generate sufficient numbers of cells for biochemical analysis, cells were expanded through 1–2 passages in PDGF + FGF-2 before replating in the presence of PDGF alone. When cells achieved approximately 50% confluence, MeHg, Pb, or paraquat was added to their medium at concentrations indicated in the text. Doses for the toxicants were chosen on the basis of dose-response curves to identify sublethal exposure levels (unpublished data), as a reflection of blood and brain toxicant levels of these compounds and, where applicable, on the basis of previous reports. All toxicant concentrations examined were confirmed to cause death of less than 5% of cells over the time course of the experiment.

For analysis of the effects of potential inhibitors of toxicant action, cells were exposed to the blocking compound of interest 1 hr before addition of toxicant. The concentrations of inhibitors used are listed as following: 0.5 μM BIM-1 (PKC inhibitor), 0.5 μM PP1/PP2 (Src family kinase inhibitors), and 10 mM NH_4Cl (lysosome inhibitor); and the concentrations of toxicants used are listed as following, except when mentioned specifically: MeHg (20 nM), Pb (1 μM), and paraquat (5 μM).

To examine the degradation of PDGFRα, O-2A/OPCs were treated with MeHg (20 nM) for different durations with or without cycloheximide (CHX; 1 μg/ml) added 1 hr before MeHg. The cells were then collected and lysed for Western blotting. For example, in the multi-toxicant analysis of Figures 8–10, for analysis of PDGFRα, O-2A/OPCs were exposed for 24 hr to MeHg (20 nM), Pb (1 μM), and paraquat (5 μM) for 24 hr in the presence of 0.5 μM bisindolylmaleimide 1 (BIM-1), 0.5 μM of PP1, 1 mM NAC, or 1 mM procysteine, which had been added 1 hr prior to toxicant addition. Cells were lysed for Western blot analysis using anti-PDGFRα (pY742) antibody. The membranes were de-probed and then re-probed with antibody against total PDGFRα and anti–β-tubulin antibody. For analysis of Fyn activity and c-Cbl phosphorylation, progenitors were exposed to MeHg (20 nM), Pb (1 μM), and paraquat (5 μM) for 3–4 hr in the presence of 0.5 μM BIM1, 0.5 μM PP1,or 1 mM NAC (each of which was added 1 hr before addition of toxicant).

Cell Transfection and Luciferase Activity Assay

Cells were deprived of PDGF-AA for 5 hr before re-exposure to PDGF-AA (10 ng/ml) for 1 hr for Western blot or 6 hr for luciferase assays of pathway activation. Transient transfection was performed using Fugene 6 (Roche, Basel, Switzerland) transfection solution according to the manufacturer's protocol. For the luciferase assay, cells seeded in 12-well plates were transfected with a reporter plasmid SRE-Luc (firefly) or NFκB-Luc (firefly) (BD-Clontech, Palo Alto, California, US) and an internal control plasmid pRLSV40-LUC. Analyses of luciferase activity were performed according to the protocol of the Dual Luciferase Assay System (Promega, Madison, Wisconsin, US), which uses an internal control of Renilla luciferase for quantification, and relative light units were measured using a luminometer.

Antibodies and Immunoblotting

Anti-phosphorylated ERK monoclonal, anti-ERK monoclonal, anti-TrkC polyclonal, anti-Fyn polyclonal, anti-EGFR polyclonal, anti–c-Met polyclonal, anti–phosphotyrosine monoclonal, and anti-PDGFRα polyclonal antibodies were obtained from Santa Cruz Biotechnology (Santa Cruz, California, US). Anti–c-Cbl monoclonal antibody was obtained from BD PharMingen (San Diego, California, US). Anti-phosphorylated Akt monoclonal and anti-Akt polyclonal antibodies were obtained from Cell Signaling Technology (Beverly, Massachusetts, US). Anti-phosphorylated PDGFRα polyclonal antibody was obtained from Biosource (Carlsbad, California, US). The cell culture samples were collected and lysed in RAPI buffer, whereas dissected tissue samples were sonicated in RAPI buffer. Samples were resolved on SDS-PAGE gels and transferred to PVDF membranes (PerkinElmer Life Science, Wellesley, Massachusetts, and US). After being blocked in 5% skim milk in PBS containing 0.1% Tween 20, membranes were incubated with a primary antibody, followed by incubation with an HRP-conjugated secondary antibody (Santa Cruz Biotechnology). Membranes were visualized using Western Blotting Luminol Reagent (Santa Cruz Biotechnology). All analyses of signaling pathway components were conducted in the presence of ligand for the receptor pathway under analysis (either PDGF-AA for PDGFRα, NT-3 or TrkC, HGF for c-Met, or EGF for EGFR).

In Vitro BrdU Incorporation Assay

Cell proliferation was assessed by BrdU incorporation and by using the mouse anti-BrdU mAb IgG1 (1:100; Sigma) to label dividing cells. Stained cells on coverslips were rinsed two times in 1 × PBS, counterstained with 4′6-diamidino-2-phenylindole (DAPI; Molecular Probes, Eugene, Oregon, US) and mounted on glass slides with

Fluoromount (Molecular Probes). Staining against surface proteins was performed on cultures of living cells or on cells fixed with 2% paraformaldehyde. Staining with intracellular antibodies was performed by permeabilizing cells with ice-cold methanol for 4 min or by using 0.5% Triton for 15 min on 2% paraformaldehyde–fixed cells. Antibody binding was detected with appropriate fluorescent dye–conjugated secondary antibodies at 10 μg/ml (Southern Biotech, Birmingham, Alabama, US) or Alexa Fluor–coupled antibodies at a concentration of 1 μg/ml (Molecular Probes), applied for 20 min. Anti-BrdU monoclonal antibody was obtained from Sigma.

Intracellular Reactive Oxygen Species Measurement and Analysis of Pb Uptake

Cells were plated in 96-well microplates and grown to about 60% confluence. Prior to treatment, cells were washed twice with Hank's buffered saline solution (HBSS), loaded with 20 μM H2DCFDA (in HBSS 100 μl/well), and incubated at 37°C for 30 min. Cells were then washed once with HBSS and growth medium to remove free probe. Then, fresh growth medium was added and a baseline fluorescence reading was taken prior to treatment. For NAC pre-treatment, NAC was added into media 1 hr before further addition of MeHg, and both compounds remained in the medium during the incubation period with H2DCFDA. Fluorescence was measured in a Wallac 1420 Victor^2 multilabel counter (PerkinElmer) using excitation and emission wavelengths of 485 nm and 535 nm, respectively, at different time courses as indicated in the figures. Results are presented as the value change from baseline by the formula $(Ft_{exp} - Ft_{base})/Ft_{base}$ normalized with the control group, where Ft_{exp} = fluorescence at any given time during the experiment in a give well and Ft_{base} = baseline fluorescence of the same well.

We further determined whether pre-treatment with NAC altered levels of intracellular Pb by analysis with the Leadmium Green AM dye (Molecular Probes), according to the manufacturer's instructions. In five separate experiments, we found no significant difference between O-2A/OPCs treated with 1 μM Pb versus [Pb + NAC] (unpaired t-test), and the values for both Pb-treated samples were 7-fold higher than control values. All of these data strongly support the hypothesis that the major effect of NAC is to antagonize cellular oxidation.

Immunoprecipitation Assay

For the co-immunoprecipitation assay, anti-c-Cbl monoclonal antibody (BD PharMingen) or anti-PDGFRα polyclonal antibody (Santa Cruz Biotechnology) was added to the pre-cleared cell lysates (250 μg of total protein), and the mixtures were gently rocked for 2 hr at 4°C. A total of 30 μl of protein A/G agarose was then added to the mixture followed by rotating at 4°C overnight. The protein A/G agarose was then spun down and washed thoroughly three times. The precipitates were resolved on an 8% SDS-PAGE gel and then were subjected to Western blot analysis using an anti-p-Tyr (for c-Cbl phosphorylation assay) or ubiquitin (for PDGFR ubiquitination assay) antibody (Santa Cruz Biotechnology).

Fyn Kinase Assay

Fyn kinase activity was quantified using the Universal Tyrosine Kinase Assay Kit (Takara, Madison, Wisconsin, and US). The O-2A/OPCs exposed to different treatments were solubilized with an equal volume of the extraction buffer provided with the kit for 15 min, and the resulting lysates were centrifuged at 13,000 × g for 15 min at 4°C; 250 μg of total cell lysates were immunoprecipitated with anti-Fyn antibody (Santa Cruz Biotechnology). Following immunoprecipitation, Fyn immune complexes were washed four times with extraction buffer, and then Fyn kinase activities of each sample were assayed using the kit according to the manufacturer's instructions.

Rho Kinase Assay

Rho kinase activity was quantified using the CycLex Rho-Kinase Assay kit (MBL International, Woburn, Massachusetts, US) as described. Cells were lysed and about 500 μg of total cell lysates were immunoprecipitated with anti-ROCK1 antibody (Sigma), and the precipitates were re-suspended with kinase reaction buffer provided in the kit. Rho kinase activities of each sample were assayed using the kit according to the manufacturer's instructions.

DNA Vector-based RNA Interference

The siRNA target sites were selected by scanning the cDNA sequence for AA dinucleotides via siRNA target finder (Ambion, Austin, Texas, US). Those 19-nucleotide segments that start with G immediately downstream of AA were recorded and then analyzed by BLAST search to eliminate any sequences with significant similarity to other genes. The siRNA inserts, containing selected 19-nucleotide coding sequences followed by a 9-nucleotide spacer and an inverted repeat of the coding sequences plus 6 Ts, were made to double-stranded DNAs with ApaI and EcoRI sites by primer extension, and then subcloned into plasmid pMSCV/U6 at the ApaI/EcoRI site. The corresponding oligonucleotides for the Fyn and c-Cbl RNAi's are listed in Table 1. Several nonfunctional siRNAs, which contain the scrambled nucleotide substitutions at the 19-nucleotide targeting sequence of the corresponding RNAi sequence, were constructed as negative controls. All of these plasmids were confirmed by complete sequencing.

Table 1. Oligonucleotide sequences for siRNAs.

Name	Oligo Sequence
Fyn-RNAi-1113-forward Fyn-RNAi-1113-reverse	5'-GTTIGCTCGACTICTIAAA TICAAGAGA TTIAAGAAGTCGAG-CAAACTI TITI-3' 5'-AATIAAAA AAGTTIGCTCGACTICTIAAA TCTCTIGAA TTIAAGAA-GTCGAGCAAACGGCC-3 '
c-Cbi-RNAi-732-forward c-Cbi-RNAi-732-reverse	5'-GTGCATCCCATCAGTICTG TICAAGAGA CAGAACTGATGGGATG-CACTI TITI-3' 5 '-AA TI AAAA AAGTGCA TCCCA TCAGTICTG TCTCTIGAA CA-GAACTGATGGGATGCACGGCC-3 '
Fyn-Scr-RNAi-1113-forward Fyn-Scr-RNAi-1113-reverse	5'-CCCGGGTTICAATATATTI CTCAAGAGA AAATATATIGAAACC-CGGGTI TITI-3' 5 '-AA TI AAAA AACCCGGGTTICAA TATA TTI TCTCTIGAG AAA TATA TIGAAACCCGGGGGCC-3 '
c-Cbi-Scr-RNAi-732-forward c-Cbi-Scr-RNAi-732-reverse	5'-TTIAAACCCGGCCCTTIGG TICAAGAGA CCAAAGGGCCGGGTTI-AAATI TITI-3' 5'-AATIAAAA AATTIAAACCCGGCCCTTIGG TCTCTIGAA CCAAA-GGGCCGGGTITAAAGGCC-3 '

Oligonucleotide sequences for the Fyn and c-Cbl RNAi's used in experiments of Figures 6 and 8. In addition, nonfunctional siRNAs, which contain the scrambled nucleotide substitutions at the 19-nucleotide targeting sequence of the corresponding RNAi sequence, were constructed as negative controls. All plasm ids were confirmed by complete sequencing.

Viral Packaging, Cell Infection, and Selection

The pJEN/neo-HA-70z-c-Cbl plasmids were generously provided by Dr. Wallace Langdon. The pBabe (puro)-HA-70z-c-Cbl plasmids were constructed by transferring the BamH1-digested HA-70z-c-Cbl from pJEN/neo-HA-70z-c-Cbl into the BamH1 digested pBabe (puro) vector. The pBabe(puro)-HA-70z-c-Cbl, pMSCV/U6-Fyn-RNAi, pMCV/U6-c-Cbl-RNAi, and the corresponding scrambled RNAi plasmids and the empty plasmids were transfected into Pheonix Ampho cells by Fugene 6 (Roche) transfection solution according to the manufacturer's protocol. Twenty-four hours after transfection, medium was changed to DMEM/F12 (SATO, but with no TH) supplemented with 10 ng/ml PDGF-AA and bFGF. Virus supernatant was collected 48 hr post-transfection, filtered through 0.45-μm filter to remove non-adherent cells and cellular debris, frozen in small aliquots on dry ice, and stored at −80°C. Twenty-four hours prior to infection, O-2AOPCs were seeded. The following day, the culture medium was aspirated and replaced with virus supernatant diluted 1:1 in the O-2A growth media. Medium was then changed into O-2A/OPC growth medium after 8 hr or overnight. Twenty-four hours after infection, the cells were collected by trypsinization and reseeded in the selective medium (growth medium + 200 ng/ml puromycin). By the next day, all noninfected cells were floating and presumably dead or dying. The infected cells were allowed to proliferate for 2 days, and then collected and re-seeded for the following experiments.

RNA Isolation and Real-time RT-PCR

Total RNA was isolated using TRIZOL reagent (Invitrogen, Carlsbad, California, US) according to the manufacturer's protocol. A total of 1 μg of RNA was subjected to reverse transcription using Superscript II (Invitrogen). The reactions were incubated at 42°C for 50 min. The FAM-labeled probe mixes for rat PDGFRα and Fyn, and the VIC-labeled GAPDH probe mix were purchased from Applied Biosystems (Foster City, California, US). For multiplex real-time PCR, reactions each containing 5 μl of 10-fold–diluted reverse transcription product, 1 μl of interest gene probe mix, 1 μl of GAPDH probe mix, and 10 μl of TaqMan Universal PCR Master Mix were performed on an iCycler iQ multicolor real-time PCR system (Bio-Rad, Hercules, California, US) and cycling condition was 50°C for 2 min and 95°C for 10 min, followed by 40 cycles of 95°C for 15 sec and 60°C for 1 min. Each sample was run in triplicate. Data were analyzed by iCycler iQ software (Bio-Rad).

Clonal Analysis

The O-2A/OPCs purified from P7 rat optic nerve were plated in poly-L-lysine–coated 25 cm^2 flasks at clonal density with DMEM medium in the presence of 10 ng/ml PDGF as previously described [59, 64, 199]. After 24 hr recovery, cells were treated with different toxicants, each for 3 days, until visual inspection and immunostaining was performed. The NAC was added 1 hr before exposure to other toxicants for NAC pretreatment, and NAC co-exists throughout the culture period. The numbers of O-2A/OPCs and oligodendrocytes in each clone were determined by counting under fluorescent microscope. The 3-dimensional graph shows the number of clones containing

O-2A/OPC cells and oligodendrocytes. Experiments were performed in triplicate in at least two independent experiments.

Animal Treatment

Six-week-old female SJL mice were treated with MeHg in their drinking water at a concentration of 100 or 250 ppb for 30–60 days prior to mating, and then throughout pregnancy and gestation. This is a level of treatment that is 75–90% below levels generally considered to be low to moderate and is below levels that have been associated with gross defects in adult or developing animals (e.g., [143-147]).

The exposure levels used in our studies were first determined as candidate exposures from the results of two different previous studies on the relationship between MeHg exposure and levels of toxicant in the brain. Studies by Weiss and colleagues [143] demonstrated that mice exposed to MeHg in their drinking water for up to 14 month have brain mercury levels roughly equivalent to that in the water. In these studies, mice exposed to MeHg in their drinking water from conception at a concentration of one part per million (ppm) had brain levels of MeHg of 1.20 mg/kg (i.e., ppm) at 14 month of age, whereas those exposed to MeHg at a concentration of 3 ppm had brain levels of 3.66 mg/kg at this age. It has also been shown, however, that mercury levels in the brain of pre-weanling animals exposed to MeHg via the mother's drinking water throughout gestation and suckling drop rapidly to one-fifth of the levels found at birth, presumably due to reduced MeHg transfer in milk [200]. As an estimated 300,000–600,000 infants in the US have blood cord mercury levels of 5.8 μg/l or more [46], and because the human brain concentrates MeHg 5–6.The 7-fold over the concentration occurring in the bloodstream, our goal was to achieve postnatal brain mercury levels of 30 ppb (i.e., ng/g) or less.

In practice, we found that exposure of female mice to MeHg in their drinking water at a concentration of 250 ppb prior to conception, and maintenance of this exposure during suckling, was associated with brain mercury levels in the offspring (examined at P14) of 50 ng/g, a fall that was precisely in agreement with predictions based on prior studies on the fall of mercury levels occurring during this period in suckling mice [200]. In offspring of dams exposed to MeHg at a concentration of 100 ppb in the drinking water, brain mercury was below the levels of detection of the Mercury Analytical Laboratory of the University of Rochester Medical Center. The exposure levels of 100 and 250 ppb are 75–90% below what has otherwise been considered to be low-dose exposure in mice.

Tissue Preparation

At the time of sacrifice, mice were anesthetized using Avertin (tribromoethanol, 250 mg/kg, 1.2% solution; Sigma) and were perfused transcardially with 4% paraformaldehyde in phosphate buffer (pH 7.4) following the removal of the blood by saline solution washing. The brains were removed and stored in 4% paraformaldehyde for 1 day, and then changed to 25% sucrose in 0.1 M phosphate buffer. Brains were cut coronally as 40 μm sections with a sliding microtome (SM/2000R; Leica, Heidelberg, Germany) and stored at −20°C in cryoprotectant solution (glycerol, ethylene glycol, and 0.1 M phosphate buffer[pH 7.4], 3:3:4 by volume). All animal experiments were conducted in accordance with National Institutes of Health guidelines for the humane use of animals.

In Vivo BrdU Incorporation Assay, BrdU Labeling, and olig2 Co-labeling for BrdU Detection

To analyze DNA synthesis *in vivo*, mice were injected with a single dose of 5-BrdU (50 mg/kg body weight), dissolved in 0.9% NaCl, filtered (0.2 μm), and applied intraperitoneally 2 hr prior to perfusion. After removal and sectioning of brains, 40 μm free-floating sections were incubated for 2 hr in 50% formamide/2× SSC (0.3 M NaCl and 0.03 M sodium citrate) at 65°C, rinsed twice for 5 min each in 2× SSC, incubated for 30 min in 2N HCl at 37°C, and rinsed for 10 min in 0.1 M boric acid (pH 8.5) at room temperature. Several rinses in TBS were followed by incubation in TBS/0.1% Triton X-100/3% donkey serum (TBS-plus) for 30 min. Sections were then incubated with monoclonal rat anti-BrdU antibody (1:2,500; Harlan Sera-lab, Loughborough, UK) and polyclonal rabbit anti-Olig2 (a generous gift from Dr. David H. Rowitch) in TBS-plus for 48 hr at 4°C. Sections were rinsed several times in TBS-plus and incubated for 1 hr with donkey anti-rat FITC and donkey anti-rabbit TRITC (Jackson ImmunoResearch Laboratories, West Grove, Pennsylvania, US). After several washes in TBS, sections were mounted on gelatin-coated glass slides using Fluoromount-G mounting solution (Southern Biotech).

Quantification of $BrdU^+$ cells was accomplished with unbiased counting methods by confocal microscopy. The BrdU immunoreactive nuclei were counted in one focal plane to avoid oversampling. In corpus callosum, $BrdU^+$ cells were counted in every sixth section (40 μm) from a coronal series between interaural AP + 5.2 mm and AP + 3.0 mm in the entire extension of the rostral and medial part of the corpus callosum. Quantitative data are presented as mean percentage normalized to control animals. Error bars represent ± the standard error of the mean.

Images, Data Processing, and Statistics

Digital images were captured using a confocal laser scanning microscope (Leica TCS SP2). Photomicrographs were processed on a Macintosh G4 and assembled with Adobe Photoshop 7.0 (Adobe Systems, Mountain View, California, US). Unpaired, two-tailed Student t-test was used for statistical analysis.

CONCLUSION

Discovering general principles underlying the effects of toxicant exposure on biological systems is one of the central challenges of toxicological research. We have discovered a previously unrecognized regulatory pathway on which chemically diverse toxicants converge, at environmentally relevant exposure levels, to disrupt the function of progenitor cells of the developing CNS. We found that the ability of low levels of methylmercury, lead, and paraquat to make progenitor cells more oxidized causes activation of an enzyme called Fyn kinase. Activated Fyn then activates another enzyme (c-Cbl) that modifies specific proteins—receptors that are required for cell division and survival—to initiate the proteins' degradation. By enhancing degradation of these receptors, their downstream signaling functions are repressed. Analysis of developmental exposure to methylmercury provided evidence that this same pathway is activated *in vivo* by environmentally relevant toxicant levels. The remarkable sensitivity of progenitor cells to low levels of toxicant exposure, and the discovery of the redox/Fyn/c-Cbl pathway as a mechanism by which small increases in oxidative status can markedly alter cell function, provide a novel and specific means by which exposure to chemically diverse toxicants might perturb normal development. In addition, the principles revealed in our studies appear likely to have broad applicability in understanding the regulation of cell function by alterations in redox balance, regardless of how they might be generated.

KEYWORDS

- **Bromodeoxyuridine**
- **Central nervous system**
- **Chemically diverse toxicants**
- **Epidermal growth factor**
- **N-acetyl-L-cysteine**
- **Oligodendrocyte-type-2 astrocyte**
- **Platelet-derived growth factor receptor-α**
- **Receptor tyrosine kinase**
- **Rho kinase activity**
- **Toxicants**

AUTHORS' CONTRIBUTIONS

Zaibo Li, Chris Pröschel, and Mark Noble conceived and designed the experiments. Zaibo Li and Tiefei Dong performed the experiments. Zaibo Li, Tiefei Dong, Chris Pröschel, and Mark Noble analyzed the data. Mark Noble wrote the chapter.

ACKNOWLEDGMENTS

It is a pleasure to acknowledge the insightful discussions with our colleagues on these studies, and in particular discussions with Margot Mayer-Pröschel, Lisa Opanashuk, Hartmut Land, and Dirk Bohmann.

COMPETING INTERESTS

The authors have declared that no competing interests exist.

Chapter 6

Estrogen-like Activity of Seafood from Chemical Contaminants

Sonia Garritano, Barbara Pinto, Marco Calderisi, Teresa Cirillo, Renata Amodio-Cocchieri, and Daniela Reali

INTRODUCTION

A wide variety of environmental pollutants occur in surface waters, including estuarine and marine waters. Many of these contaminants are recognized as endocrine disrupting chemicals (EDCs) which can adversely affect the male and female reproductive system by binding the estrogen receptor and exhibiting hormone-like activities. In this study the estrogenic activity of extracts of edible marine organisms for human consumption from the Mediterranean Sea was assayed.

Marine organisms were collected in two different areas of the Mediterranean Sea. The estrogenic activity of tissues was assessed using an *in vitro* yeast reporter gene assay (*S. cerevisiae* RMY 326 ER-ERE). Concentrations of polychlorinated biphenyls (PCBs) (congeners 28, 52, 101, 118, 138, 153, 180) in fish tissue was also evaluated.

Thirty-eight percent of extracts showed a hormone-like activity higher than 10% of the activity elicited by 10 nM 17b-estradiol (E2) used as control.

Total PCB concentrations ranged from 0.002 up to 1.785 ng/g wet weight (w wt). Chemical analyses detected different levels of contamination among the species collected in the two areas, with the ones collected in the Adriatic Sea showing concentrations significantly higher than those collected in the Tyrrhenian Sea ($p < 0.01$).

The more frequent combination of chemicals in the samples that showed higher estrogenic activity was PCB 28, PCB 101, PCB 153, PCB 180.

The content of PCBs and estrogenic activity did not reveal any significant correlation.

A broad range of environmental pollutants take place in surface waters, including estuarine and marine waters. Among of them are recognized as EDCs and include substances of very different chemical property such as therapeutic agents, alkylphenols, dioxins, pesticides, plasticizers, and surfactants. Of particular concern are the PCBs because of their ubiquity and their lipophilic and persistent nature (they tend to accumulate within the fat and tissue of animals and humans). These compounds are widely spread in the environment despite most countries have banned their production. They continue to be detected in the ecosystem including marine habitat because of deliberate or accidental dumping or through disposal of goods containing them. There is evidence that these chemicals can adversely affect the male and female reproductive system by binding the estrogen receptor and exhibiting estrogenic or anti-estrogenic activities [1-3]. The susceptibility of target tissues is related to the stage of development, the immune status of the individual and the cumulative exposure dose. Twelve PCBs congeners, PCBs 77, 81, 126, and 169 (non-ortho PCBs), PCBs 105, 114, 118, 123, 156, 157, 167, and 189 (ortho PCBs), were identified by the WHO as having dioxin-like properties, and seven congeners (PCBs 28, 52, 101, 118, 153, 138, 180) were identified by the International Council for the Exploration of the Seas (ICES) as markers of the degree of contamination. These compounds are the second greatest cause of fish advisory, according to United States Environmental Protection Agency (USEPA) [4]. The main source of exposure to PCBs for humans is represented by food, specifically of animal origin [5-7]. In epidemiological studies, PCBs have been associated with immunotoxicity [8] and neurobehavioral deficits have been reported in children prenatally exposed to PCBs and through mother's milk [9] or ascribed to the consumption of PCB contaminated food including fish accumulating these substances directly from the surrounding environment [10-12].

In order to evaluate the exposure of edible species of the Mediterranean Sea, the estrogenic activity of extracts of edible marine organisms from two areas of the Mediterranean Sea was assayed and largely consumed in the Italian diet to complex mixtures of xenobiotics that may exhibit estrogenic activities. We also evaluated the concentration of ICES-7 PCBs considered as target compounds in marine pollution studies.

MATERIALS AND METHODS

Sampling

Fish, crustaceans, and cephalopods were collected directly from professional fishing in two areas of the Mediterranean Sea, the Adriatic Sea within 40 miles S-E from the Pescara port, and Tyrrhenian Sea within 50 miles S-W from the Naples port, respectively, from June to July, 2004. The organisms collected in the Adriatic Sea were Blue-mouth (*Helicolenus dactylopterus*), Broad-tail shortfin squid (*Illex coindetii*), Red mullet (*Mullus barbatus*), European hake (*Merluccius merluccius*), Fork beard (*Phycis phycis*), Deepwater rose shrimp (*Parapenaeus longirostris*), Atlantic mackerel (*Scomber scombrus*); in Tyrrhenian Sea were Common squid (*Loligo vulgaris*), Red mullet (*Mullus barbatus*), Common gray mullet (*Mugil cephalus*), Fork beard (*Phycis phycis*), Blue whiting (*Micromesistius poutassou*), Common octopus (*Octopus vulgaris*), Gilt-head seabream (*Sparus aurata*), Common cuttlefish (*Sepia officinalis*), Atlantic mackerel (*Scomber scombrus*). These organisms, that represent different trophic positions in the marine environment, were selected because they are abundant, widely distributed in the Mediterranean area and available all over the year and also because they represent a largely consumed seafood. The specimens, all of commercial size, were wrapped in aluminum foil, then immediately refrigerated and transported to the laboratory.

Analytical Sample Preparation

Each species was classified; the length and weight of each specimen were measured and recorded. The specimens with weight > 200 g were individually analyzed, those with weights < 200 g were pooled, obtaining 20 analytical samples from the Adriatic Sea and 22 from the Tyrrhenian Sea. The edible part of the marine organisms was selected, homogenized, and subsequently lyophilized.

Fat Extraction

Fat was cold-extracted from lyophilized tissues with petroleum ether/acetone (1:1, v/v). The extract was passed through a glass tube packed with anhydrous sodium sulfate and then evaporated by rotavapor (40°C and low pressure) and the lipid residue was weighed.

Chemical Analysis

The cleanup of fat extracts (50 mg) was carried out on Extrelut-NT3/Extrelut-NT1 cartridges (Merck KGaA—Darmstadt, Germany) with the addition of 0.36 g of C-18 Isolute (40–60 mesh Merck KGaA Darmstad, Germany) and eluted with acetonitrile. The extracts were concentrated under vacuum at 40°C, cleaned up by column adsorption-chromatography on Florisil (60/100 mesh—Supelco Bellefonte, PA USA) activated at 130°C for 2 hr and eluted with 30 ml n-hexane added in 5 ml aliquots. The eluate was concentrated to a small volume (<1 ml) by evaporation at room temperature under a flow of N_2, and 1 ml isooctane was added as a keeper. Seven PCB congeners (IUPAC nn. 28, 52, 101, 118, 153, 138, and 180) were detected in seafood according to the analytical method of Italian Public Health laboratories [13].

For PCBs quantification, samples were injected into a capillary gas chromatographer with electron capture detector (GC-ECD) with Temperature program 60°C for 2 min, increasing of 10°C/min to 170°C stay for 2 min, increasing of 2°C/min to 210°C, increasing for 10°C/min to 260°C. The gas chromatography mass spectrometry (GC-MS) was used for their confirm. The internal standard solution (PCB 209) was added to the extract before injecting. The evaluation of PCB concentrations in the samples was carried out by comparison with a calibration curve obtained by a pool of the seven ICES PCB congeners: PCB 28 (2, 4, 4′ tri-chlorobiphenyls), PCB 52 (2, 2′, 5, 5′ tetra-chlorobiphenyls), PCB 101 (2, 4, 5, 2′, 5′ penta-chlorobiphenyls), PCB 118 (2, 4, 5, 3′, 4′ penta-chlorobiphenyls), PCB 138 (2, 2′3, 4, 4′, 5′ hexa-chlorobiphenyls), PCB 153 (2, 2′, 4.4′, 5, 5′ hexa-chlorobiphenyls), and PCB 180 (2, 2′, 3, 4, 4′, 5, 5′ hepta-chlorobiphenyls). All of the compounds (95–99% pure) were purchased from Dr. Ehrenstorfer (GmbH, Augsburg, Germany). In the analytical conditions applied, the detection limits were: 0.002 ng/g w wt for the PCBs nn 180 and 138; 0.003 ng/g (w wt) for the PCB n 153; 0.005 ng/g (w wt) for the PCBs nn 52, 101, 118, and 0.008 ng/g (w wt) for the PCB n 28. The mean recovery obtained by PCB standard spiked samples was 70 ± 7%. Total PCB levels were calculated as the sum ($\sum$) of all the seven determined congeners.

Yeast Strain

Estrogenic activity of organic extracts of seafood (200 g w wt) and standard PCBs was tested by *S. cerevisiae* yeast strain (RMY326 ER-ERE) containing the human estrogen receptor (hERa) and a Xenopus laevis vitellogenin estrogen-responsive element (ERE) linked to a reporter gene lac Z encoding for the enzyme β-galactosidase.

Plasmid pG/ER(G) was used as the yeast expression vector for ERα and PUCΔSS-ERE as its β-galactosidase reporter plasmid [14, 15]. The sensitivity and specificity of the yeast strain were previously assessed using 17b-estradiol (E2), diethylstilbestrol (DES), and other natural and synthetic chemicals [16, 17].

MEDIUM

A synthetic drop-out selective medium lacking uracil and thryptophane to maintain plasmid selection was prepared by adding 0.67 g yeast nitrogen base, 2% glucose, 10 ml a stock aminoacids solution (30 mg L-isoleucine, 150 mg L-valine, 20 mg L-arginine-HCl, 30 mg L-lysine-HCl, 20 mg L-methionine, 50 mg L-phenylalanine, 200 mg L-threonine, 30 mg L-tyrosine in 100 ml water), 1 ml a stock L-hystidine-HCl solution (200 mg in 100 ml water), 1 ml a stock L-leucine solution (1 gr in 100 ml water) and an adenine hemisulfate solution (200 mg in 100 ml water) to 90 ml water endotoxin-free cell cultures (Sigma). Bacto-Agar (DIFCO) (3 g) was used for the solid media.

Yeast Assay

Yeast cultures were incubated at 28°C for 7 hr by continuously shaking on an orbital shaker (210 rpm) in 1 ml of selective medium. The cultures were then diluted in fresh medium to an optical density of 0.1 (OD_{600} nm) and incubated at 30°C for 17 hr (overnight) in the absence or presence of 17b-estradiol (positive control), solvent (negative control), pure chemicals, and organic extracts. Dimethylsulphoxide (DMSO) was used as solvent. Solutions of standard PCB congeners and extracts were evaporated under a gentle flow of nitrogen and the pellet was resuspended in 10 μl DMSO.

The PCB standards were tested at 5 μg/ml. Congeners 138 and 153 were assayed at 0.05 μg/ml because at higher concentrations strongly inhibited yeast cells growth. The samples were added to the yeast culture so that the concentration of solvent DMSO did not exceed 1% (v/v).

β-galactosidase Assay

Yeast cells were harvested by centrifugation and the pellet was resuspended in 1 ml of Z-buffer (60 mM $Na_2HPO_4{\cdot}7H_2O$, 40 mM $NaH_2HPO_4{\cdot}H_2O$, 10 mM KCl, 1 mM $MgSO_4{\cdot}7H_2O$, and 35 mM 2-mercaptoethanol, pH 7.0). After centrifugation, the pellet was resuspended in 150 μl of Z-buffer. The cells were permeabilized by adding 50 μl dichloromethane, 20 μl 0.1% SDS, 5–50 μl resuspended cells (out of 150 μl) plus Z-buffer for a total of 150 μl including cells, followed by vortexing for 10 sec. The enzymatic reaction was started by adding 700 μl 2 mg/ml o-nitrophenyl β-D-galactopyranoside (ONPG) to the Z-buffer and incubating at 30°C for 5–10 min. The reaction was terminated by the addition of 500 μl 1 M Na_2CO_3 and the absorbance at 420 nm (OD_{420}) of the sample was measured. The β-gal activity was normalized to the number of cells assayed and expressed as Miller units using the following formula [18]:

$$\beta\text{-gal units (M.U.)} = (1{,}000 \times OD_{420})/(t \times V \times OD_{600})$$

t = length of incubation (min) V = volume of culture used in the assay (ml)

The β-galactosidase activity induction elicited by estradiol, the reference estrogen, showed a sigmoid shape (Figure 1) and adequately fitted a linear dose-response relationship after Log transformation of Miller Units.

The β-galactosidase activity of the samples was expressed as a percentage of the activity obtained with 10 nM E2 (positive control) [19].

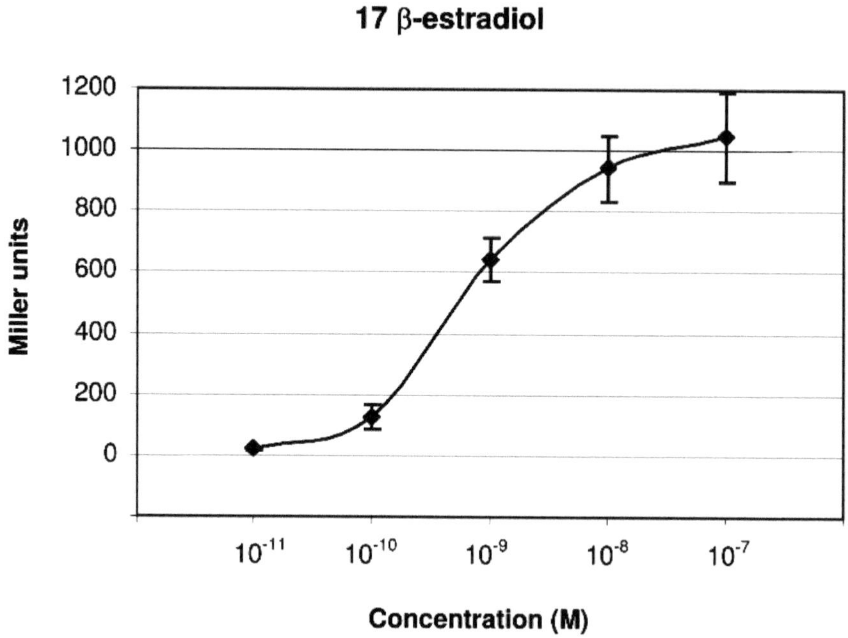

Figure 1. Dose-response curve of E2 concentrations. Data represent the mean ± S.D. of 16 independent experiments.

Statistical Analysis

Chemo-analytical and biological results were processed using principal components analysis (PCA) [20, 21] and partial least square regression (PLS) [22]. The PCA is a rotation of the original data in order to orientate the first new axis in the direction of the maximum explained variance. The second new axis will be oriented perpendicularly to the first, in order to maximize the residual variance, and so on until all the information of the system is explained. Using PCA is possible to:

1. estimate correlation between variables
2. display objects (finding outliers, clusters, ...)
3. summarize the "information" of a system
4. reduce dataset dimension
5. find mean aspect of the system

The PLS regression is a biased regression method that allow to obtain very stable models and it can be used also when the ratio cases/predictors is smaller then one and when there are predictors that are strongly correlated.

This method uses predictors PCs and responses to find the couple that as got the best correlation and goes on using other PCs until there is still usable "information." At the end the number of PLS components to be used is the one that let maximize the explained variance in prediction.

Expected estrogenicity according to the content of individual PCBs measured in each sample was calculated from the chemical data for each sample by using the principle of concentration additivity and relative potencies of the various chemicals as determined with the yeast estrogen screen.

The use of Log transformation of PCB and β-galactosidase activity values normalized their distribution and stabilized the variance allowing the use of parametric methods. The Statgraphics Plus statistical package was used (Magnugistic, Rockville MD, USA).

RESULTS

PCBs Assessment

Twenty samples of marine organisms from the Adriatic Sea and 22 from the Tyrrhenian Sea, mainly belonging to species, *M. barbatus*, *S. scombrus*, *P. phycis*, *M. merluccius*, and *O. vulgaris* were analyzed for the presence of seven reference PCBs, namely PCB 28, PCB 52, PBC 101, PCB 118, PCB 138, PCB 153, and PCB 180. They were chosen because considered to be suitable indicators of industrial marine pollution by International Agencies. Forty-one percent of 294 chemical determinations gave negative results (under the detection limit). Total PCB concentrations ranged from 0.002 up to 1.785 ng/g w wt. Species from the Adriatic Sea were more contaminated than those collected in the Tyrrhenian Sea (Table 1) and the analysis of variance showed that this difference was statistically significant (p = 0.002) (Figure 2).

Table 1. The PCB concentrations (ng/g wet weight) determined in marine species collected in Adriatic and Tyrrhenian Sea.

Adriatic Sea samples	Fat (g%)	PCB 28	PCB 52	PCB 101	PCB 118	PCB 138	PCB 153	PCB 180	PCB
1. H. dactilopterus	0.3	ND[a]	ND	ND	ND	73	8	n.d	81
2. I. coindetii	0.5	ND	ND	ND	ND	37	42	109	188
3. F. coindetii	0.5	ND	ND	ND	ND	26	41	72	139
4. M. barbatus	1.7	ND	ND	ND	76	184	264	208	732
5. M. barbatus	1.4	ND	ND	15	ND	123	168	137	443
6. M. merluccius	0.4	46	4	7	10	21	27	11	126
7. M. merfuccius	0.8	41	4	10	12	37	48	52	204
8. M. merluccius	0.9	44	5	18	23	69	82	29	270
9. M. merfuccius	0.4	ND	39	14	ND	56	66	21	196
10. M. merluccius	0.4	ND	ND	ND	ND	ND	9	ND	9

Table 1. *(Continued)*

Adriatic Sea samples	Fat (g%)	PCB 28	PCB 52	PCB 101	PCB 118	PCB 138	PCB 153	PCB 180	PCB
11. P. phycis	0.6	15	ND	ND	55	ND	ND	ND	70
12. P. phycis	0.6	ND	ND	ND	ND	6	7	12	25
13. P. phycis	0.8	ND	1	ND	10	ND	6	ND	17
14. P. phycis	0.5	ND	1	ND	8	ND	5	ND	14
15. P. phycis	0.7	ND	ND	ND	ND	6	8	12	26
16. P. fongirostris	0.4	ND	2	ND	ND	2	4	8	16
17. S. scombrus	0.5	244	ND	135	ND	497	196	92	1164
18. S. scombrus	0.6	ND	ND	ND	ND	ND	78	400	478
19. S. scombrus	0.6	1469	191	ND	ND	ND	ND	125	1785
20. S. scombrus	0.7	ND	ND	ND	ND	320	311	6	637
Tyrrhenian Sea samples	**Fat (g%)**	**PCB 28**	**PCB 52**	**PCB 1 0 1**	**PCB 118**	**PCB 138**	**PCB 153**	**PCB 180**	**PCB**
1. L. vulgaris	0.7	12	ND[a]	ND	ND	ND	ND	ND	12
2. L. vulgaris	0.6	18	ND	ND	ND	1	ND	ND	19
3. M. barbatus	1.3	5	ND	ND	1	ND	5	2	13
4. M. barbatus	1.2	ND	ND	ND	ND	ND	1	1	2
5. M. barbatus	1.4	10	ND	15	ND	ND	ND	1	26
6. M. cephalus	0.6	7	1	1	ND	ND	ND	ND	9
7. P. phycis	0.5	6	ND	1	3	ND	6	2	18
8. P.phycis	0.7	17	6	ND	ND	ND	2	ND	25
9. P. phycis	1.5	ND	ND	1	ND	ND	2	4	7
10. P. phycis	0.8	3	ND	ND	ND	ND	3	ND	6
11. M. poutassou	0.4	ND	ND	13	9	1	2	1	26
12. M. poutassou	0.4	8	3	27	25	15	ND	ND	78
13. 0. vulgaris	0.6	2	ND	1	ND	ND	ND	ND	3
14. 0. vulgaris	0.7	5	ND	ND	ND	ND	ND	ND	5
15. 0. vulgaris	0.6	17	ND	2	3	ND	1	1	24
16. 0. vulgaris	0.5	5	ND	ND	ND	ND	ND	ND	5
17. S. au rata	1.2	1	2	11	16	ND	ND	ND	30
18. S. officina/is	0.6	ND	ND	ND	ND	ND	ND	3	3
1 9. S. officina/is	0.6	4	ND	ND	ND	ND	3	2	9
20. S. scombrus	0.4	4	ND	3	ND	3	ND	ND	10
21. S. scombrus	0.3	7	ND	3	10	ND	3	2	25
22. S. scombrus	0.8	5	ND	16	ND	4	6	3	34

[a]ND detection limit

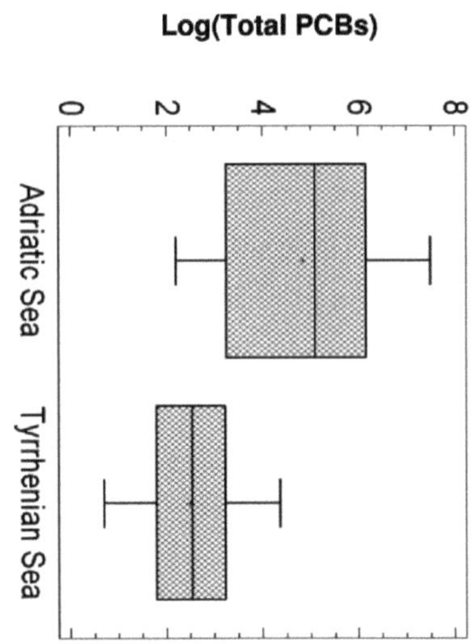

Figure 2. Differences in total PCBs content (Log) in samples from the Adriatic and the Tyrrhenian Sea.

Individual PCB congeners distribution also differed depending on the area of fishing (Figure 3). The species collected in the Adriatic Sea were more often contaminated with hexa- and heptachlorobiphenyls, and this is in accord to studies from other authors [23], whereas samples collected in the Tyrrhenian Sea were mostly contaminated by PCB 28.

Multivariate statistical analysis (PCA) showed different groups of correlation between the PCBs in the two datasets. The first two principal components are able to explain 64% of variance in the Adriatic dataset, and 63% of variance in the Thyrrenian dataset (Figures 4 and 5).

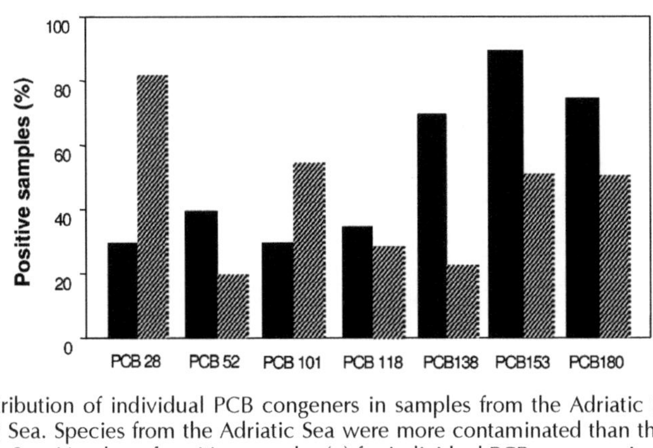

Figure 3. Distribution of individual PCB congeners in samples from the Adriatic Sea (■) and the Tyrrhenian (■) Sea. Species from the Adriatic Sea were more contaminated than those collected in the Tyrrhenian Sea. Number of positive samples (n) for individual PCB congener in the Adriatic Sea: PCB 28 (n = 6), PCB 52 (n = 8), PCB 101 (n = 6), PCB 118 (n = 7), PCB 138 (n = 14), PCB 153 (n = 18), PCB 180 (n = 15). Number of positive samples (n) for individual PCB congener in the Tyrrhenian Sea: PCB 28 (n = 18), PCB 52 (n = 4), PCB 101 (n = 12), PCB 118 (n = 7), PCB 138 (n = 5), PCB 153 (n = 11), PCB 180 (n = 11).

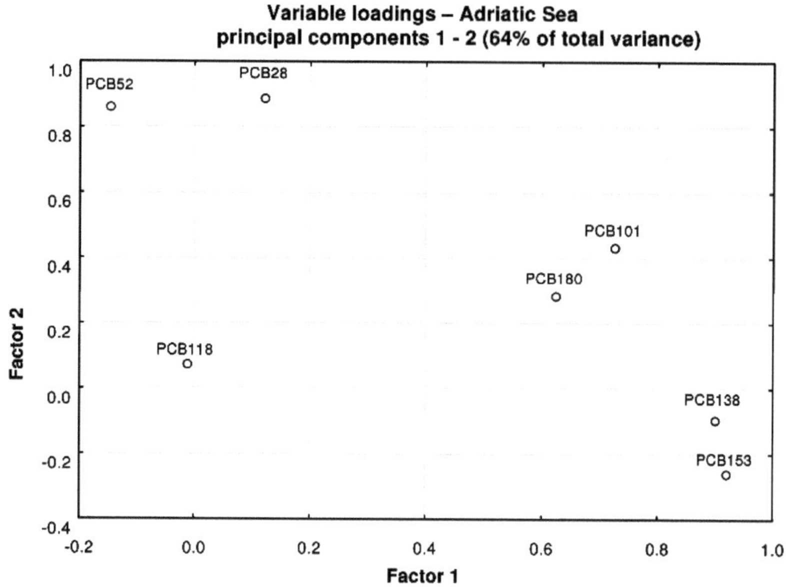

Figure 4. Principal components analysis loading plot of PCBs contamination in samples collected in the Adriatic Sea.

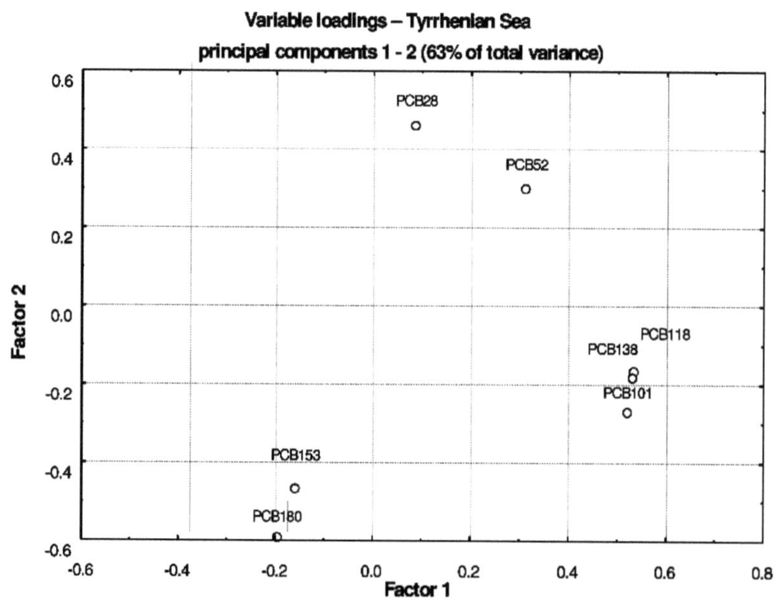

Figure 5. Principal components analysis loading plot of PCBs contamination in samples collected in the Tyrrhenian Sea.

Taking into account the species, the multifactor analysis of variance showed that the difference in PCBs contamination observed in the two seas is mainly due to the PCBs content of *S. scombrus* and *M. barbatus*. These species showed for samples collected in the Adriatic Sea, higher concentrations of total PCBs (Table 1), accordingly to other authors [24].

β-galactosidase Activity

The β-galactosidase activity elicited by seafood extracts is reported in Figures 6 and 7. Thirty-eight percent of seafood samples showed ER-mediated responses higher than 10% E2. The greatest response measured was 42.96% of the activity elicited by the natural hormone. Tyrrhenian samples were more frequently positive than Adriatic ones (50% vs. 25%), although the analysis of variance did not denote any significant differences between the agonistic activity ($p = 0.13$). *S. scombrus*, *O. vulgaris*, *P. phycis*, *M. barbatus* were the more frequently inducing species in Tyrrhenian samples, while for the Adriatic samples the estrogen-like activity was mainly due to *M. merluccius*.

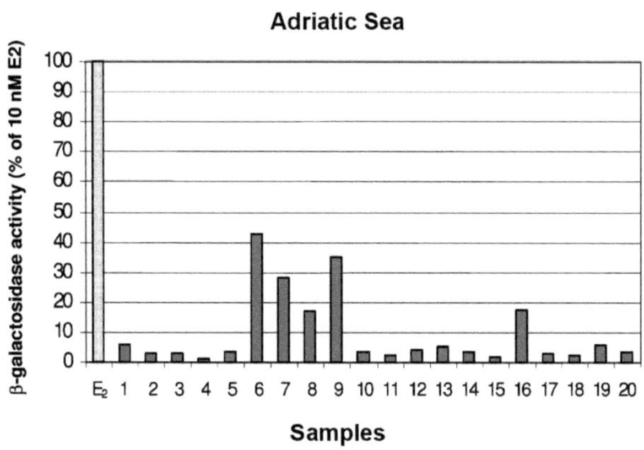

Figure 6. The β-galactosidase activity induced by tissue extracts of seafood from the Adriatic Sea. Results are expressed as percent activity induced by 10 nM E2.

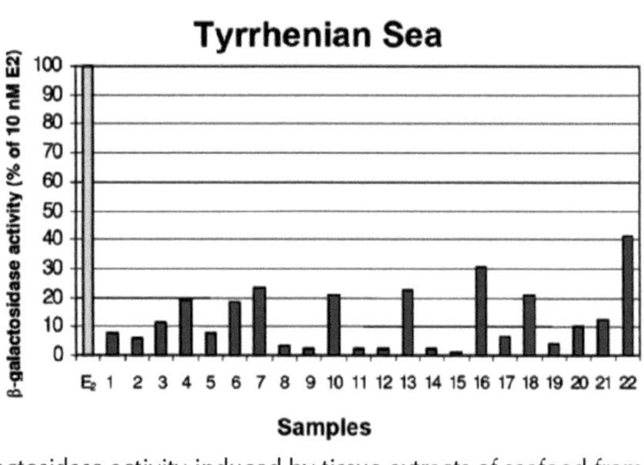

Figure 7. The β-galactosidase activity induced by tissue extracts of seafood from the Tyrrhenian Sea. Results are expressed as percent activity induced by 10 nM E2.

PCBs and β-galactosidase Activity Correlation

The content of total PCBs and β-galactosidase activity did not show any significant correlation (a negative borderline correlation was observed, $p = 0.07$).

The PLS analysis showed that in samples with estrogenic activity higher than 10% E2, the most observed combination of these contaminants was PCB 28, PCB 101, PCB 153, PCB 180.

However, total PCBs concentration explained only 33% and 15.77 % of variance, respectively in samples from the Tyrrhenian and the Adriatic Sea.

To better investigate these results, the β-galactosidase activity of individual PCBs was assessed in the yeast assay. The results are showed in Table 2. The β-galactosidase activity of PCB standards ranged from 18.35 % (PCB 101) up to 85.88 % (PCB 118), with PCB 118 resulting the most estrogenic in the yeast assay. This congener belongs to the class of 12 PCBs identified by the WHO as "dioxin-like" because of their toxicity and certain features of their structure which make them similar to 2, 3, 7, 8-tetrachlorodibenzo-p-dioxin (2, 3, 7, 8-TCDD).

Table 2. The β-galactosidase activity induced by PCB congeners (standard solutions).

Congeners	Concentrations µg/ml[a]	β-galactosidase activity (percent of 10 nM E2)
PCB 28	5.0	31 .10 ± 6.58
PCB 52	5.0	31 .96 ± 7.82
PCB 101	5.0	18.35 ± 4.26
PCB 118	5.0	85.88 ± 19.79
PCB 138	0.05	20.95 ± 7.82
PCB 153	0.05	26.37 ± 5.63
PCB 180	5.0	22.92 ± 8.67

[a]Congeners 138 and 153 were assayed at 0.05 µg/ml because at higher concentrations they strongly inhibited yeast cells growth

The predicted response according to the individual content of PCBs measured in each seafood sample was calculated. Comparison between predicted and measured enzymatic activity showed a statistically significant difference depending on the two areas. In the Tyrrhenian Sea enzymatic activity measured in the samples was higher than the expected activity, while an inverse correlation was found in the Adriatic Sea. The biological effects of PCBs are often similar to (although less potent than) those of TCDD by activating the aryl hydrocarbon (Ah) receptor. Additionally, some PCBs or mixtures of PCBs exhibit agonistic activity, whereas others are actually antiestrogenic [2, 25, 26]. Thus, the estrogenic activity of each congener was determined but the prediction of the effect of the same congener *in vivo* may be extremely difficult, depending on the interactions in a complex environmental mixture.

In this study we analyzed the content of PCBs because of their well-documented ability to influence the endocrine system. However, fish tissue may contain a mixture of several environmental compounds other than PCBs interfering with the endocrine system due to the widespread contamination of surface waters with scarcely treated urban and industrial waste that could have additive, synergistic or antagonistic effects. This may account for the final biological activity observed in the samples. Actually, the USEPA estimates there are more than 87,000 of potential EDCs.

Nevertheless, detecting so many chemicals would take an unreasonable investment of time and resources, so it is necessary to develop screening programs using short term bioassays to assess the risk of exposure for biota to EDCs through the environment and diet.

Finally, we underline that endogenous hormones could interfere in the estrogen-like activity elicited by animal organic extracts, marine organisms included as recently pointed out by some authors [27]. It is known that in marine organisms, estrogen level may vary depending on differences in species, sex, age, life cycle, and season and we tested a raw fat extract in which endogenous hormones are still present. Thirteen species with different habitat and reproductive periods were analyzed, and for three species (*M. barbatus*, *P. phycis*, *S. scombrus*) collected both in the Adriatic Sea and in the Tyrrhenian Sea in the same sampling season, statistically significant differences (Fisher's exact test: $p < 0.004$) in β-galactosidase activity were observed when considered on the whole (Table 3). This result may indicate the role of the aquatic environment in bioaccumulation of xenoestrogens.

Table 3. Comparison of the estrogenic activity in the same species from the two habitats.

Species	Adriatic Sea		Tyrrhenian Sea		
	Number of samples	Number of positive samples	Number of samples	Number of positive samples	Fisher's exact test
M. barbatus	2	-	3	2	p > 0.3
P. phycis	5	-	4	2	p > 0. 16
S. scombrus	4	-	3	2	p > 0. 14
Total	11	0	10	6	p < 0.004

CONCLUSION

Most studies have been focused on the evaluation of the content of environmental contaminants such as PCBs and related persistent organic pollutants (POPs) into tissues of fish and other aquatic organisms [5, 7, 23, 28-30] or on reproductive effects due to *in vivo* exposure in natural environment [31, 32] or to specific sources of pollution [33-35].

In our monitoring we detected a generally low PCBs content in most seafood samples and alone they cannot justify the estrogenicity of the extracts. The approach proposed in this work, namely to measure the overall estrogenicity of chemicals each presented at low concentrations, may suggest the probable intake of estrogen-like chemicals for humans.

Then, a useful application of the yeast assay could be aimed to direct chemical analyses to only biologically active samples as a first monitoring level. This bioassay may provide a useful integration to chemical approach, and could be used to identify edible seafood exposed to estrogenic organic chemicals, depending on geographical natural habitat. As yet, however, there is a paucity of analytical data on extracts of edible marine organisms exhibiting estrogenic activity. Fish products may represent an important dietary source of environmental contaminants with endocrine activity to humans, particularly when they represent a relevant part of food intake [11, 36]. Many compounds may be present in the environment in trace amounts, but have high biological activity. It is important to assess health risk for biota and the level exposure to environmental contaminants. The foetus and the new-borns in humans are particularly vulnerable to pollutants exposure due to transplacental and lactational transfer of maternal burdens at critical periods of development [37]. The scientific evidence demonstrated a link between chronic exposure to low concentrations of chemicals through the environment or the food-chain and reproductive animal health [31, 34, 38]. Subtle health effects have been documented in certain Arctic populations exposed to a variety of contaminants present in the food chain (in traditional foods), particularly mercury and PCBs and the greatest concern is for fetal and neonatal development [37, 39, 40]. The possibility that bio-accumulative properties of persistent organic chemicals with hormone-like activity and the chronic low level exposure may contribute to overall breast cancer risk in women, as well as reproductive and developmental effects in humans [10, 41] has heavy implications for the prevention of these diseases in western countries.

KEYWORDS

- **β-galactosidase**
- **Endocrine disrupting chemicals**
- **Mediterranean Sea**
- **Partial least square regression**
- **Polychlorinated biphenyls**
- **Principal components analysis**

AUTHORS' CONTRIBUTIONS

Sonia Garritano and Barbara Pinto carried out the biological analyses and helped to draft the manuscript, Teresa Cirillo collected the samples and performed the chemical analyses, Marco Calderisi provided statistical data analysis, Renata Amodio-Cocchieri participated in the conceiving of the study, Daniela Reali participated in the design of the study and in coordination and led the writing of the manuscript. All authors read and approved the final manuscript.

ACKNOWLEDGMENTS

This research was partially supported by the Italian Ministry of the Environment.

COMPETING INTERESTS

The author(s) declare that they have no competing interests.

Chapter 7

Microbial Contamination and Chemical Toxicity

Jose Mendoza, James Botsford, Jose Hernandez, Anna Montoya, Roswitha Saenz, Adrian Valles, Alejandro Vazquez, and Maria Alvarez

INTRODUCTION

The Rio Grande River is the natural boundary between US and Mexico from El Paso, TX to Brownsville, TX, and is one of the major water resources of the area. Agriculture, farming, maquiladora industry, domestic activities, as well as differences in disposal regulations and enforcement increase the contamination potential of water supplies along the border region. Therefore, continuous and accurate assessment of the quality of water supplies is of paramount importance. The objectives of this study were to monitor water quality of the Rio Grande and to determine if any correlations exist between fecal coliforms, *E. coli*, chemical toxicity as determined by Botsford's assay, *H. pylori* presence, and environmental parameters. Seven sites along a 112 km segment of the Rio Grande from Sunland Park, NM to Fort Hancock, TX were sampled on a monthly basis between January, 2000 and December, 2002.

The results showed great variability in the number of fecal coliforms, and *E. coli* on a month-to-month basis. Fecal coliforms ranged between 0 and 10^6 CFU/100 ml while *E. coli* ranged between 6 and > 2419 MPN. *H. pylori* showed positive detection for all the sites at different times. Toxicity ranged between 0 and 94% of inhibition capacity (IC). Since values above 50% are considered to be toxic, most of the sites displayed significant chemical toxicity at different times of the year. No significant correlations were observed between microbial indicators and chemical toxicity.

The results of the present study indicate that the 112 km segment of the Rio Grande river from Sunland Park, NM to Fort Hancock, TX exceeds the standards for contact recreation water on a continuous basis. In addition, the presence of chemical toxicity in most sites along the 112 km segment indicates that water quality is an area of concern for the bi-national region. The presence of *H. pylori* adds to the potential health hazards of the Rio Grande. Since no significant correlation was observed between the presence of *H. pylori* antigens and the two indicators of fecal contamination, we can conclude that fecal indicators cannot be used to detect the presence of *H. pylori* reliably in surface water.

El Paso, TX and Ciudad Juarez, Mexico comprise the largest metropolitan area of the bi-national region with a semi-arid environment receiving an average of 17.7 cm of rain per year. The Rio Grande/Rio Bravo is the major watershed of this bi-national region. The major groundwater reservoirs of the area are the Hueco Bolson and the Mesilla Bolson. The river serves as an important natural resource for industry, agriculture, domestic water supply, recreation, and wildlife habitat for both countries [1].

Unfortunately, the Rio Grande is also a reservoir for infectious micro-organisms and toxic pollutants [2]. A variety of activities contributing to the chemical and microbial contamination of water supplies have been identified and include improperly installed and maintained septic systems, landfills, injection wells, land application of waste, irrigation, runoff, animal feed lots, and so forth [3]. It is estimated that at the present rate of consumption groundwater supplies will be depleted in approximately 20 years.

The presence of "colonias" (unincorporated and economically disadvantaged communities) with inadequate wastewater disposal methods, the application of untreated or improperly treated sewage for disposal or irrigation purposes, the numerous maquiladoras (international industry in Mexico), and differences in disposal regulations between US and Mexico result in a high probability of anthropogenic activities being responsible for the contamination of Rio Grande water supplies. Although, groundwater has traditionally been considered a safe source of drinking water, more than half of the reported waterborne disease outbreaks have been linked to contaminated groundwater [4].

Infectious diseases including cholera, amoebiasis, hepatitis A, salmonellosis, shigellosis, giardiasis, ascariasis, and other intestinal infections are not uncommon in the border region. The Texas Department of Health showed that hepatitis A, salmonellosis, dysentery, cholera, and other diseases occur at much higher rates in colonias than in Texas as a whole [2]. The occurrence of infectious diseases is associated with conditions prevalent in border counties, that is, potentially contaminated water from shallow wells in colonias, poor hygiene, and low socioeconomic status. Results from previous toxic chemicals studies by the Texas Commission on Environmental Quality (TCEQ), the International Boundary and Water Commission (IBWC) and the United States Environmental Protection Agency (USEPA) ranked several Rio Grande sites as areas of concern [1, 5]. The main pollutants found in water were arsenic, copper, nickel, chloride, unionized ammonia, and phenolic compounds. The USEPA has recently included Helicobacter pylori on the Contaminant Candidate List (CCL) (62FR 52193). The CCL identifies contaminants which are not currently regulated but known or anticipated to occur in public water systems. Little is known about the mode of transmission of *H. pylori*. A waterborne transmission route has been proposed since this microorganism has been found in surface water, groundwater and drinking water [6-9]. The occurrence and persistence of *H. pylori* in border water supplies has not been established. However, *H. pylori* antibodies were detected in 21% of children between the ages of 4–7 in a study of 365 primary school children conducted in an area of El Paso where half of the population does not drink piped water and 86% use septic tanks [10]. In order to fully understand the risk factors that promote *H. pylori* infections, it is extremely important to determine if this bacterium is present in border water supplies. Several studies have indicated that microorganisms can be found in the environment in a "viable, non culturable state". While these microorganisms cannot be cultured in regular culture media, their genomes remain viable, and given the right conditions, they can become infectious [11, 12]. Viruses have been reported to be in a similar state and to be activated and become infectious under certain environmental conditions [13]. It has been suggested that *H. pylori* can be found in a viable, non-culturable, metabolically active state in water supplies [8, 12]. Information on the occurrence of *H. pylori* in border water supplies will provide valuable information on the route of transmission and survival of this infectious agent.

Agrochemicals, pesticides, heavy metals, arsenic, and polychlorinated biphenyls (PCBs) (presumably from illegal dumping, agriculture and maquiladora activities) have been detected in the river and may be associated with fish deformities, leukaemia, and congenital malformations in humans. Fecal coliforms were also identified as an area of concern in three of the 19 segments of the Rio Grande Basin [2]. Two of these sites were located in the El Paso/Ciudad Juarez region. In order to determine the extent of chemical contamination of the river, it is important to develop and utilize assays that are appropriate for field work. Botsford's toxicity assay is inexpensive, rapid, and can be conducted with minimal training [14]. It is a novel assay that provides values comparable to the fat head minnow (*Pimephales promelas*) assay, and two commercial assays (Microtex and Polytox) that use bacteria as indicator organisms. Botsford's assay has been used to detect toxicity due to the presence of several inorganic and organic chemicals. It uses the ability of *Rhizobium meliloti* to reduce MTT (3-(4,5-dimethylthiazol-2-yl)-2,5-diphenyltetrazolium bromide) under non toxic conditions [14].

Water quality of the river is one of the most important concerns facing communities that are dependent on the river for drinking water, agriculture, and watershed. Microbial and chemical contamination must be monitored continuously to determine the condition of the water in order to contain the spread of diseases and to eliminate non-point sources of contamination.

River Flow

The demands on the water of the Rio Grande river have changed in the last 2 years (2001–2003) due to the drought conditions that the area is experiencing. Agriculture as well as residential "drought-condition" limits has been in place for the last several years. To conserve water and to control pollution, most of the water is now being diverted into concrete-lined canals leaving part of the river sites between El Paso, TX and Fort Hancock, TX with low or no flowing water. The concrete lined-canals are used for irrigation purposes. The decreased flow in some of the sample sites has resulted in observable changes in the microbial population and chemical composition of the water. The seven sites extend over 112 km distances that vary in terms of the potential sources of contamination. Table 1 and Figure 1 describe the seven surface water sites with the potential sources of contamination. The least impacted in terms of flow are Sites 1 and 2 since they receive water from New Mexico and are before the diversion dams that distribute water between Mexico and the USA. Two secondary-treatment Waste Water Treatment Plants (WWTPs) in New Mexico discharge their effluent into the river approximately 3 km upstream from Site 1. The standards for fecal coliforms in WWTP's effluent is 1,000 cfu/100 ml (monthly average) for New Mexico, while in Texas the standard for fecal coliform is 200 cfu/100 ml. The Sunland Park Horse Race Track in New Mexico is almost adjacent to Site 1 and wild avian species frequent this area in large quantities. Conditions between Site 1 and Site 2 are influenced by street runoff, the Montoya agricultural drain, and other industrial discharge. Water treated for drinking purposes in the Canal Treatment Plant is taken from a concrete-lined canal approximately 5 km downstream from Site 2. Sites 3 and 5 are located around the most populated areas of the city and are likely to be affected by anthropogenic activity. These sites are in a concrete-lined section of the river and have only a trickle of flow due to the diversion of the river above Site 3. Two international vehicle and pedestrian bridges are located close to this concrete-lined section of the river. The flow in Site 3 consists mostly of leakage from the International Dam. Street runoff and municipal streams also contribute to the flow in Site 3. The Haskell Waste Water Treatment Plant (which is a tertiary treatment plant) returned most of its treated effluent into the river between Sites 3 and 5 before the Americas canal was completed. Site 5 has same flow as Site 3, except when the Americas Canal cannot handle effluent from Haskell WWTP. Since the completion of this canal, most of this treated wastewater is returned to the Americas canal, which then empties into the Riverside Canal. Water that is treated in the Jonathan Rogers Water Treatment Plant for municipal use is taken from the Riverside canal at this point. Both of the drinking water plants are closed during non-irrigational seasons. Site 6 is located close to this point and some of the overflow, which the Riverside Canal cannot handle, is diverted back into the river. The majority of the year this site is stagnant or has insignificant flow. Since the water in the Americas canal comes from a mixture of river water and treated water, Site 6 shows differences from Site 5 in microbial contamination and chemical toxicity. For most of this study, the 17.6 km natural segment between Site 5 and Site 6 has been completely dry. The conditions at Site 7 are unique since this area is far from populated areas and water taken from the river at American and International diversion dams merges from both sides of the border. Most of the flow at Site 7 is due to a return gate from the Mexican side, which includes partially treated wastewater (primary treatment only) along with some gates from irrigation canals in the US side of the river.

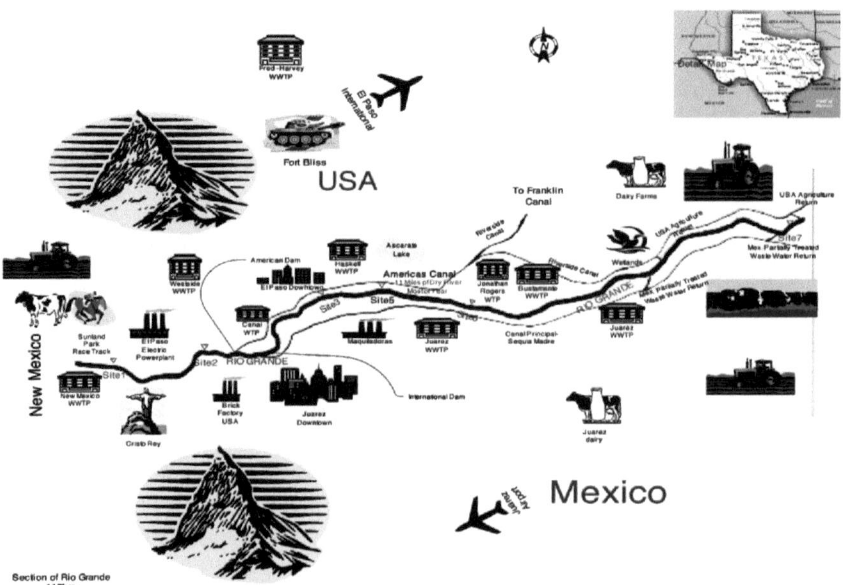

Figure 1. Sampling locations in the Rio Grande basin.

Table 1. Sample site descriptions.

Site 1. Rio Grande at Sunland Park bridge near Texas/ New Mexico state line	A horse race track is located upstream and El Paso Electric is downstream [influenced by urban and agricultural runoff]
Site 2. Rio Grande at Courchesne Bridge in El Paso	This site is influenced by flows coming from Elephant Butte Dam. The use of water for irrigation upstream contributes large volumes of irrigation return flow and agricultural runoff. It also receives urban runoff and wastewater discharges from Anthony, Canutillo, and El Paso
Site 3. Rio Grande 2.4 Km upstream of Haskell street [WWTP]	Vehicle traffic is heavy and the area is also affected by urban runoff
Site 4. El Paso Haskell Street WWTP outfall	Wastewater is discharged to the concrete lined/ channel part of the river
Site 5. Rio Grande 1.3 Km downstream from Haskell WWTP	Vehicle traffic is heavy and the area is affected by urban runoff
Site 6. Rio Grande at Riverside Diversion Dam	Influenced by urban runoff, wastewater discharges from the Haskell Street WWTP and runoff from industrial facilities on both sides of the border
Site 7. Rio Grande at Alamo Grade Control structure 9.7 Km upstream of Fort Hancock Port of Entry	Receives large amounts of wastewater from domestic and industrial sources

Microbial

Fecal coliform counts for Sites 1–3 (Figure 2) showed values ranging between 0 and 1.9×10^5 CFU/100 ml for Site 1, from 1.3×10^2 to 2.9×10^5 for Site 2, and from 0 to 3.7×10^5 for Site 3. Site 4 counts were from 0 to 27 CFU/100 mL, and most counts were below 10 CFU/100 mL (these results were not plotted). The Site 4 sample is from an effluent sampling faucet located inside Haskell WWTP and was not taken directly from the river. Fecal coliforms results for Sites 5–7 (Figure 3) ranged between 0 and 1.9×10^5 CFU/100 ml for Site 5, between 0 and 1×10^5 for Site 6 and between 0 and 3×10^6 for Site 7. *E. coli* most probable number (MPN) determined with the IDDEX Colilert system for all sites are shown in (Figure 4). Values ranged between 6 and 2,419, which is the upper detection limit. A high percentage of the samples were at the upper detection limit. Site 4 was not analyzed for MPN of *E. coli*. Table 2 shows the results for *H. pylori* determinations using the HpSA Antigen test. Out of 31 months, Site1 tested positive for *H. pylori* 17 times, Site 2 (13 times), Sites 3 and 5 (17 times), Site 4 (15 out of 24 samples) data not shown, Site 6 (14 times), and Site 7 (16 times).

Chemical Toxicity

Results of Chemical toxicity for Sites 1–7 are shown in Figure 5. Values above 50% are considered toxic. Chemical toxicity ranged between 0 and 94% for Site 1, between 0 and 91% for Site 2, and between 0 and 92% for Site 3. Values for Site 4, out of the 24 samples, the highest chemical toxicity registered was 87% on three occasions, these values were not plotted. Chemical toxicity ranged between 0 and 94% for Sites 5 and 6, and between 0 and 90% for Site 7.

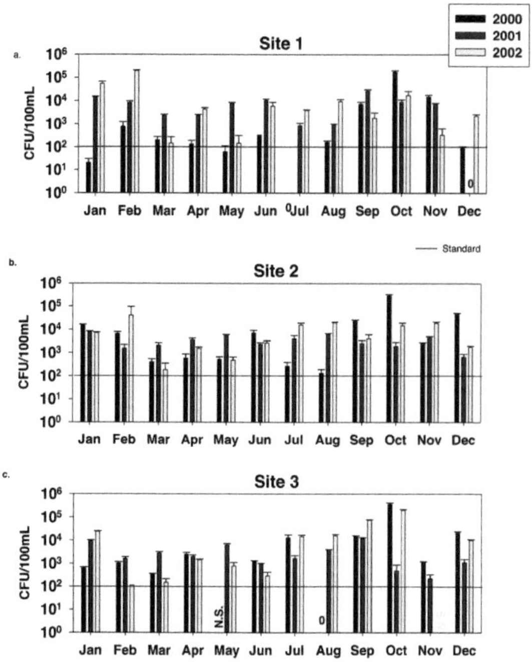

Figure 2. Fecal coliform bacteria on M-FC agar media for Sites 1–3.

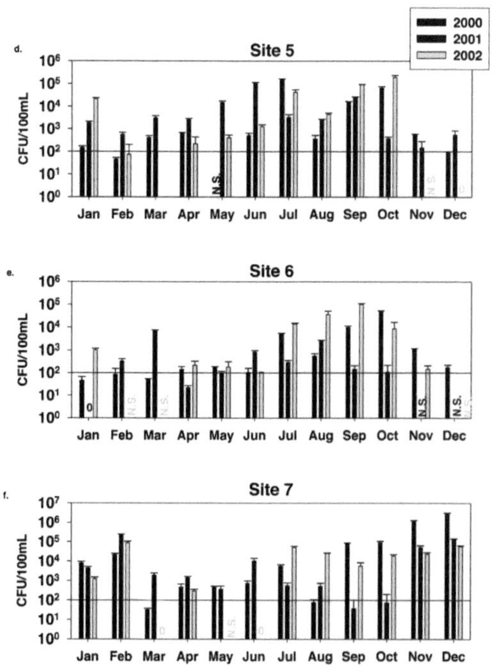

Figure 3. Fecal coliform bacteria on M-FC agar media for Sites 5–7.

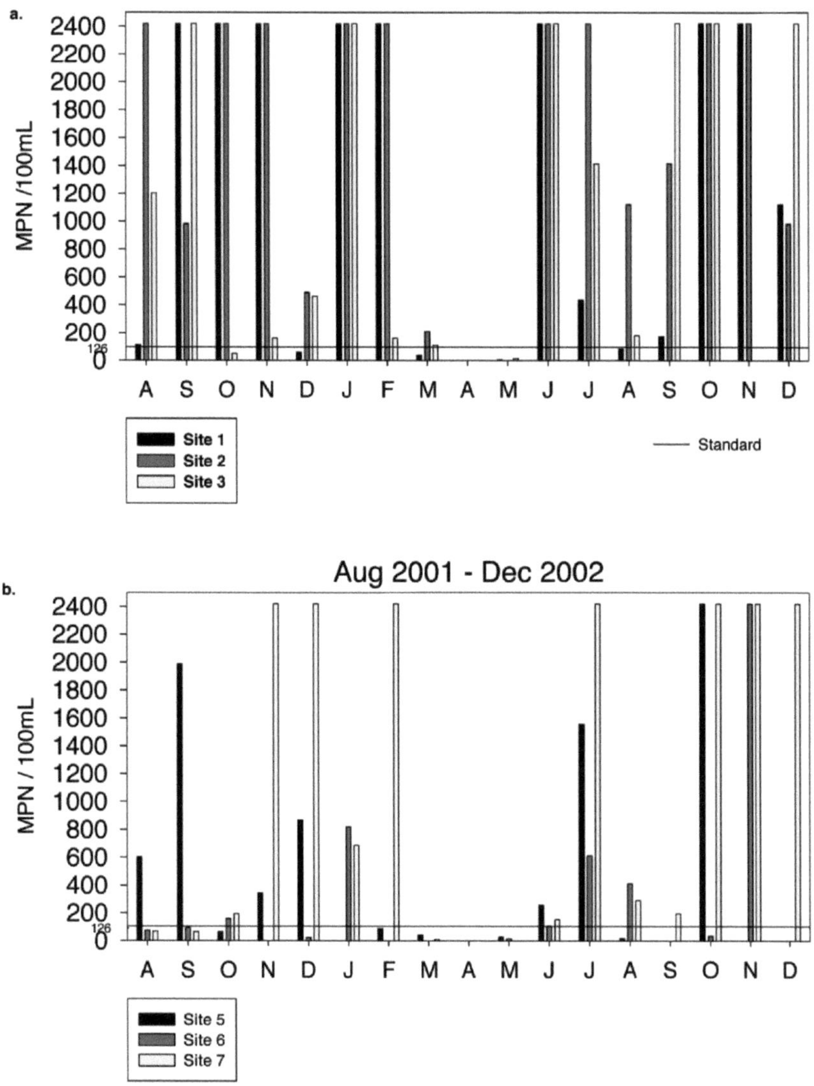

Figure 4. The *E. coli* most probable number (MPN) using the IDEXX Colilert assay for August, 2001–December, 2002.

Table 2. The *H. pylori* HpSa ELISA test results.

	Site 1	Site 2	Site 3	Site 5	Site 6	Site 7
Jun-00		+			+	
Jul-00						
Aug-00	+	+	+	+		

Table 2. *(Continued)*

	Site 1	Site 2	Site 3	Site 5	Site 6	Site 7
Sep-00	+	+	+	+	+	+
Oct-00						+
Nov-00						
Dec-00			+	+	+	+
Jan-01	+	+	+	+	+	+
Feb-0						
Mar-01						
Apr-01	+	+				
May-01	+	+	+	+	+	+
Jun-01						
Jul-01						
Aug-01	+	+	+	+	+	+
Sep-01	+	+	+	+	+	+
Oct-01	+	+	+	+	+	+
Nov-01	N/D	N/D	N/D	N/D	N/D	N/D
Dec-01			+			
Jan-02	+		+	+	+	+
Feb-02	+	+	+	+	N/D	+
Mar-02	+	+	+	+	+	+
Apr-02	+	N/D	+	+	+	+
May-02	+	+	+	+	+	+
Jun-02	+		+	+		+
Jul-02	+		+	+	+	+
Aug-02						
Sep-02	+	+	+	+	+	+
Oct-02	+					+
Nov-02						
Dec-02				+		

Note: Squares with + indicate detection of H. pylori antigen.

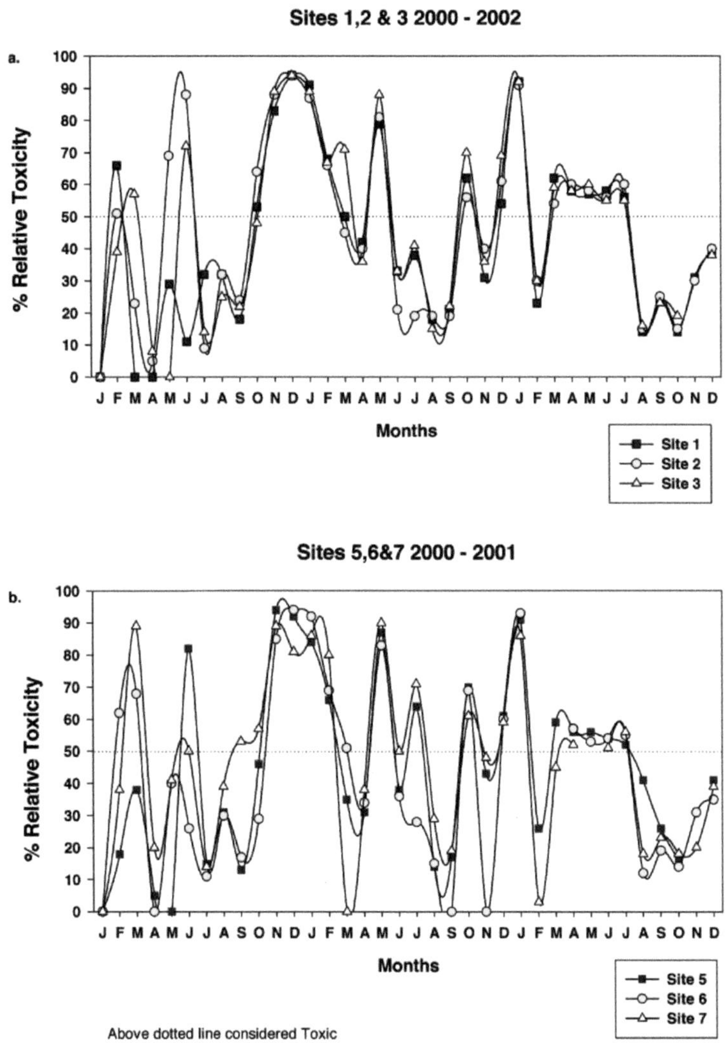

Figure 5. Relative chemical toxicity utilizing the Botsford assay.

DISCUSSION

Microbial

The TCEQ's standard for fecal coliforms on contact recreation areas is 200 CFUs/100 ml (Chapter 307 Texas Surface Water Quality Standards, Appendix A). Although, four of the seven tested sites do not support contact recreation activity because of the low flow most of the year, this standard was used for comparative purposes. During the 3-year evaluation period, Site 1 exceeded the limit 75% of the time, Site 2 94.4% of the time, and Site 3 88.6% of the time. Sunland Park and Santa Teresa WWTPs located between New Mexico and Texas discharge their effluent before Site 1 and are potential contributors of contaminants to Sites 1 and 2. The Montoya agricultural drain is likely to impact Sites 2 and 3. The flow in Site 3 consists mostly of leakage from the International Dam and municipal streams. Site 5 exceeded 77.1% of the time, Site 6 exceeded only 42.9% of the time, and Site 7 exceeded the limit 80.6% of the time. Sites 1, 2, and 3 had fecal coliform counts around 10^3 CFU/100 ml most of the time. Site 5 had counts between 10^2 and 10^4 most of the time. In Site 6, the counts were between 10^2 and 10^3 CFU/100 most of the time. It was noted at Site 6, throughout the course of this study, that the counts remain close or below the standard. As mentioned before, Site 6 receives the overflow from the Riverside canal, which contains a chlorinated effluent from the Haskell WWTP and industrial discharges. This site has been stagnant the majority of the time. This may be the reason for the lower bacterial counts observed. Site 7 had fecal coliform counts of 10^4 the majority of the time. It was at this site that we observed the highest counts reaching 10^6 CFU/100 ml in November–December, 2000. Extremely high fecal coliform counts were only observed in the first year of this study. During 2001, two wastewater treatment plants became operational in Mexico. The Mexican wastewater treatment plants only provide primary treatment. However, drought conditions have forced most crops to be irrigated with waters that would normally be returned to the river. The lower fecal coliform counts observed at Site 7 during the last 2 years may be the results of the WWTPs reducing the microbial contaminants combined with drought-condition irrigation practices.

The *E. coli* is the most specific test for fecal contamination. As the numbers of *E. coli* increase, there is a statistically greater risk that people using the river will experience gastro-intestinal illness. The TCEQ standard for this organism is 126 CFU/100 ml for multiple samples (Chapter 307 Texas Surface Water Quality Standards). *E. coli* counts during the last 17 months of assessment (Figure 4) revealed consistent trends. Site 1 exceeded the limit 62.5% of the time, Site 2 exceeded 87.5% of the time, Site 3 exceeded 68.8% of the time, Site 5 exceeded 43.8% of the time, Site 6 exceeded 31.2% of the time, and Site 7 exceeded the limit 68.8% of the time. These values agree with the results for fecal coliform counts, that is, Site 2 exceeds the limit with the highest percentage, and Site 6 has the lowest percentage of time where the limit is exceeded. When river flow is diverted for irrigation beginning in middle of February, all river sites reveal lower *E. coli* counts. From August, 2001 to October, 2001 lower counts (around the standard or lower) were recorded in Site 6 and 7. Site 1, 2, and 3 reached the upper limit of the assay at least once on this period while Sites 5, 6, 7 remained very low. This could be the result of treated effluent entering the Rio Grande between Sites 3 and 5, which happen as a result of the Canal of the Americas not been able to handle all the effluent from Haskell WWTP. Precipitation at the El Paso International Airport for August, 2001 was 4.3 cm, which supports the reasoning for this variation. However, *E. coli* counts appear to be influenced by rainfall, runoff and chemical discharges into the river including those used in the waste water treatment process.

Chemical Toxicity

Botsford's Chemical Toxicity assay was standardized using a variety of toxic chemicals as previously reported [14]. According to this method, any samples showing greater than 50% inhibition are considered toxic (IC50) (Figure 5). Using this standard, Site 1 displayed significant toxicity 45.7% of the time, Site 2 was toxic 48.6% of the time, Site 3 was toxic 45.7% of the time, Site 5 was toxic 44.1% of the time, Site 6 was toxic 48.5% of the time, and Site 7 was toxic 50% of the time. Botsford's assay provides an initial assessment of the relative toxicity of a sample but does not identify the toxic chemicals present. It is not clear at this point if it is the synergistic effect of several chemical compounds that contribute to the toxicity displayed using Botsford's assay. Further detailed analyses of samples will be needed in order to fully establish the value of using Botsford's assay for surface water testing.

Statistical Analyses

Analyses were performed to quantify relationships between biological variables and chemical toxicity, and to measure the strength of relationships between *H. pylori* and fecal contamination indicators such as fecal coliforms and *E. coli*. The assumption was that toxic chemicals present in the river may have an adverse effect on the bacterial population including fecal coliforms, *E. coli* and *H. pylori*. The All-Possible regression procedures using r^2 and Stepwise were first used to find any significant relationship between independent variables such as chemical toxicity and *H. pylori* versus biological variables including fecal coliforms and *E. coli*. By-site and overall analyses were performed and no relationships of significant levels were observed (data not shown). Further analyses were performed using the same procedures between the same independent variables and environmental parameters such as pH, precipitation, specific conductivity, and DO. Chemical toxicity showed a significant relationship between temperature of sample with an $r^2 = 0.1615$, $p < 0.001$ (data not shown).

To analyze any relationships between data from all parameters that were measured in the study, the Pearson-Moment correlation procedure was used. Tables 3, 4, 5, and 6 were generated with the results from the analyses. First, an overall analysis was performed using data from all sites. As expected, there was a highly significant relationship between fecal coliforms and *E. coli* (shown in Table 3). Chemical toxicity shows a significant relationship with specific conductivity and a negative relationship with temperature of sample (shown in Table 3). Non-point contamination events and the decreased river flow in the winter months could be contributing to these results. The same analysis was performed by-site between chemical toxicity, *H. pylori*, Fecal coliforms, and physiochemical parameters. The same strong relationship was observed between *E. coli* and fecal coliforms, except for Site 6 which did not show a strong relationship between *E. coli* and fecal coliforms (shown in Table 4). As mentioned earlier, this site has unique hydrologic characteristics, which could temporarily affect the microbial population. No significant relationships were observed between the presence of *H. pylori* antigen and indicators of fecal contamination including fecal coliforms and *E. coli*. Differences in survival rates between *H. pylori* and fecal coliforms may explain the lack of correlation. Therefore, the traditional indicators of fecal contamination cannot be used to detect the presence of *H. pylori*.

Table 3. Overall Pearson Product-Moment coefficients for correlation analysis between biological and environmental parameters.

	Fecal Coliform	Total Coliform	*E. coli*	*H. pylori*	Specific Conductivity	pH	Temperature	Dissolved Oxygen
Chemical Toxicity	0.0601 0.3915 05	0.1637 0.0208** 208	-0.1001 0.3769 80	-0.0454 0.581 150	0.2076 0.0093** 156	-0.0599 .4576 156	-0.4046 0.0001** 167	0.0634 0.4197 164
Fecal Coliform		0.53 0.0001** 205	0.6631 0.0001** 80	-0.0782 0.341 150	0.0247 0.7595 156	0.0271 0.7364 156	0.0558 0.4734 167	-0.1368 0.0806 164
Total Coliform			0.7502 0.0001** 80	0.0686 0.4039 150	0.0653 0.4179 156	0.0273 0.7352 156	-0.2156 0.0051** 167	0.0168 0.8304 164
E. coli				-0.0863 0.4707 72	0.2314 0.1134 48	-0.3289 0.0224** 48	-0.01625 0.9054 56	-0.3455 0.0079** 58
H. pylori					0.0269 0.7769 113	0.1627 0.085 113	0.0291 0.7456 126	-0.0802 0.3775 123
Specific Conductivity						-0.3106 0.0001 ** 157	-0.4432 0.0001 ** 146	0.1106 0.1822 147
pH							0.1713 0.0387** 146	0.221 0.0071 ** 47
Temperature								-0.4047 0.0001 ** 162

Values are Pearson Correlation Coefficients/Probability IRI under Ho: Rho= 0/Number Observations.** Probability of 95% of greater level of significance.

Table 4. Pearson Product-Moment coefficients for correlation analysis by site between fecal coliforms and biological and environmental parameters.

	Site I	Site 2	Site 3	Site 5	Site 6	Site 7
Chemical Toxicity	0.2158 0.213 35	0.1242 0.4772 35	0.1972 0.2634 34	-0.0175 0.9217 34	-0.2365 0.1851 33	0.0644 0.7174 34
Total Coliform	0.5844 0.0002** 35	0.3486 0.0401 ** 35	0.51221 0.002** 34	0.393 0.0215** 34	0.5991 0.0002** 33	0.5132 0.0019** 34
E. coli	0.6779 0.0055** 15	0.7753 0.0011 ** 14	0.61 0.0205** 14	0.6001 0.0391 ** 12	0.3386 0.3083 11	0.8466 0.0001 ** 14
H. pylori	-0.0065 0.9752 25	-0.3487 0.0875 25	-0.1 881 0.3679 25	0.00052 0.998 26	0.0623 0.7724 24	-0.1777 0.3952 25
Specific Conductivity	0.0943 0.6468 26	0.2294 0.2497 27	0.0607 0.7633 27	-0.4797 0.0113** 27	-0.4706 0.0234** 23	0.03 0.8843 26
pH	0.3704 0.0624 26	-0.1124 0.5764 27	-0.2436 0.2206 27	0. 1451 0.47 27	0.0527 0.811 23	0.0942 0.6469 26
Temperature	-0.1215 0.5714 24	-0.0054 0.9771 30	0.0882 0.6429 30	0.4589 0.0107** 30	0.4981 0.0113** 25	-0.2452 0.2083 28

Values are Pearson Correlation Coefficients/Probability IRI under Ho: Rho= 0/Number of observations.** Probability of 95% of greater level of significance.

Table 5. Pearson Product-Moment coefficients for correlation analysis by site between chemical toxicity and biological and environmental parameters.

	Site I	Site 2	Site 3	Site 5	Site 6	Site 7
Fecal Coliform	0.21584	0.1242	0.1972	-0.0175	-0.2365	0.0644
	0.213	0.4772	0.2634	0.9217	0.1851	0.7174
	35	35	34	34	33	34
Total Coliform	0.2666	0.0337	0.1474	0.039	0.2719	0.1182
	0.12151	0.8473	0.4053	0.8263	0.1257	0.5053
	35	35	34	34	33	34
E. coli	-0.0747	-0.0766	-0.213	-0.4612	0.1376	-0.0093
	0.7912	0.7945	0.4646	0.1312	0.6866	0.9748
	15	14	14	12	11	14
H. pylori	-0.1733	-0.2405	-0.2019	-0.0709	0.2174	0.022
	0.4074	0.2468	0.333	0.7307	0.3078	0.9167
	25	25	25	26	24	25
Specific Conductivity	0.2434	0.391	0.2393	0.1371	0.3778	-0.0375
	0.2307	0.0437**	0.2293	0.1371	0.0758	0.8554
	26	27	27	27	23	26
pH	-0.0796	-0.0303	-0.07382	0.0665	0.0928	-0.342
	0.6988	0.8807	0.7144	0.7417	0.6734	0.0872
	26	27	27	27	23	26
Temperature	-0.1215	-0.3753	-0.4489	-0.217	-0.5488	-0.1965
	0.5714	0.0409**	0.0128**	0.2492	0.0045**	0.3162
	24	30	30	30	25	28

Values are Pearson Correlation Coefficients/Probability IRI under Ho: Rho= 0/Number of observations. ** Probability of 95% of greater level of significance.

Table 6. Pearson Product-Moment coefficients for correlation analysis by site between *H. pylori*, and biological and environmental parameters.

	Site I	Site 2	Site 3	Site 5	Site 6	Site 7
Fecal Coliform	-0.0065	-0.2405	0.1972	-0.0175	-0.2365	0.0644
	0.9752	0.4772	0.2634	0.9217	0.1851	0.7174
	25	35	34	34	33	34
Total Coliform	0.2666	0.0337	0.1474	0.039	0.2719	0.1182
	0.12151	0.8473	0.4053	0.8263	0.1257	0.5053
	35	35	34	34	33	34
E. coli	-0.0747	-0.0766	-0.213	-0.4612	0.1376	-0.0093
	0.7912	0.7945	0.4646	0.1312	0.6866	0.9748
	15	14	14	12	11	14
H. pylori	-0.1733	-0.2405	-0.2019	-0.0709	0.2174	0.022
	0.4074	0.2468	0.333	0.7307	0.3078	0.9167
	25	25	25	26	24	25
Specific Conductivity	0.2434	0.391	0.2393	0.1371	0.3778	-0.0375
	0.2307	0.0437**	0.2293	0.1371	0.0758	0.8554
	26	27	27	27	23	26
pH	-0.0796	-0.0303	-0.07382	0.0665	0.0928	-0.342
	0.6988	0.8807	0.7144	0.7417	0.6734	0.0872
	26	27	27	27	23	26
Temperature	-0.1215	-0.3753	-0.4489	-0.217	-0.5488	-0.1965
	0.5714	0.0409**	0.0128**	0.2492	0.0045**	0.3162
	24	30	30	30	25	28

Values are Pearson Correlation Coefficients/Probability IRI under Ho: Rho= 0/Number of observations. ** Probability of 95% of greater level of significance.

Correlation between Chemical and Microbial Contamination, *H. Pylori*, and Biological Indicators

Analyses were performed to quantify relationships between biological variables and chemical toxicity, and to measure the strength of relationships between *H. pylori* and fecal contamination indicators such as fecal coliforms and *E. coli*. The All Possible regression procedure using r^2 and Stepwise was first used to find any significant relationship between independent variables chemical toxicity, and *H. pylori* and biological variables. By site and overall analyses were performed and no relationships of significant level were observed (data not shown). Further analyses were performed using the same procedure between same independent variables and physicochemical parameters. Chemical toxicity showed a relationship between temperatures of samples of $r^2 = 0.1615$, $p < 0.001$ (data not shown).

The results of this comprehensive three-year river monitoring effort confirm what previous isolated studies have suggested, that is, that the Rio Grande is heavily contaminated with bacteria of fecal origin. Since, several areas of the Rio Grande within this segment are used for recreational purposes by individuals from both sides of the border, the public health implications need to be addressed. Studies have been initiated in our laboratory to determine the sources of fecal contaminants using antibiotic resistance analyses (ARA), genotyping and ribotyping. Future studies will include more detailed chemical analyses of water samples showing a high degree of chemical toxicity.

MATERIALS AND METHODS

Seven Rio Grande sites described in Table 1 and Figure 1 were sampled. Sampling was done on a monthly basis for a period of 36 months. All samples were collected and processed according to methods described in the Study Methods section of the Second Phase Bi-national Study [1] and in Standard Methods for the Examination of Water and Wastewater [15]. Samples were collected midstream by submerging sterile 1 l plastic cubitainers to a depth of 30 cm and opening the container underneath till full with sample then closing underwater. Environmental parameters were taken at each site using a Hydro Lab Quanta Multi Parameter Analyser. Samples were kept on ice until delivered to the laboratories.

Chemical Toxicity

Botsford's assay was adapted and used to test for toxicity in water samples [14]. Rhizobium meliloti was grown in a chemically-defined medium (CDM) supplemented with 0.1% casamino acids with 1% mannitol as the carbon source. Bacterial cultures were incubated at 30°C overnight and centrifuged at 10,000 × g for 10 min. After washing the cells once with 0.01 M KH_2PO_4 buffer, pH 7.5, the cells were resuspended in 0.01 M phosphate buffer, pH 7.5 to an absorbance value between 0.31 and 0.38 at 550 nm. Assays were conducted by combining 0.2 ml of 0.1 M Tris-HCl buffer, pH 7.5 and 2.1 ml of test water sample. Nanopure-quality water was used as a negative control. Water samples were filtered using a 0.2 µm syringe filter to remove sediments and other organisms that could interfere with absorbance reading. The diluted bacterial suspension was then added to make a final volume of 3.3 ml and the initial absorbance reading was taken using a spectrophotometer at a fixed wavelength of 550 nm. The 100 μl of a 3 mM solution of MTT (3-(4,5-dimethyl-thiazol-2-yl)-2,5-diphenyltetrazolium bromide) was added and the mixture was incubated at 30°C for 20 min. To prevent the high concentrations of minerals in samples from interfering with the assay, 10 μl of EDTA 5 mM was added to the mixture before incubation. The difference in absorbance between the initial reading and the absorbance after 20 minute incubation (ΔA) was used in the calculation of MTT reduction. The difference between ΔA of river water samples and ΔA of pure-water controls determined the degree of inhibition of MTT reduction by potential toxic compounds in the water. Samples that show an inhibition capacity of 50% (IC50) or more indicate potential toxicity as determined by standardization using toxic chemicals [14].

Microbial

The procedure described in Standard Methods for the Examination of Water and Waste Water [15] and modifications described by Elmund et al. [16] was used. Fecal coliforms were detected using the MF technique, incubating the filters placed on sterile absorbent pads saturated with M-FC broth for fecal coliforms. Tests for *H. pylori* were done using the Premier Platinum HpSA (EIA) *H. pylori* assay, which detects *H. pylori* antigens (Meridian Diagnostics, Inc.). For increased sensitivity, 10 ml samples were concentrated 100-fold by ultrafiltration. The Premier Platinum HpSA test utilizes polyclonal antibodies absorbed to micro wells. Concentrated samples and a peroxidase conjugated polyclonal antibody were added to the wells and incubated for one hour at room temperature. A wash was performed to remove unbound material. Substrate was added and incubated for ten minutes at room temperature. Color developed in the presence of bound enzyme. Stop solution was added and the results were interpreted spectrophotometrically. The IDEXX®, using the Quanti-Tray®, assay was used to determine *E. coli* MPN. This assay has an upper detection limit of 2419.2. IDEXX Colilert® medium was added to the samples, sealed in Quanti-Tray, incubated for 24 hr at 37°C, and then quantified under a 365 nm ultraviolet light. *E. coli* uses the enzyme called β-glucuronidase to metabolize β-D-glucoronide leaving 4-methyl-umbelliferyl (MUG) to glow under ultraviolet light. Data collection using this assay began in August, 2001.

Statistical Analyses

Statistical analyses of data were done in collaboration with Dr. Melchor Ortiz, Professor of Biometry at the U.T. Houston School of Public Health using SAS 6.12 analysis software. All-Possible regressions using Stepwise regression and r^2 procedures and the Pearson Product-moment correlation procedures were used to analyze data on bacterial indicators, *H. pylori*, chemical toxicity and physical/chemical parameter. The Pearson Product-Moment correlation coefficient measures the strength of the linear relationship between two variables. For response variables X and Y, it is denoted as r and computed as shown to the right. If there is an exact linear relationship between two variables, the correlation is 1 or –1, depending on whether the variables are positively or negatively related. If there is no linear relationship, the correlation tends toward zero.

$$r = \frac{\sum (X - \bar{X})(Y - \bar{Y})}{\sqrt{\sum (X - \bar{X})^2} \sqrt{\sum (Y - \bar{Y})^2}}$$

CONCLUSION

The results of the present study indicate that the 112 km segment of the Rio Grande river from Sunland Park, NM to Fort Hancock, TX exceeds fecal coliforms standards for contact recreation water on a continuous basis. In addition, the presence of chemical toxicity in most sites along the 112 km segment as detected by Botsford's assay, indicate that water quality is an area of concern for the bi-national region. The presence of *H. pylori* adds to the potential health hazards of the Rio Grande. Since no significant correlation was observed between the presence of *H. pylori* antigens and the two indicators of fecal contamination, we can conclude that fecal indicators cannot be used to detect the presence of *H. pylori* antigens reliably in surface water. Also, no correlation was found between Botsford's chemical toxicity assay and fecal indicators indicating that the toxic chemicals present in the river are having a differential effect on the bacterial population. The river is a dynamic system where biological and chemical components may interact in a complex synergistic or antagonistic manner. The results of this study indicate that no simple conclusions can be drawn by studying a single indicator or parameter.

KEYWORDS

- **Colonias**
- ***E. Coli***
- **Fecal coliforms**
- **Leukaemia**

AUTHORS' CONTRIBUTIONS

Jose Mendoza participated in the design of experiments and worked on sample collection, fecal coliforms, chemical toxicity assays, and statistical analyses. James Botsford developed and supervised chemical toxicity assays. Jose Hernandez worked on *H. pylori* assay. Anna Montoya worked on sample collection, fecal coliforms, and chemical toxicity assays. Roswitha Saenz participated in the design of the study and worked on fecal coliforms. Adrian Valles worked on sample collection, fecal coliforms, and chemical toxicity assays. Adrian Valles participated in the design of the study and supervised sampling and fecal coliform enumeration. Maria Alvarez conceived the idea for this study and participated in all aspects of its design, implementation and coordination. All authors read and approved the final manuscript.

ACKNOWLEDGMENTS

This study was supported in part by the MBRS-RISE Grant (Number-R25 GM60424-01) to El Paso Community College, The Paso Del Norte Health Foundation's Center for Border Health Research, and The International Boundary and Water Commission's Clean Rivers Program-US Section.

Chapter 8

Heavy Metal Tolerance in *Stenotrophomonas maltophilia*

Delphine Pages, Jerome Rose, Sandrine Conrod, Stephane Cuine, Patrick Carrier, Thierry Heulin, and Wafa Achouak

INTRODUCTION

Stenotrophomonas maltophilia is an aerobic, non-fermentative Gram-negative bacterium widespread in the environment. *Stenotrophomonas maltophilia* Sm777 exhibits innate resistance to multiple antimicrobial agents. Furthermore, this bacterium tolerates high levels (0.1–50 mM) of various toxic metals, such as Cd, Pb, Co, Zn, Hg, Ag, selenite, tellurite, and uranyl. *Stenotrophomonas maltophilia* Sm777 was able to grow in the presence of 50 mM selenite and 25 mM tellurite and to reduce them to elemental selenium (Se^0) and tellurium (Te^0) respectively. Transmission electron microscopy (TEM) and energy dispersive X-ray (EDX) analysis showed cytoplasmic nanometer-sized electron-dense Se^0 granules and Te^0 crystals. Moreover, this bacterium can withstand up to 2 mM $CdCl_2$ and accumulate this metal up to 4% of its biomass. The analysis of soluble thiols in response to 10 different metals showed eightfold increase of the intracellular pool of cysteine only in response to cadmium. Measurements by Cd K-edge EXAFS spectroscopy indicated the formation of Cd-S clusters in strain Sm777. Cysteine is likely to be involved in Cd tolerance and in Cd-S clusters formation. Our data suggest that besides high tolerance to antibiotics by efflux mechanisms, *S. maltophilia* Sm777 has developed at least two different mechanisms to overcome metal toxicity, reduction of oxyanions to non-toxic elemental ions and detoxification of Cd into CdS.

The *S. maltophilia* is a non-fermentative, aerobic Gram-negative bacterium prevalent in the environment, which constitutes one of the dominant rhizosphere inhabitant, frequently isolated from the rhizosphere of wheat, oat, cucumber, maize, oilseed rapè, and potato [1-4]. *Stenotrophomonas maltophilia* shows plant growth-promoting activity as well as antagonistic properties against plant pathogens. It is currently being studied for its biological control of plant pathogens and was therefore utilized for the development of biopesticides [5]. *Stenotrophomonas maltophilia* is also able to degrade xenobiotic compounds [6, 7], to detoxify high molecular weight polycyclic aromatic hydrocarbons [8], possessing therefore a potential for soil decontamination (bioremediation). This bacterium was also increasingly described as an important nosocomial pathogen in debilitated and immunodeficient patients [9, 10], as well as associated with a broad spectrum of clinical syndromes, for example bacteraemia, endocarditis, respiratory tract infections [11]. *Stenotrophomonas maltophilia* displays intrinsic resistance to many antibiotics, making selection of optimal therapy difficult. The mechanisms underlying this multiresistance to drugs seem to result from a combination of reduced permeability [12], and expression of efflux pumps. Two RND efflux systems have been identified, SmeABC [13] and SmeDEF [14, 15].

Considering on one hand the rhizospheric origin of various opportunistic pathogens [16] including *S. maltophilia* and, on the other hand, the description of horizontal gene transfers in the rhizosphere [17], the tolerance of this bacterium to a wide range of toxic oxianions and metals must be addressed.

In the present study, we evidenced the tolerance of the strain Sm777 that belongs to *S. maltophilia* species, to very high concentrations of various toxic metals, especially cadmium, selenium, and tellurium, involving two different tolerance mechanisms.

RESULTS AND DISCUSSION

The strain Sm777 was isolated as a culture contaminant associated to Pseudomonas strains and was revealed in a contest of heavy metal tolerance studies. This rod-shaped bacterium was persistent in cultures containing a high concentration of cadmium, and was identified as a *Stenotrophomonas maltophilia* by 16S rDNA sequencing. The sequence analysis (using the BLAST database of the National Center for Biotechnology Information; [http://www.ncbi.nlm.nih.gov]) showed that strain Sm777 matched 99.5% with 16S rDNA of the *S. maltophilia* LMG 958T (accession n° DQ469587).

Minimal Inhibitory Concentrations (MICs) of Drugs and Heavy Metals

Stenotrophomonas maltophilia Sm777 was able to grow during 16 hr in the presence of 500 µM $CdCl_2$, 20 mM tellurite or 50 mM selenite without any significant increase of the lag phase. It is worth noting that strain Sm777 also grew to a high density (10^9 cfu.ml^{-1}) in the presence of high concentrations of other heavy metals (0.1 mM $CoCl_2$, 5 mM $CuSO_4$, 4 mM $ZnSO_4$, 10 mM $NiSO_4$, 0.05 mM $HgCl_2$, 0.02 mM $AgNO_3$, >1 mM uranyl, and 5 mM $Pb(NO_3)_2$). Moreover, this bacterium was resistant to a wide range of antibiotics, such as kanamycin (50 µg.ml^{-1}), gentamycin (100 µg.ml^{-1}), tetracycline (50 µg.ml^{-1}), and 50 µg.ml^{-1} of nalidixic acid. These may suggest that strain Sm777 overproduces some multidrug resistance (MDR) efflux pumps that are known to be involved in bacterial resistance to a wide range of compounds by extruding antibiotics and other toxic compounds.

Oxianions Reduction

To verify the hypothesis of over-expression of efflux systems to get ride of drugs and heavy metals, we analyzed and localized the elemental composition of bacteria grown in the presence of tellurite and selenite, by using EDX pectroscopy in conjunction with TEM or environmental scanning electron microscopy (ESEM).

The chemical microanalysis (TEM-EDX) of reddish colonies of strain Sm777 grown in the presence of selenite revealed cytoplasmic electron-dense Se^0 granules (Figure 1A). No detectable extracellular particles were observed. The intracellular Se^0 granules strongly suggest that selenite tolerance of strain Sm777 is not related to an efficient efflux system. On the contrary, a *S. maltophilia* strain isolated from a seleniferous agricultural drainage pond sediment was shown to transform selenate and selenite and to form spherical extracellular deposits consisting of Se [18]. The TEM-EDX observations of black colonies of strain Sm777 grown in the presence of tellurite revealed the presence of Te^0 crystals in the cytoplasm and proved that tellurite was taken up by the cells and was reduced into tellurium in the intracellular compartment (Figure 1B).

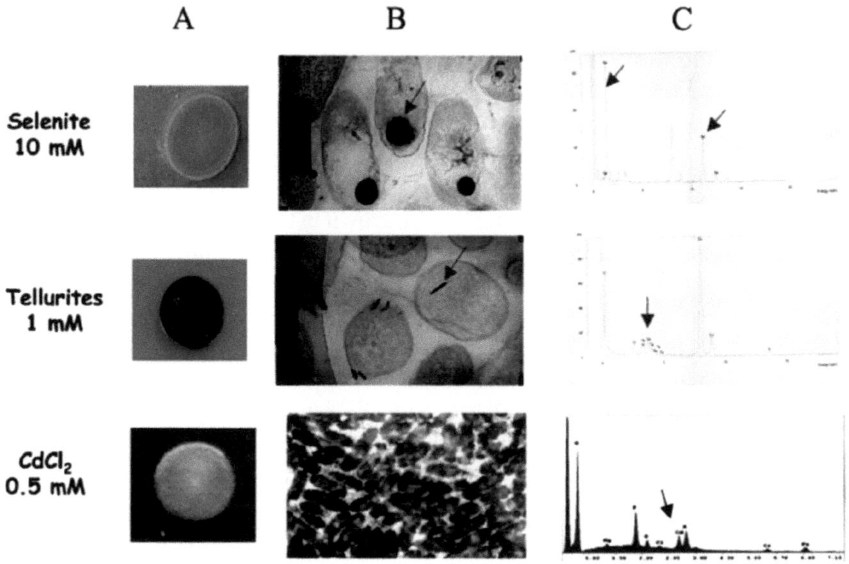

Figure 1. The ESEM-EDX and TEM-EDX observations. Microscopic observations and representative energy-dispersive X-ray spectra of electron-dense particles of *S. maltophilia* Sm777 cells grown in 10 fold-diluted TSB medium solidified with 15 g.l–1 agar, and supplemented with metals. (A and B) Colony shape, TEM-EDX micrographs and spectra of *S. maltophilia* Sm777 grown in the presence of selenite (10 mM) and tellurite (1 mM). (C) Colony shape, ESEM-EDX observation and analysis of cells grown in the presence of $CdCl_2$ (500 μM). Arrows on micrographs indicate the presence of intracellularly localized electron-dense particles of Se and Te, and arrows on spectra indicate metal-specific peak detected.

Active efflux of the metal is a frequently utilized strategy to produce tolerance by lowering the intracellular concentration to subtoxic levels. However, our data showing intracellular nanometer-sized particles of elemental selenium or tellurium, suggest that MDR efflux pumps probably do not mediate the heavy metal tolerance mechanism in strain Sm777 since tellurite and selenite-tolerance was associated to an intracellular reduction of these oxyanions and then by their accumulation.

Tolerance of *S. Maltophilia* to Cadmium

The ESEM observations coupled to EDX analysis of strain Sm777 grown in the presence of 500 μM $CdCl_2$ revealed the presence of Cd associated to bacterial cells, but did not allow localizing it exactly (Figure 1C). The bacterial Cd content was determined by ICP-AES as previously described [19]. This analysis revealed an accumulation of Cd strongly associated with the bacterial cell wall or incorporated into cells. Hence, this strain was able to accumulate Cd representing up to 4% of its dry mass. The presence of a cluster of genes from Gram-positive bacteria involved in both antibiotic and heavy metal resistance has been described in *S. maltophilia* D457R [20]. This cluster contains genes encoding a macrolide phosphotransferase (*mphBM*) and a cadmium efflux determinant (*cadA*). This study indicated a lateral gene transfer between Gram-positive and Gram-negative bacteria. The role of these genes in heavy metal tolerance of *S. maltophilia* has not been clearly evidenced yet.

Cysteine Accumulation in Response to Cadmium

The role of thiol compounds in the protection against heavy metals is well known [21]. Moreover, the chemical sequestration of Cd is thought to occur by coordination of cysteine thiolate groups. For that reason, we determined the concentration of soluble thiol compounds of strain Sm777 cells in response to Cd. We noticed an increase of intracellular cysteine pool when bacteria were grown in the presence of 500 μM $CdCl_2$ (Figure 2). Unlike other bacteria or yeast, no modification of gluthatione content was observed [21]. Moreover, no modification of the intracellular pool of cysteine was observed in response to the following metals: $NiSO_4$, $CuSO_4$, $Pb(NO_3)_2$, $ZnSO_4$, $CoCl_2$, $HgCl_2$, $AgNO_3$, tellurite, and selenite. The increase of intracellular pool of cysteine might reduce the bioavailability of Cd.

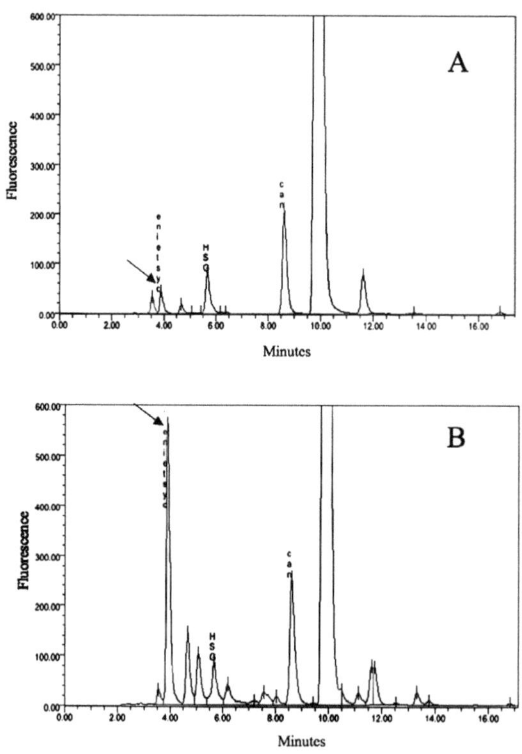

Figure 2. Soluble thiols analysis. The HPLC analysis of nonprotein thiols in *S. maltophilia* Sm777 grown in TSB/10 without (A) or supplemented (B) with 500 μM of $CdCl_2$. The arrow indicates cysteine peack. N-acetyl-L-cysteine (NAC) was used as an internal standard.

Park and Imlay [22] have shown that high levels of intracellular cysteine promote oxidative DNA damage by driving the Fenton reaction. They actually found that when cysteine homeostasis is disrupted, intracellular cysteine acts as an adventitious reductant of free iron and thereby promotes oxidative DNA damage.

The toxic effect of Cd is mainly mediated by its high degree of reactivity with S, O, and N atoms in biomolecules. Cysteine promotes an oxidative stress in cells, however it also protects against Cd toxicity probably by chelating Cd. The resulting metal thiolate complex formation may neutralize the toxicity of heavy metal. To deal with this dilemma, increasing the intracellular cysteine pool, bacterial cells are potentially exposed to an oxidative stress, but these cysteine residues may be stabilized by formation of Cd-cysteine complex decreasing that way the amount of free Cd and free cysteine.

Formation of CdS Particles

When strain Sm777 was grown under aerobic conditions on solid media containing 500 μM $CdCl_2$, it formed yellow colonies (Figure 1C). This observation suggested that bacterial cells may have transformed the Cd(II) into CdS as previously reported for *Klebsiella pneumoniae* [23], and for *Klebsiella planticola* [24]. To test this hypothesis, we used Cd K-edge EXAFS spectroscopy to probe the detailed coordination environment of the metal. The EXAFS spectrum was adjusted using different atomic neighbors around Cd. The nature, number, and distances of atoms surrounding Cd in the sample are detailed in Table 1 and the calculated and experimental EXAFS curves are compared in Figure 3A. The EXAFS modeling indicated that the first coordination sphere of Cd was composed of four sulfur atoms at 2.50 and 2.64 Å and confirmed the formation of CdS compounds. The EXAFS calculations also indicated the presence of Cd in the second coordination sphere at 3.42 and 3.68 Å. These Cd–Cd contributions indicated that CdS_4 tetrahedra present in the cell bond to form Cd-S-Cd clusters. The low number of Cd atoms around each Cd (1.3) suggested that the size of the Cd-S-Cd clusters is small and can be a mixture of Cd dimers and trimers as illustrated in Figure 3B. Thus, Cd-S clusters are formed in the cells and the low coordination number for the Cd–Cd contributions suggests that the product is less crystalline than the CdS reference compound. However, it is not possible to conclude whether these Cd-S clusters are surrounded by poly-thiols molecules or not.

The mechanism underlying the formation of CdS by strain Sm777 remains unclear; it is obvious that strain Sm777 formed CdS under aerobic conditions, whereas the formation of CdS in *Clostridium thermoaceticum* is mediated by the production of H_2S under stringent reductive conditions [25]. The aerobic sulfide production and Cd precipitation by *Escherichia coli* was possible by over-expression of the *Treponema denticola* cysteine desulfhydrase gene which product converts cysteine to sulfide under aerobic conditions. However, Cd precipitation as CdS was effective only when cysteine was added to the growth medium [26], whereas the production of CdS by strain Sm777 did not require any exogenous supply of cysteine. The high increase of intracellular pool of cysteine suggests that the bacterium reorients its metabolism to the production of cysteine that might be converted to sulfide used for CdS formation. Cysteine is able to form high-affinity metal ligand clusters and to promote the formation of CdS particles.

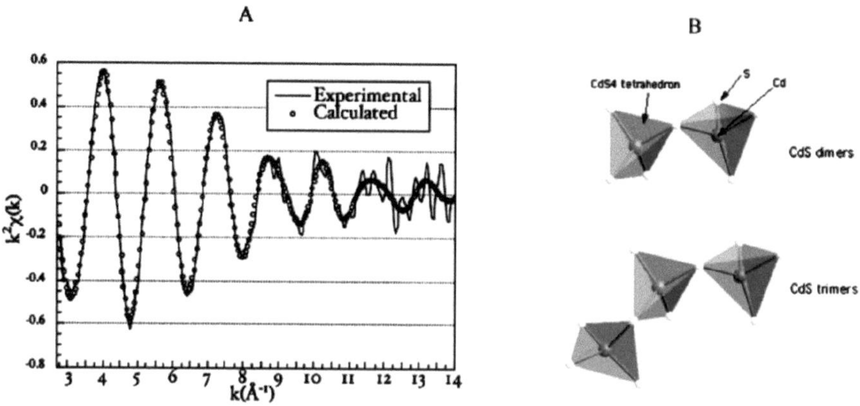

Figure 3. Extended X-ray absorption fine structure (EXAFS) spectroscopy. (A) Comparison between experimental and calculated EXAFS of *S. maltophilia* Sm777 strain cells. (B) Modeling of Cd dimers and trimers.

Table 1. Cd atomic environment.

Atomic pair	Interatomic distance R (A)	Debye-Waller parameter (A)	Number of atoms	Residue
Cd↔S	2.50	0.090	3.1	0.02
Cd↔S	2.64	0.100	0.9	
Cd↔Cd	3.42	0.092	0.4	
Cd↔Cd	3.68	0.110	0.9	

Structural parameters of the Cd atomic environment derived from EXAFS modeling of the Cd Kedge EXAFS spectrum of *S. maltophilia* Sm777 bacterial cells.

Alonso and colleagues [20] showed that a *Stenotrophomonas* strain has acquired a cluster of antibiotic and heavy metal resistance genes from Gram positive bacteria. Most of these genes are homologues of genes previously found on *Staphylococcus aureus* plasmids. In the present study, we evidenced the high tolerance to various heavy metals by *S. maltophilia* Sm777. To our knowledge, this is the first report indicating the high ability of a member of this species to tolerate and to detoxify several heavy metals. This bacterial species is also described as an opportunistic pathogen responsible for nosocomial infections. The severity of these infections is due to the virulence factors of the bacteria and to their occurrence in debilitated patients in whom invasive devices are used. To get more insight in the different mechanisms of heavy metals tolerance, and to identify pathogenesis related genes, it would be of great interest to perform a genome analysis and functional genomic studies of this species.

MATERIALS AND METHODS

Growth Conditions

Stenotrophomonas maltophilia Sm777 was grown aerobically in an incubating shaker at 30°C in tenfold diluted tryptic soy broth (TSB/10) (DIFCO Laboratories, Detroit,

USA). For growth on plates, media were solidified with 15 $g.l^{-1}$ Bacto-agar (DIFCO Laboratories, Detroit, USA).

Determination of Metals and Antibiotics Maximum Tolerance Concentrations

To determine the maximal tolerated concentration (MTCs) for different heavy metals, bacteria were grown on 10 ml of TSB/10 in the presence of different concentrations of different metals, $CdCl_2$, $NiSO_4$, $CuSO_4$, $Pb(NO_3)_2$, $ZnSO_4$, $CoCl_2$, $HgCl_2$, uranyl acetate, and $AgNO_3$, at 30°C under shaking. The MTCs corresponded to the highest concentration of each metal at which growth was still observed [27]. The MTCs for the four antibiotics, kanamycin, gentamycin, nalidixic acid, and tetracycline were also determined, and are expressed in $\mu g.ml^{-1}$. Experiments were performed in triplicate for each condition.

Analysis of Cadmium Accumulation

To determine the Cd content of bacterial cells grown in TSB/10 supplemented with 500 μM $CdCl_2$ for 48 hr cells were harvested, rinsed three times using TSB/10 and dried at 55°C for 24 hr. Following addition of 5 ml HNO_3 (70%), mineralization was carried out in a microwave oven (Mars X; CEM Corp., Matthews, NC). Metal content was determined using an inductively coupled plasma atomic emission spectrometry (ICP-OES device; Varian); standard solutions were supplied by Merck.

Soluble Thiols Analysis

Cells were harvested and rinsed with TSB/10 and stored at −80°C until analysis. Non-protein thiols were extracted by disruption of cells by sonication of 5–7 mg of frozen bacteria in 0.5–0.7 ml of extraction buffer (6.3 mM diethylenetriamine pentaacetic acid (DTPA)–0.1% (vol/vol) trifluoroacetic acid). Thirty microliters of 100 μM N-acetyl-L-cysteine was added as an internal standard. The homogenate was centrifuged at 10,000× g for 15 min at 4°C (Centromix 1236 V, Rotor 20RT) and the supernatant was filtered (0.22 μm). The derivatization procedure was modified from Rijstenbil and Wijnholds [28]. Filtered extracts (125 μl) were mixed with 225 μl of reaction buffer (0.2 M 4-(2-hydroxy-ethyl)-piperazine-1-propanesulfonicacid pH 8.2 containing 6.3 mM DTPA) and 5 μl of 25 mM monobromobimane dissolved in acetonitrile. Following 15 min of incubation in the dark at room temperature, the reaction was stopped by adding 150 μl of 1 M methane sulfonic acid. The samples were stored at 4°C in the dark until high-performance liquid chromatography (HPLC) analysis. The bimane derivatives were separated on a reversed-phase Nova-Pak C18 analytical column (pore size, 60 Å; particle size, 4 μm; dimensions, 3.9 by 300 mm; Waters catalog no. 11695) using two eluents (0.1% (vol/vol) trifluoroacetic acid in water and acetonitrile) at a flow rate of 1 $ml.min^{-1}$. Fluorescence was monitored by a Waters 464 detector ($\lambda_{excitation}$ = 380 nm; $\lambda_{emission}$ = 470 nm). Calibration curves of glutathione were used in all measurements. Cysteine, GSH and -glutamylcysteine (-EC) (from Sigma) were used as standard.

ESEM-EDX AND TEM-EDX OBSERVATIONS

For TEM, bacterial cells were harvested from TSA/10 plates containing tellurite (1 mM) or selenite (10 mM). Cells were then fixed in 2.5% glutaraldehyde and postfixed with osmium tetroxide in sodium cacodylate buffer. Dehydration was performed in ethanol and inclusion in epoxy resin. Ultrathin sections were made using a Reichert ultramicrotome. Electron micrographs and chemical microanalyses were obtained with a Jeol (Tokyo, Japan) 100CX transmission electron microscope coupled with an EDX spectrometer. The ESEM microscope coupled with an EDX spectrometer observations were realized on colonies grown on TSA/10 containing 500 μM $CdCl_2$.

Extended X-ray Absorption Fine Structure (EXAFS) Spectroscopy

The Cd K-edge X-ray absorption spectroscopy (XAS) experiments were carried out at the European Synchrotron Radiation Facility (ESRF, Grenoble-France) on the FAME (BM30-b) beamline with Si (220) monochromator crystals using the fluorescence detection mode. The storage ring was operated at 6 GeV with a current of 200 mA. The XAS spectra were scanned from 100 eV below to 800 eV above the Cd K-edge. The pre-edge part was extracted from the X-ray absorption near edge structure (XANES) region (extended from 26,600 eV to 26,650 eV). The XANES spectra intensity was normalized by fitting the photoelectric background above the absorption edge with a second order polynomial function. The EXAFS data reduction was done using a series of programs developed by Michalowicz [29] based on standard procedures [30]. The extracted EXAFS was k^2 weighted (with k = wave vector) to enhance the high-k region and Fourier transformed over the k range 2.4 to 14–15 $Å^{-1}$, to R space using a kaiser apodization window with t = 2.5. The resulting pseudo-radial distribution functions (RDFs) are uncorrected for phase shift leading to a shift of the peaks by 0.3–0.4 Å. Separate peaks in the RDF corresponding to successive shells of neighboring atoms around Cd were isolated by back-Fourier transformation (BFT) for single or multiple shell analysis. The analysis of partial c(k) was based upon the curved wave EXAFS formalism [31] in the single scattering approximation. Curve fitting was performed with a non linear least-square procedure, and phase ($f_{backscattere}r(k)$, $d_{central\ atom}(k)$) and amplitude ($|f_{backscatterer}(q, k, R)|$) functions used were calculated with FEFF8 [32]. Phase and amplitude functions of Cd-S and Cd-Cd atomic pairs were tested on reference compounds ($Cd(OH)_2$, CdS).

KEYWORDS

- Cysteine
- Cytoplasmic
- Gluthatione
- *Stenotrophomonas maltophilia*

AUTHORS' CONTRIBUTIONS

Conceived and designed the experiments: Wafa Achouak. Performed the experiments: Delphine Pages and Sandrine Conrod. Analyzed the data: Thierry Heulin. Contributed reagents/materials/analysis tools: Sandrine Conrod, Patrick Carrier, and Jerome Rose. Wrote the chapter: Wafa Achouak.

ACKNOWLEDGMENTS

We are grateful to M. Lesourd for the contribution to TEM observations and EDX analysis, to I. Felines for the contribution to ESEM observations and to A de Groot for critical reading of the manuscript.

Chapter 9

Chemical Defense in Marine Biofilm Bacteria

Carsten Matz, Jeremy S. Webb, Peter J. Schupp, Shui Yen Phang, Anahit Penesyan, Suhelen Egan, Peter Steinberg, and Staffan Kjelleberg

INTRODUCTION

Many plants and animals are defended from predation or herbivory by inhibitory secondary metabolites, which in the marine environment are very common among sessile organisms. Among bacteria, where there is the greatest metabolic potential, little is known about chemical defenses against bacterivorous consumers. An emerging hypothesis is that sessile bacterial communities organized as biofilms serve as bacterial refuge from predation. By testing growth and survival of two common bacterivorous nanoflagellates, we find evidence that chemically-mediated resistance against protozoan predators is common among biofilm populations in a diverse set of marine bacteria. Using bioassay-guided chemical and genetic analysis, we identified one of the most effective antiprotozoal compounds as violacein, an alkaloid that we demonstrate is produced predominately within biofilm cells. Nanomolar concentrations of violacein inhibit protozoan feeding by inducing a conserved eukaryotic cell death program. Such biofilm-specific chemical defenses could contribute to the successful persistence of biofilm bacteria in various environments and provide the ecological and evolutionary context for a number of eukaryote-targeting bacterial metabolites. Predators are potent agents of mortality and natural selection in biological communities. Plants and animals synthesize a broad range of secondary metabolites that are deterrent or toxic to their consumers, thus functioning as defense compounds. Such chemicals are often common in sessile eukaryotic organisms such as marine sponges and corals, seaweeds and terrestrial plants [1-4], which lack escape or avoidance mechanisms. However, chemically-mediated antipredator defenses of bacteria and their ecological and evolutionary consequences remain a greatly understudied field. Particularly, the increasing number of biologically active compounds isolated from marine bacteria raises the question of their ecological functions [5].

In many aquatic ecosystems, a common mode of life for bacteria is in biofilms, that is sessile high-density consortia of cells glued together by an exopolymeric matrix. While bacteria in the plankton typically occur at concentrations of 10^5–10^7 cells ml^{-1}, cell densities in bacterial biofilms tend to be two or three orders of magnitude higher. The ability of biofilm bacteria to achieve high densities has been ascribed to various factors, including greater resistance and/or tolerance to exogenous stresses or the accumulation of nutrients at surfaces [6, 7]. However, an alternative hypothesis for the prevalence of biofilms is that they serve as a protective niche against predation [8, 9], allowing for higher densities of cells than in planktonic environments.

Predation by phagotrophic protists, the protozoa, constitutes a major mortality factor for both planktonic and biofilm bacteria [10-12]. Aggregations of bacterial cells are rapidly detected and grazed by protozoa [13, 14], which in turn can drive shifts in bacterial communities from edible to grazing-resistant species or ecotypes [15]. In the plankton, size-selective protozoan grazing typically favors less edible ultramicrobacteria (cell volume < 0.1 μM^3) or elongated, filamentous cells [16-18]. Bacterial swimming speeds of >30 $\mu M\ s^{-1}$, particularly in combination with small cell size, can also significantly reduce prey capture and handling probabilities of bacterivorous flagellates [19].

Although bacteria-protist interactions in biofilms remain largely unexplored, bacterial morphotypes found in grazing-exposed freshwater biofilms do not appear to be comprised of the grazing-resistant phenotypes reported for bacterioplankton [12]. Escape by swimming by biofilm bacteria would eliminate the biofilm, and the formation of filamentous, less edible cell morphologies may be limited by strong competition for nutrients and space at the typically high cell densities found in biofilms. Thus, by analogy to sessile macroorganisms, we might expect that chemical defenses would be an important mode of defense against predators in the high cell density, clonal consortia that comprise biofilms. As a corollary, we might further expect—as suggested by the existing literature—that planktonic populations would rely more heavily on alternative defenses such as cell morphology or escape, with chemical defenses less prevalent (though we note the increasing evidence for chemical defenses in marine eukaryotic plankton [20]).

In this study, we for the first time compared the presence and efficacy of chemical defenses in biofilms versus planktonic populations of marine bacteria. We found a much greater prevalence of chemical defenses in the biofilm mode of life. We then went on to investigate the nature and mode of action of one such defense which is highly effective against a wide range of ecologically important protozoan predators. By contrasting the occurrence of chemical defense in biofilm and plankton populations, we tested the potential adaptive advantage for bacteria growing as biofilms when exposed to protozoan grazers.

RESULTS AND DISCUSSION

Our first hypothesis, that evidence for chemical defenses would be more prevalent in biofilm versus planktonic populations of bacteria, was tested using members of the well-characterized biofilm community found on the surface of the marine macroalga *Ulva australis* [21-25]. Clone libraries of the *16S ribosomal RNA* gene have previously uncovered considerable bacterial diversity in this community, including representatives of the five main bacterial groups, Bacteriodetes, Planctomycetes, α-, δ-, and γ-proteobacteria [25]. Besides some host-specific taxa [25], the culturable fraction of the bacterial community associated with *U. australis* is comprised of bacterial species found also in the planktonic community [21-23]. A diverse set of bacterial strains from this community were screened for their effect on protozoan predators when the bacteria were grown either as a biofilm or planktonically. Protozoa generally show a strict niche separation into suspension and surface feeding species, thus lacking species that are capable of efficiently utilizing both suspended and attached prey. To test for biofilm versus plankton effects, we used an experimental system contrasting two niche-specific predatory flagellates which are typically found in coastal ecosystems [26]. The surface feeder *Rhynchomonas nasuta* and the suspension feeder *Cafeteria roenbergensis* have repeatedly been isolated from coastal microbial communities and are among the 20 most commonly reported species of marine heterotrophic flagellates [27]. Despite their contrasting feeding modes, these two predators show high similarity in cellular features that are relevant for the comparison of feeding experiments, for example cell size, feeding rates, and susceptibility to bacteria producing biologically active compounds [26, 28].

Our screen revealed that planktonic populations of the 30 bacterial strains tested were widely edible, as evidenced by a moderate to high increase in cell numbers of the suspension feeder *C. roenbergensis* (180-fold increase on average, Figure 1A). In 27 bacterial strains (90%), we found a significant increase in predator numbers relative to the non-food control treatment ($P < 0.01$). Cell yield values, a measure describing the number of bacteria consumed to produce one flagellate cell, were typically 10–30 bacteria flagellate^{-1}, indicating high food quality for planktonic cells of 22 strains (73%). Grazing of *C. roenbergensis* resulted in a significant reduction of planktonic cell numbers of all 30 bacterial strains relative to the predator-free control ($P < 0.001$).

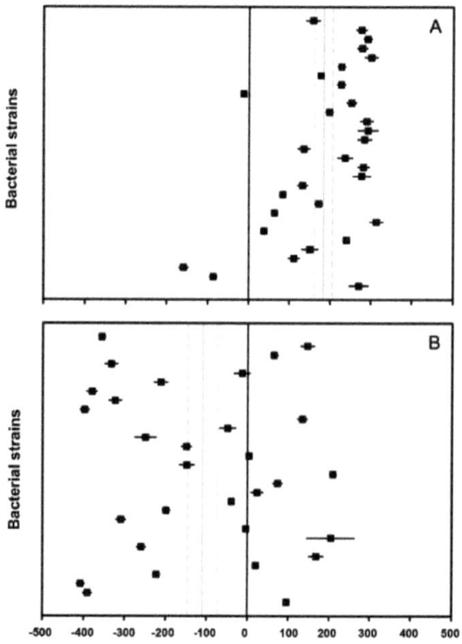

Figure 1. Grazing resistance and antiprotozoal activity in biofilm bacteria. Change in relative abundance of the protozoan predators *C. roenbergensis* feeding on planktonic populations (A) and *R. nasuta* feeding on biofilm populations (B) of bacteria isolated from the marine green alga *U. australis*. Bacteria were considered to be edible or toxic if predator numbers showed a significant increase or decrease relative to the non-food control treatment. Data are means ± SD of four replicates. Dashed lines represent total means ± SE of 30 strains.

In contrast, the majority of biofilm populations were resistant or toxic to the surface feeder *R. nasuta* (Figure 1B). Biofilms of 17 strains (57%) caused a significant reduction in predator numbers relative to the non-food control treatment ($P < 0.001$), which concurred with relatively stable biofilm biomasses. Only 11 strains were significantly reduced by *R. nasuta* ($P < 0.01$). Numbers of *R. nasuta* remained unchanged in biofilms of five strains, while eight bacterial strains supported growth of the flagellate predator ($P < 0.01$). Nineteen out of the 22 strains that were resistant or toxic to the surface feeder (when grown as biofilms) showed at the same time high susceptibility to the suspension feeding flagellate (when grown planktonically). Biofilms have recently been proposed to serve as refuge from protozoan grazing for bacterial pathogens such as *Vibrio cholerae* [26]. The current dataset expands this concept to non-pathogenic bacteria and across a broader range of bacterial taxa, suggesting that biofilm-specific resistance against predation is more widespread than previously thought.

Our previous studies have shown that the formation of less edible biofilm structures, such as microcolonies and exopolymers, may slow down flagellate growth but still leads to a several 100 fold increase in flagellate numbers [26, 29, 30]. Moreover, constant flagellate numbers in non-food control treatments of our present study indicate that the reduction in predator numbers was not a result of starvation. Thus, the substantial decline of protozoan numbers strongly suggests that grazing resistance was not based on the physical inaccessibility of biofilm cells, or poor nutritive value, but rather chemical interference with predator viability. To confirm this for a more restricted set of bacteria, we investigated in more detail biofilms of two γ-proteobacteria, *Pseudoalteromonas tunicata* and *Microbulbifer* sp., which showed the strongest negative effects against the flagellate consumers.

To identify the putative protozoa-active compound produced by *P. tunicata*, we chose a dual approach of chemical extraction and transposon mutagenesis. Antiprotozoal activity of biofilm extracts was evaluated by monitoring the total number and the number of active flagellate cells in cultures of *C. roenbergensis* and *R. nasuta* growing on heat-killed *P. aeruginosa*. Bioassay-guided chromatographic fractionation of crude biofilm material followed by nuclear magnetic resonance (NMR)-spectroscopic analysis revealed that the antiprotozoal activity of *P. tunicata* is elicited by the purple pigment violacein, an L-tryptophan-derived alkaloid consisting of three structural units: 5-hydroxyindole, 2-pyrrolidone, and oxindole (Figure 2A). Antiprotozoal activity was also detected for the violacein derivative deoxyviolacein, which is produced at 3-fold lower quantities. Violacein was first described from cellular extracts of the β-proteobacterium *Chromobacterium violaceum* [31]. Analysis of the *P. tunicata* genome sequence revealed a 8 kb region encoding a putative biosynthetic gene cluster consisting of five open reading frames (Genbank accession numbers ZP01133841–ZP0113384), each with high (>40%) predicted amino acid sequence identity to the violacein operon *vioABCDE* of *C. violaceum* [32, 33].

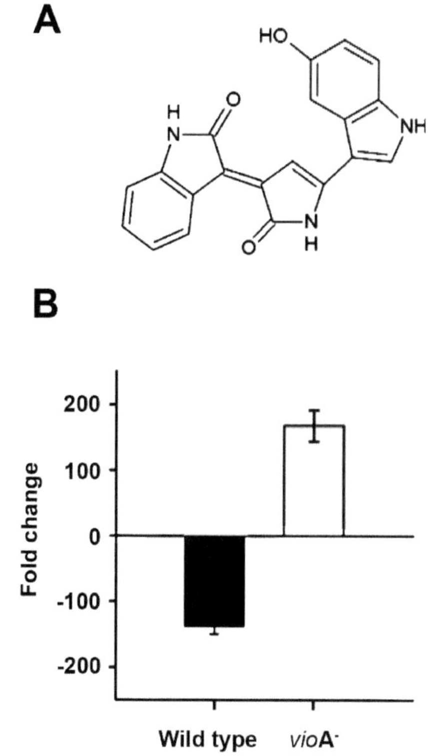

Figure 2. Defensive metabolite violacein. (A) The molecular structure of violacein, the antiprotozoal constituent of *P. tunicata* cells. (B) Mortality and growth of *R. nasuta* on biofilms of wild-type *P. tunicata* and the violacein-deficient *vioA* mutant. Data are means ± SD of four replicates.

The role of violacein in the killing of protozoan predators by *P. tunicata* was supported through the analysis of a *vioA* mutant, defective in violacein biosynthesis. While biofilms of the *P. tunicata* wild type killed the flagellate *R. nasuta* within 24 hr, the *vioA* mutant was completely grazed resulting in significantly higher predator numbers compared to the wild type and the non-food control treatment ($P < 0.001$, Figure 2B). Similar effects of *P. tunicata* wild type and violacein-negative mutant as well as of purified violacein were found against a broad spectrum of protozoan predators, including amoebae, ciliates, and flagellates. The PCR-amplification of the putative *vioA* gene and biochemical analysis of cell extracts revealed that violacein is synthesized by another two species found in *U. australis* biofilms, *Microbulbifer* sp., and *Pseudoalteromonas ulvae*. Comparing the isolated strains with the known violacein-producers *C. violaceum*, *Janthinobacterium lividum*, and *Pseudoalteromonas luteoviolacea* revealed that, despite spanning several genera, all six species share a niche preference for sessile microbial communities, such as those found in biofilms, sediments, and soils [21, 34-36]. The putative biofilm prevalence of violacein-producers may find support in the fact that the violacein gene cluster has not been detected in the currently largest metagenomics sequence database of pelagic ocean waters [37, 38].

To test whether violacein is produced predominantly by biofilm populations, we analyzed five violacein-producing bacteria, *P. tunicata*, *P. ulvae*, *P. luteoviolacea*, *Microbulbifer* sp., and *C. violaceum*, for their violacein biosynthesis rates in biofilms versus planktonic cells. While all five species grew to considerable cell densities in plankton ($\geq 2 \times 10^7$ cells ml^{-1}) and biofilm (OD490 nm ≥ 0.2), cell extracts from biofilms contained 3 (for *P. ulvae*) to 59 (for *Microbulbifer* sp.) times more violacein per protein biomass than the corresponding plankton extracts (Figure 3). Differences between biofilm and plankton extracts were significant for all five species ($P < 0.001$).

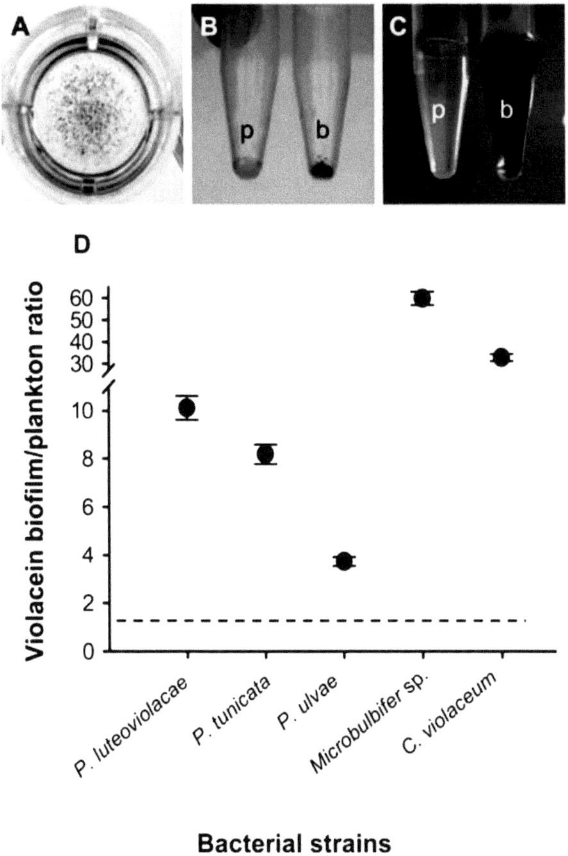

Figure 3. Biofilm-enhanced production of violacein. (A) Formation of violacein-containing biofilms by *Microbulbifer* sp. at well bottom. (B) Cell pellets of planktonic (p) and biofilm (b) bacteria. (C) Butanol extracts from planktonic (p) and biofilm (b) bacteria. (D) Relative violacein content of biofilm and planktonic populations of five bacterial species. Dashed line indicates an equal distribution. Data are means ± SD of four replicates.

In nutrient addition experiments, we examined the influence of cell density on the relative violacein content in biofilms versus planktonic bacteria. Although population densities of both biofilm and plankton were positively correlated with the nutrient concentrations applied, the violacein production by planktonic cells remained minimal at low to intermediate nutrient levels while the violacein content in biofilm populations was relatively high and increased continually. However, the relative differences in violacein production between biofilm and planktonic bacteria became significantly smaller with increasing cell densities at high nutrient concentrations ($P < 0.01$). It appears that the localized high cell densities in biofilms allow for higher violacein content particularly in environments of low productivity. The central role of population density for the biofilm-enhanced production of violacein is intriguing, given that violacein biosynthesis in *C. violaceum* is regulated by quorum sensing [39]. Assuming biofilm and cell density dependent gene regulation for the five violacein-producing species, our data may illustrate the adaptive advantage of clonal growth and cellular cooperation as found in biofilms for the synthesis of antipredator compounds. As with quorum sensing compounds themselves, it may only be adaptive for bacteria to produce violacein when they are in sufficiently high densities (e.g., in biofilms) to produce concentrations that are active (in this case, inhibitory against predators).

To further investigate the effectiveness of violacein-mediated defense, we analyzed concentration-dependent effects on feeding, growth and viability of the flagellate *R. nasuta*, the ciliate *Tetrahymena* sp. and the amoeba *Acanthamoeba castellanii*. In these experiments, we tested (i) mono-species biofilms of *P. tunicata*, *P. ulvae*, *P. luteoviolacea*, and *Microbulbifer* sp. and (ii) purified violacein at concentrations from 50 to 0.1 μM. Although biofilms were grown under low nutrient conditions, all three protozoan predators showed 100% mortality within 24 hr when feeding on biofilms of all four violacein-producing species. Chemical extractions from cell pellets and cell-free culture supernatants revealed that violacein is stored intracellularly. Sub-cellular fractionation of *P. tunicata* cells further suggest that violacein is associated with the outer membrane and accumulates in the periplasm. It is possible that periplasmic storage aids to increase the defense efficiency by minimizing the loss of compounds into seawater, avoiding autotoxicity and ensuring direct contact with a potential consumer. Our findings of cell bound storage may also implicate a primary ecological function of violacein in antipredator defense rather than in competitor inhibition [40].

When exposed to micro- and submicromolar concentrations of pure violacein, protozoan cells showed subtle morphological changes that are indicative of stress. While cell numbers of *A. castellanii*, for example, remained stable over 24 hr, up to 85% of cells exhibited the inactive rounded cell type (Figures 4A and B). This state of feeding inactivity was followed by a sharp decline of *A. castellanii* cell numbers about 26 hr after the onset of feeding inhibition. Within hours 100% of *A. castellanii* cells were lysed in the presence of 1 μM violacein, but not in stationary phase cultures of *A. castellanii* without violacein exposure (Figure 4A). Similar to the killing of planktonic flagellates by *C. violaceum* [41], microscopic analysis revealed that the uptake of a single violacein-producing bacterium can cause lysis of *A. castellanii* within less than 1 hr. Even though individual cells are ingested and do not survive to reproduce, the remaining clonal prey population may benefit from reduced grazing pressure.

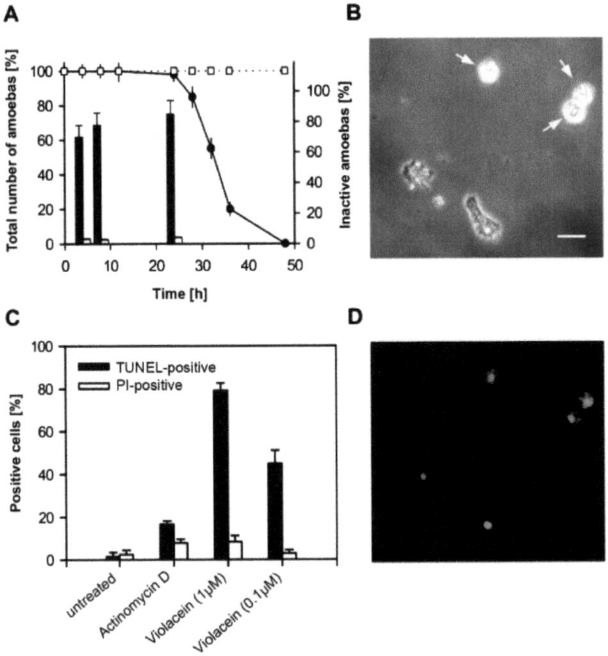

Figure 4. Induction of predator death program. (A and B) Inactivation and (C and D) apoptosis-like changes in the amoeba *A. castellanii* upon exposure to violacein. (A) Relative number of total amoebae and inactive cells in the absence (open symbols) or presence of 1 µM violacein (filled symbols). (B) Rounded cell morphology indicates inactive/stressed amoeba. Scale bar, 20 µm. (C) The DNA degradation and maintenance of membrane integrity in *A. castellanii* after 24 hr as revealed by TUNEL and propidium iodide (PI) staining. Positive control is treatment with actinomycin D. (D) Bright fluorescence indicates fragmentation of nuclear DNA in *A. castellanii* upon 24-hours-exposure to 1 µM violacein. Data are means ± SD.

The catastrophic decline in population density suggests the induction of an autolytic process in protozoa feeding on a violacein-producing biofilm. It is known that unicellular eukaryotes, such as protists and fungi, undergo autocatalytic cell death in response to environmental stress, which is analogous to programmed cell death or apoptosis in multicellular organisms [42]. To test whether violacein triggers specifically an apoptosis-like cell death program in protozoan predators, we used morphological and biochemical markers characteristic for apoptosis. Detection of DNA fragmentation by using the TUNEL assay revealed that after 24 hr of incubation 79% of *A. castellanii* cells in the 1 µM violacein treatment and 44% of cells in the 0.1 µM violacein treatment were TUNEL-positive (Figures 4C and D). Untreated *A. castellanii* remained below 3% while the positive control using 10 µM of the apoptosis inducer actinomycin D had about 17% of positive cells. Simultaneous tests for the maintenance of cell membrane integrity with propidium iodide (PI) revealed intact cell membranes despite the degenerative processes inside *A. castellanii* cells (Figures 4C and D). Analysis with annexin-V antibodies, however, revealed that violacein, similar to actinomycin D, induces the exposure of phosphatidyl serine on the outer layer of the cell membrane in about 20%

of *A. castellanii* cells. Using colorimetric assays for caspase-like protease activity, we also observed an increase in caspase-3-like activity in *A. castellanii* when exposed to violacein. Taken together, these findings suggest that the release of low amounts of violacein from ingested prey bacteria induces an apoptosis-like cell death mechanism in protozoan predators, which not only leads to feeding inactivity but to the effective decline in predator cell densities. Interestingly, violacein isolated from *C. violaceum* has recently been described to induce apoptosis in mammalian cell lines and thus has been proposed as possible novel therapeutic agent for anticancer treatments [43, 44]. Although the molecular target of violacein in the eukaryotic cell remains unidentified, the induction of an apoptosis-like cell death mechanism suggest that an ancient eukaryotic cell process may be the natural target of violacein-mediated defense.

In contrast to predator-prey interactions among plants and animals, the study of antipredator defenses in bacteria is still in its infancy, partly because predation has not been generally recognized by microbiologists as an important fitness determinant for bacteria. However, predation is likely to be particularly intense on "sessile" biofilms, and thus it is something of a conundrum in the microbial sciences how bacteria can build up and persist in such high numbers on surfaces despite the immense consumption rates of their natural predators. One answer to this conundrum is suggested by our results, which introduce chemically mediated resistance to protozoan grazing as a potentially key mechanism underlying grazer resistance in biofilms. We further identify the molecular basis of one biofilm-associated chemical defense, violacein. Assuming that effective chemical defenses are a widespread phenomenon in bacterial communities and generally promoted by the biofilm life form, bacterial antipredator compounds could significantly affect the stability of biofilm communities, the functioning of microbial food webs, and the coupling of energy and nutrient fluxes. Furthermore, we argue that our findings of biofilm chemical defenses may provide a new perspective on the causes of biofilm persistence in infectious diseases as phagocytic predation by protozoa shares fundamental cellular mechanisms with host immune cells.

The long coevolution of bacteria and protists and the considerable metabolic capabilities found in bacteria raise the question of how sophisticated bacteria-protist systems are—an interaction that may have shaped adaptation and diversity at the bacterial-eukaryotic interface more than we currently appreciate.

MATERIALS AND METHODS

Strains and Culture Conditions

Bacterial strains were isolated from the surface of the green alga *Ulva australis*, collected from the rocky intertidal zone near Sydney, Australia. Unless otherwise stated all bacterial isolates were routinely grown at 25°C on Vaatanen nine-salt solution (VNSS) marine medium [45]. Protozoan predators included common representatives of the three major feeding types: the flagellates *Rhynchomonas nasuta* and *Cafeteria roenbergensis* [26], the ciliates *Tetrahymena* sp. and *Euplotes* sp. (isolated by M. Weitere and V.C.L. Meyer, respectively), and the amoebae *Acanthamoeba castellanii* ATCC 30234 and *A. polyphaga* ATCC 30872. Cultures of *R. nasuta*, *C. roenbergensis*,

Tetrahymena sp., *A. castellanii*, and *A. polyphaga* are axenic and were maintained as described previously [22, 43].

Grazing Bioassay

Biofilm-specific protection against protozoan grazing was tested in a bioassay employing the common marine flagellates, *R. nasuta* and *C. roenbergensis*, as described elsewhere [26]. Bacterial overnight cultures were diluted to 10^5 cells ml^{-1} in carbon-reduced VNSS medium (0.5 g/l peptone, 0.25 g/l yeast, 0.25 g/l glucose), transferred into 24-well tissue culture plates and grown at 20°C with shaking to give an $OD_{490\ nm}$ of 0.2, as determined by a crystal violet staining assay [26]. For the comparison of plankton versus biofilm persistence, planktonic bacteria were separated from surface-associated cells after 24 hr incubation by transferring the planktonic phase of each well to a new plate. Cell numbers of planktonic bacteria were adjusted to a concentration of 1×10^7 ml^{-1}. Subsequently, the suspension-feeder *C. roenbergensis* was added to the planktonic population while the surface-feeder *R. nasuta* was introduced to the biofilm population (both at a final concentration of 1×10^3 ml^{-1}). Numbers of flagellates and planktonic bacteria, and biofilm biomass were followed over 4 days. In the biofilm assay, surface-associated *R. nasuta* was quantified by inverted confocal microscopy and bacterial biomass by a crystal violet staining assay. In the plankton wells, cell numbers of *C. roenbergensis* and suspended bacteria were determined by epifluorescence microscopy following formalin fixation (2%) and DAPI staining. Generally each treatment was run in replicate wells of four. To examine survival of *Tetrahymena* sp. and *A. castellanii* feeding on violacein-producing biofilms of *P. tunicata*, *P. ulvae*, *P. luteoviolacea*, and *Microbulbifer* sp., experiments were run analogously to the *R. nasuta* design with an inoculum size of 500 amoebae/ml.

Chemical and Genetic Analysis of Antiprotozoal Activity

Freeze-dried cell material of *P. tunicata* was extracted exhaustively with methanol. The active methanol fraction was then partitioned sequentially by flash column chromatography eluting with solvent systems of increasing polarity to yield 29 fractions (hexane, ethyl acetate, dichloromethane, and methanol). These fractions were evaluated for antiprotozoal activity in a 24-well-plate flagellate bioassay, containing *C. roenbergensis* and *R. nasuta* (at a final concentration of 2×10^3 cells ml^{-1}) supplemented with heat-killed *Pseudomonas aeruginosa* prey. Fractions and crude extract were diluted in NSS medium to achieve the desired concentrations in a 500 μl final volume per well. Total flagellate numbers and the number of active cells were monitored by direct inspection with an inverted light microscope and compared to the solvent-only control treatment. The active fractions were further purified by semi-preparative high-performance liquid chromatography (HPLC) (Phenomenex, silica gel column, 5 μ, 250 × 10 mm). Separation was achieved by applying a linear gradient of hexane/ethyl acetate (from 20:80 to 15:85) over 35 min. The purified violet pigment was analyzed using a spectrophotometer (DU 640, Beckman), and ^1H and ^{13}C NMR spectroscopy (Bruker DMX 500 MHz). To support the role of violacein in the killing of protozoan predators a *P. tunicata* transposon mutant library was generated using mini-Tn10 transposon [47] and screened for mutants that were defective in violacein production.

The DNA sequence of the gene disrupted by the transposon insertion in the violacein negative mutant was obtained using a panhandle PCR method with transposon specific primers, Tn10C (5′-GCTGATTGACGGGACGGCG-3′) and Tn10D (5′-CCTCGAG-CAAGACGTTTCCCG-3′) as described [47]. Panhandle PCR products were then sequenced using transposon specific primers (Tn10C and Tn10D) via a primer walking strategy. The DNA sequences were compared to those in the Genbank data-base and additional sequence information was obtained by analysis of the draft genome sequence for *P. tunicata* using a BLAST-search algorithm and gene comparisons made using the integrated microbial genomes system (IMGs) [48]. For the specific comparison of *P. tunicata* wild type and *vioA* mutant, biofilms of both strains (OD490 nm = 0.2) were exposed to *R. nasuta* (final concentration of 1×10^3 ml^{-1}) as described above for the grazing bioassay. Numbers of flagellates and biofilm biomass were followed over 4 days and compared to the non-food control treatment. Each treatment was run in replicate wells of four.

Violacein Production Kinetics

Overnight cultures of *P. tunicata, P. ulvae, P. luteoviolacea, Microbulbifer* sp., and *C. violaceum* were diluted to 10^5 CFU/ml in VNSS medium of four different carbon concentration (100%, 50%, 25%, 12.5%). Biofilms were grown in semi-continuous culture on 24-well microtiter plates for 48 hr. Suspended cells were harvested by centrifugation (14,000 rpm, 15 min); washed biofilms were harvested mechanically from the well bottom. Violacein was extracted from cell pellets by a combined treatment of sodium dodecyl sulfate (5% final concentration) and water saturated butanol and quantified photometrically as described elsewhere [49]. Simultaneously, pellets of planktonic and biofilm cells were sampled for the colorimetric quantitation of total protein based on the bicinchoninic acid (BCA) protein assay (Pierce, Rockford, IL). Bacterial overnight cultures were further used for sub-cellular fractionation according to [50].

Analysis of Protozoan Fitness

Cultures of *A. castellanii* attached to glass coverslips were exposed to different concentrations of purified violacein (50 μM, 10 μM, 1 μM, 0.1 μM). Numerical and morphological changes in *A. castellanii* populations were monitored over at least 24 hr by determining the number of cells with regular shape (= active cells) and rounded shape (= inactive cells). The apoptosis inducer actinomycin D (ICN Biomedicals, Aurora, USA) was used as positive control treatment as described elsewhere [51]. Hallmarks characteristic for eukaryotic cell death programs include DNA fragmentation, cell membrane integrity, phosphatidyl serine exposure, and caspase activity [42, 46]. The DNA fragmentation was detected by using terminal deoxynucleotidyl transferase-mediated labeling of dUTP nick ends (Roche, Switzerland). The integrity of the amoeba cell membrane was checked by adding PI. Phosphatidyl serine asymmetry was analyzed by measuring FITC-conjugated annexin V binding to the cell membrane (Sigma, St Louis, USA). The intracellular activity of caspases was determined using the caspase-3 intracellular activity assay kit I (Calbiochem, Germany). All samples were analyzed with the Leica DMLB epifluorescent microscope. A minimum of 100 cells per sample was examined. Experiments were run in triplicate wells and repeated twice.

KEYWORDS

- **Antiprotozoal activity**
- **Bacterivorous**
- **Biofilms**
- ***Microbulbifer***
- **γ-proteobacteria**

AUTHORS' CONTRIBUTIONS

Conceived and designed the experiments: Carsten Matz and Jeremy S. Webb. Performed the experiments: Carsten Matz and Shui Yen Phang. Analyzed the data: Carsten Matz, Jeremy S. Webb, Peter J. Schupp, Suhelen Egan, Peter Steinberg, and Staffan Kjelleberg. Contributed reagents/materials/analysis tools: Peter J. Schupp, Anahit Penesyan, and Suhelen Egan. Wrote the chapter: Carsten Matz, Suhelen Egan, Peter Steinberg, and Staffan Kjelleberg.

ACKNOWLEDGMENTS

We thank our colleagues at the Centre for Marine Bio-Innovation for discussions.

Chapter 10

Reverse and Conventional Chemical Ecology in *Culex* Mosquitoes

Walter S. Leal, RosBngela M. R. Barbosa, Wei Xu, Yuko Ishida, Zainulabeuddin Syed, Nicolas Latte, Angela M. Chen, Tania I. Morgan, Anthony J. Cornel, and André Furtado

INTRODUCTION

Synthetic mosquito oviposition attractants are sorely needed for surveillance and control programs for *Culex* species, which are major vectors of pathogens causing various human diseases, including filariasis, encephalitis, and West Nile encephalomyelitis. We employed novel and conventional chemical ecology approaches to identify potential attractants, which were demonstrated in field tests to be effective for monitoring populations of *Cx. p. quinquefasciatus* in human dwellings. Immunohistochemistry (IHC) studies showed that an odorant-binding protein (OBP) from this species, CquiOBP1, is expressed in trichoid sensilla on the antennae, including short, sharp-tipped trichoid sensilla type, which house an olfactory receptor neuron sensitive to a previously identified mosquito oviposition pheromone (MOP), 6-acetoxy-5-hexadecanolide. The CquiOBP1 exists in monomeric and dimeric forms. Monomeric CquiOBP1 bound MOP in a pH-dependent manner, with a change in secondary structure apparently related to the loss of binding at low pH. The pheromone antipode showed higher affinity than the natural stereoisomer. By using both CquiOBP1 as a molecular target in binding assays and gas chromatography-electroantennographic detection (GC-EAD), we identified nonanal, trimethylamine (TMA), and skatole as test compounds. Extensive field evaluations in Recife, Brazil, a region with high populations of *Cx. p. quinquefasciatus*, showed that a combination of TMA (0.9 μg/l) and nonanal (0.15 ng/μl) is equivalent in attraction to the currently used infusion-based lure, and superior in that the offensive smell of infusions was eliminated in the newly developed synthetic mixture.

Mosquitoes in the genus *Culex* are major vectors of pathogens causing human diseases throughout the world, including *Wulchereria bancrofti* and arboviruses, such as, St. Louis encephalitis (SLE), Japanese encephalitis (JE), Venezuelan equine encephalitis (VEE), Western equine encephalitis (WEE), and West Nile Virus (WNV) [1]. Surveillance and control programs have reduced the threat from endemic SLE in California and now mitigate the impact of WNV invasion [2]. Monitoring mosquito populations and mosquito-borne virus activity are the cornerstones of surveillance programs. Although CDC-style CO_2 traps can be very effective tools for monitoring mosquito populations, the majority of mosquitoes collected in these traps have not taken a bloodmeal [3]. In theory sampling with gravid traps should represent a more efficient surveillance tool because they collect proportionately more parous and gravid females that have taken bloodmeals and thus have had the opportunity to become horizontally infected [4, 5]. One of the major disadvantages of the gravid traps, however, is the use of cumbersome, infusion-based attractants whose offensive smell hinders population monitoring in human dwellings. This prompted us to undertake a multi-disciplinary approach to explore the development of user-friendly, chemically-based lures for gravid females in the *Culex pipiens* complex. First, we examined the expression pattern of an OBP, CquiOBP1, which was first isolated from the Southern House mosquito, *Cx. p. quinquefasciatus* [6] and later determined to be homologous to a female antennae-specific OBP from *Cx. tarsalis* and identical to OBPs from *Cx. p. pipiens* and *Cx. p. molestus* (Ishida and Leal, EU723597 and EU723598). Having observed that CquiOBP1 binds to a previously identified MOP in a pH-dependent manner, we used CquiOBP1 as a molecular target to identify a candidate compound. Then, we identified electrophysiologically-active compounds from rabbit chow infusion and conducted extensive field tests to develop a simple and convenient synthetic attractant lure for trapping gravid females of the Southern house mosquito.

RESULTS AND DISCUSSION

Expression of CquiOBP1

Recombinant CquiOBP1 was prepared by a perisplamic expression system, which is known to generate properly folded, functional OBPs [7]. Purification by a combination of ion-exchange chromatography and gel filtration generated samples of high purity, as indicated by liquid chromatography-electrospray ionization mass spectrometry (LC-ESI/MS) analysis (Figure 1). The MS data also suggest that all six cysteine residues in CquiOBP1 are linked to form three disulfide bonds (observed, 14,479 Da; calculated, 14,486 Da or 14,480 with disulfide bridges). Formation of three disulfide bridges have also been shown in a homologous protein [8], AgamOBP1, an OBP cloned from the malaria mosquito, *Anopheles gambiae* sensu stricto [9]. During gel filtration separations, recombinant CquiOBP1 samples were isolated in monomeric and dimeric forms (Figure 2), but the dimer slowly dissociated into the corresponding

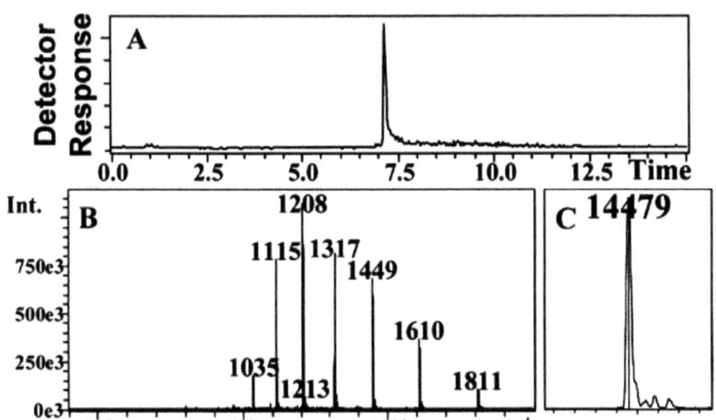

Figure 1. Mass spectral data for purified recombinant CquiOBP1. (A) The HPLC separation, (B) mass spectrum of CquiOBP1 peak, and (C) deconvolution data indicating a molecular mass of 14,479 Da.

monomer. Because we did not observe dimerization of the isolated monomer and considering that rCquiOBP1 monomer migrated in native gel as the native protein (data not shown), we used only the monomeric form of the protein in subsequent studies.

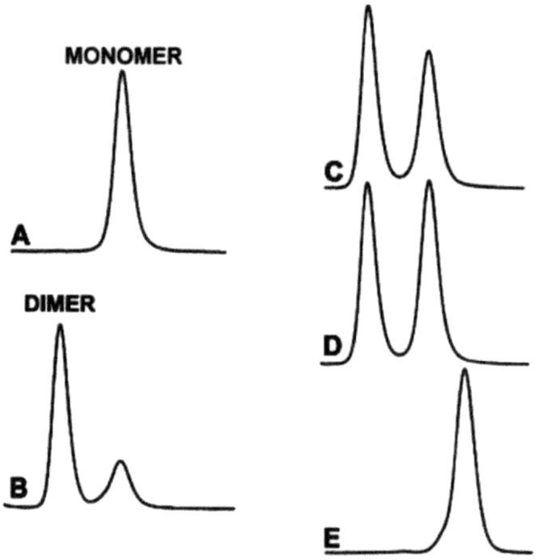

Figure 2. Gel filtration elution profiles of CquiOBP1. (A) Monomeric and (B) dimeric form of CquiOBP1. The dimer was isolated with a minor peak of the monomer. The dimer dissociates into monomer as indicated by the increase in the second peak (C) after 1 hr at room temperature and (D) overnight at 4°C. The dimeric form is also dissociated with organic solvent. (E) Sample D analyzed with acetonitrile in the mobile phase.

Localization of CquiOBP1 in Olfactory Sensilla of *Cx. p. Quinquefasciatus*

The CquiOBP1 was the first OBP isolated from any mosquito species [6]. Since then homologous proteins have been isolated and/or cloned from the malaria mosquito, *A. gambiae* s. s. [9], *Cx. tarsalis* [10], *Cx. Pipiens*, and *Cx. molestus* (Ishida and Leal, EU723597 and EU723598), and the yellow fever mosquito, *Aedes aegypti* [11, 12]. Although CquiOBP1 was demonstrated to be expressed specifically in female antennae of the Southern house mosquito, OBPs from mosquitoes are yet to be mapped in specific olfactory tissues.

The olfactory sensilla in *Culex* mosquitoes are morphologically comparable to those in *A. aegypti* [13]. The antennae are endowed with three types of trichoid (single-walled multiporous) sensilla and one type of grooved (double-walled multiporous) peg sensilla, whereas the maxillary palps house only (single-walled multiporous) peg sensilla. The trichoid sensilla are further classified into sharp-tipped (A1), long (A1-I), or short (A1-II), and blunt-tipped (A2), whereas the grooved pegs are designated as A3. Single-sensillum recordings indicate that MOP is detected by an olfactory receptor neuron with large spike amplitude (Figure 3A) in the short, sharp-tipped trichoid (A1-II) sensilla. Although these sensilla are morphologically indistinguishable, we identified two types of A1-II sensilla involved in MOP reception, a more frequently encountered type which responded with nearly the same sensitivity to both enantiomers of MOP (Figure 3A) and a more rare type that responded only to the compound with the same stereochemistry as the natural pheromone, (5R,6S)-MOP [14-16], and was silent to its antipode (data not shown). Both types of sensilla housed also a small spike amplitude neuron, which was very sensitive to skatole, a compound previously identified as a mosquito oviposition attractant [17, 18]. Although we found olfactory receptor neurons sensitive to (5S,6R)-MOP, behavioral and field evaluations indicated that the non-natural stereoisomer is neither an attractant nor repellent [14-16, 19-22].

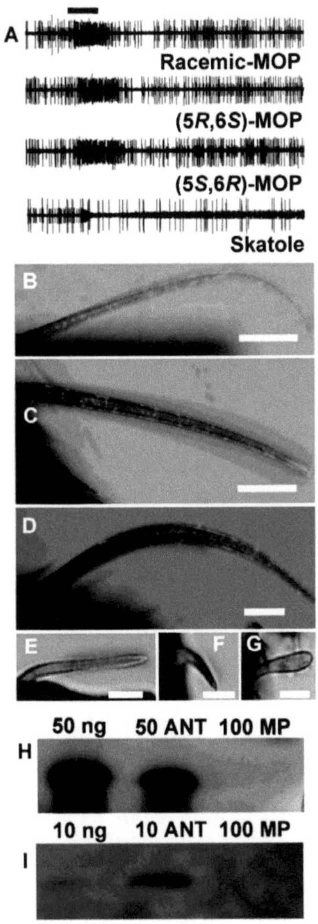

Figure 3. Oviposition pheromone reception. (A) Single sensillum recordings from short, sharp-tipped trichoid sensilla on the antennae of female *Cx. quinquefasciatus*. Response of a neuron to both the natural stereoisomer, (5R,6S)-MOP, and its antipode. The sensilla housed a second olfactory receptor neuron, characterized by a smaller spike amplitude, which was very sensitive to skatole. Bar denotes stimulus duration, 500 ms. Immunohistochemical localization of CquiOBP1 in the trichoid (B) long, sharp-tipped sensilla, (C) long, sharp-tipped sensilla with high density labeling at the excised tip, (D) MOP-detecting short, sharp-tipped sensilla, and (E) blunt-tipped sensilla on the antennae of female *Cx. quinquefasciatus*. The CquiOBP1 was not detected in the (F) grooved peg sensilla on the antennae, and (G) the peg sensilla on maxillary palps. Scale bars, B, C: 10 μm, others, 5 μm. (H) Western blot analysis of protein extracted from olfactory tissues compared to recombinant CquiOBP1. (I) Same analysis as in H, but with 5x lower amounts of rCquiOBP1 and antennal extract. The ANT, antenna-equivalent; MP, maxillary palp-equivalent. While signal was detected from 10 antenna-equivalent, no signal was observed from extracts of 100 maxillary palp-equivalent.

The IHC with affinity purified CquiOBP1-specific antibody provided data on the precise localization of CquiOBP1 in *Cx. p. quinquefasciatus* olfactory tissues. Labeling indicated that CquiOBP1 is expressed in most types of sensilla on the antennae, but not in the grooved pegs on the maxillary palps (Figure 3B–G). Fluorescence signal showed the presence of CquiOBP1 in the long, sharp-tipped sensilla (A1-I) (Figure 3B), the short, sharp-tipped (A1-II) sensilla (Figure 3D), and blunt-tipped (A2) sensilla (Figure 3E). Sensilla fractured or cut during preparations showed high density labeling at the exposed site suggesting that these incisions allowed more penetration of antibody (Figure 3C), whereas sensilla from control treatments showed no labeling (data not shown). Density of labeling in the sensillar lymph of grooved pegs (A3) sensilla (Figure 3F) was below the detection limit. We concluded that CquiOBP1 is not expressed in the A3 sensilla, but a caveat is the possibility that the double-walled structures of these sensilla filter out fluorescence signal. No density labeling was observed in the single-walled multiporous peg sensilla (Figure 3G) on the maxillary palps, but in this case we were able to unambiguously demonstrate that CquiOBP1 is not expressed in the maxillary palps. Western blot analyses (Figures 3H and I) confirmed that, as opposed to antennae, the maxillary palps express no detectable amounts of CquiOBP1. Previously, we have shown that the grooved peg sensilla on the maxillary palps are more than CO_2 detectors and house olfactory receptor neurons, which are highly sensitive to 1-octen-3-ol, and various plant-derived compounds [23]. Therefore, CquiOBP1 is unlikely to be involved in the transport of any of these odorants in the sensillar lymph of the peg sensilla on the maxillary palps. On the other hand, expression of CquiOBP1 in all types of antennal trichoid sensilla suggests that this olfactory protein may be involved in the detection of MOP and other semiochemicals.

pH-dependent Binding of MOP to CquiOBP1

Given the high level of CquiOBP1 expression in antennae, we reasoned that it might be involved in the detection of MOP and other oviposition attractants. Indeed, MOP binds to CquiOBP1 with apparently high affinity at pH 7 (Figure 4A). Since moth pheromone-binding proteins undergo pH-dependent conformational changes [7, 24] that leads to decreased binding affinity at low pH [25, 26], we examined the effect of pH on the ability of CquiOBP1 to bind MOP. At low pH, the amount of ligand recovered by incubation with protein was not significantly different from that detected in buffer only (Figure 4A) (Wilcoxon–Mann–Whitney unpaired rank sum test, $P > 0.05$) thus suggesting that at low pH MOP does not bind to CquiOBP1, or binds with significantly reduced affinity. To examine further the pH-dependent binding of CquiOBP1 to MOP, we tested binding at high and low pH by fluorescence using N-phenyl-1-naphthylamine (NPN) as a fluorescent reporter [27]. The MOP replaced NPN at pH 7 (Figure 5A), but not at pH 5 (Figure 5B) thus confirming that binding affinity is lost or is very weak at low pH. The molecular basis for loss of binding affinity of pheromones to pheromone-binding proteins at low pH has been elucidated in moths [7, 24, 28-30]. Protonation of acidic residues in the C-terminus of the silkworm pheromone-binding protein at low pH triggers the formation of an additional α-helix [29, 31], which occupies the binding cavity [28]. By contrast, the structure of AgamOBP1 [8], a protein homologous to CquiOBP1, lacks a corresponding C-terminal helix that is found in the moth protein. Therefore, CquiOBP1 may have a different mechanism for pH-dependent odorant binding. Circular dichroism data suggest that the helical content of CquiOBP1 is reduced at low pH (Figure 6) thus implying possible unwinding of helical structure(s) at low pH.

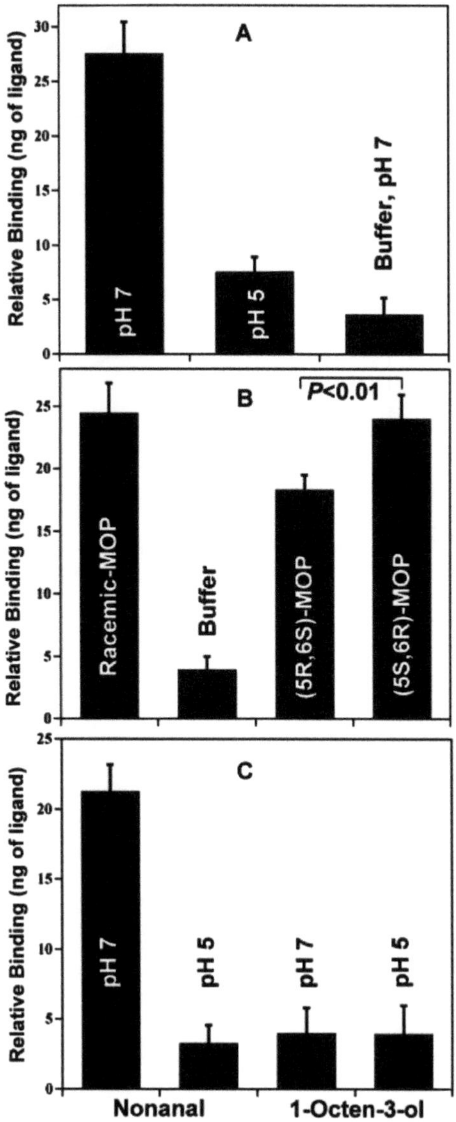

Figure 4. Binding of test ligands to antennae-specific CquiOBP1. (A) The pH-dependent binding of racemic MOP to CquiOBP1. (B) Binding of enantiomers compared to racemic MOP. The non-natural stereoisomer, (5S,6R)-MOP showed significantly higher affinity for CquiOBP1 than the natural pheromone. (C) Nonanal bound to CquiOBP1 with high affinity at high but not low pH, whereas 1-octen-3-ol did not bind the protein at high or low pH.

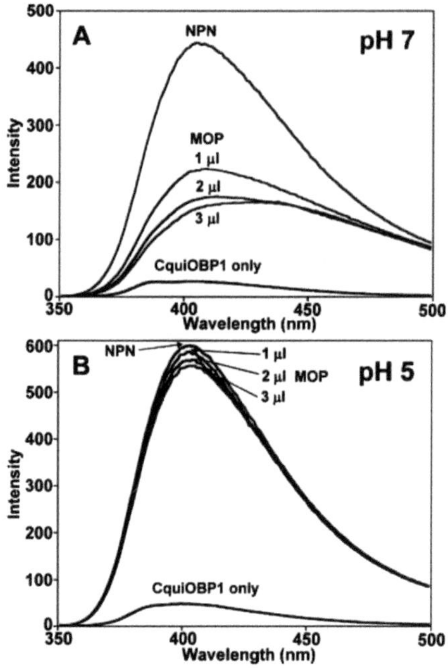

Figure 5. Competitive binding of MOP to CquiOBP1. Fluorescence emission spectra of CquiOBP1 alone (15 µg/ml; black), in the presence of NPN (2 µl, 3.2 mM), and after titrating with increasing amounts of MOP (1–3 µl, 3.2 mM). (A) Replacement of fluorescent reporter by MOP is indicated by decrease of emission, and suggests competitive binding. (B) Excitation of the fluorescent reporter was not changed with addition of MOP thus indicating no MOP binding at low pH.

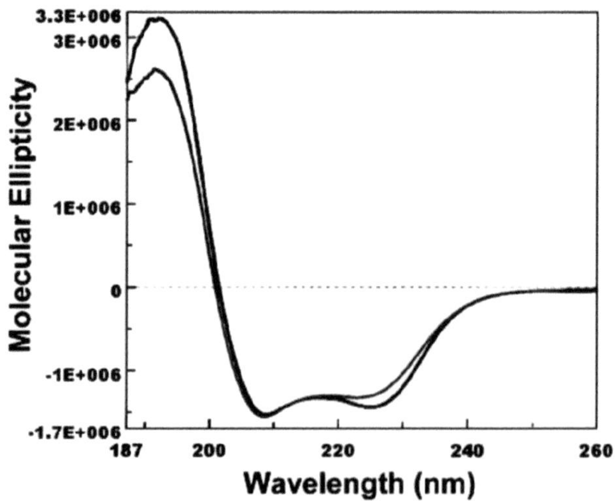

Figure 6. Far-UV circular dichroism spectra of CquiOBP1 at pH 6.5 and pH 5. The helical-rich protein underwent unwinding of α-helix at low pH as indicated by the change in the intensity of the second minima.

Next, we examined if CquiOBP1 could discriminate MOP enantiomers. The stereochemistry of the natural product isolated from eggs of *Culex pipiens* fatigans (= *Cx. p. quinquefasciatus*) has been determined to be (5R,6S)-6-acetoxy-5-hexadecanolide, ((5R,6S)-MOP) [16]. We isolated the oviposition pheromone from the Southern house mosquito, analyzed the extract by GC with a chiral column and confirmed that *Cx. p. quinquefasciatus* produces (5R,6S)-MOP (Figure 7C) thus ruling out possible geographical or other variations. Binding assays showed that both stereoisomers of MOP bound to CquiOBP1 at pH 7 (Figure 4B), but not at pH 5 (data not shown). Intriguingly, the antipode of the natural compound, (5S,6R)-MOP showed significantly higher binding affinity than the natural pheromone (Wilcoxon–Mann–Whitney unpaired rank sum test, N = 15, $P < 0.01$) (Figure 4B). Because this assay [26] is sensitive to the amount of ligand incubated, it is worth mentioning that the amounts of the enantiomers were adjusted by GC to measure binding to analytically equal amounts of the two enantiomers. This chiral mismatch of MOP and CquiOBP1 suggests that odorant receptor(s) may play a more significant role than CquiOBP1 in the chiral discrimination observed by olfactory receptor neurons in one type of A1-II sensilla (data not shown). An inability to discriminate enantiomers has also been observed in the pheromone-binding protein from the Japanese beetle [32].

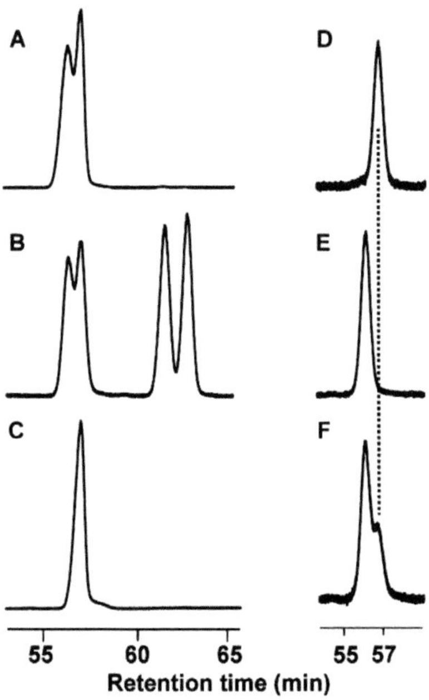

Figure 7. Resolution of MOP stereoisomers on a chiral column. (A) Partially resolved enantiomers of MOP. (B) Random mixture of stereoisomers showing two clusters of well separated stereoisomers: erythro- and threo-MOP. (C) The pheromone isolated from egg rafts of *Cx. quinquefasciatus* showed the same retention time as the second peak in the erythro-MOP cluster, and is thus confirmed to be (5R,6S)-MOP. (D) The natural stereochemistry and (E) the configuration of the antipode were retained upon binding. (F) Competitive binding with the two enantiomers showed that CquiOBP1 has a higher affinity for the non-natural stereoisomer (first peak) than for natural stereoisomer, (5R,6S)-MOP (the second peak).

We determined the stereochemistry of ligands recovered in these binding assays by GC. It was not surprising that both the natural stereoisomer and its antipode retained their absolute configuration. When synthetic (5R,6S)-MOP was incubated with CquiOBP1 the pheromone with the same configuration as the natural product was recovered from the bound protein (Figure 7D). Likewise, the antipode retained its configuration and was recovered as (5S,6R)-MOP (Figure 7E). Interestingly, a racemic mixture of MOP extracted from bound-CquiOBP1 had a much higher proportion of the non-natural stereoisomer (Figure 7F). In these enantiocompetitive assays, despite the protein being incubated with a racemic mixture slightly richer in the natural stereoisomer (Figure 7A) more (5S,6R)-MOP was still recovered from the bound protein. Inversion of configuration can be ruled out on the basis of experiments with one enantiomer at a time. Taken together, these data suggest that CquiOBP1 has a higher affinity for (5S,6R)-MOP than the enantiomer with the same absolute configuration as the oviposition pheromone.

Prospecting for Oviposition Attractants

Since the MOP is detected only by antennae and CquiOBP1 is likely involved MOP reception (see above), we used this molecular target in a "reverse chemical ecology" approach to identify candidate compound(s) for subsequent field tests. We measured binding of CquiOBP1 to a few compounds known/inferred to be mosquito attractants. While CquiOBP1 bound to nonanal with apparently high affinity, we detected no binding to 1-octen-3-ol (Figure 4C), 1-octyn-3-ol, (R)-4-isopropenyl-1-methylcyclohexene (D-limonene), and oxidized D-limonene (data not shown). Attempts to determine binding of CquiOBP1 to skatole and cresols were unsuccessful, because high levels of non specific binding generated high backgrounds after incubation even with buffer alone. The fact that 1-octen-3-ol is detected with high sensitivity by olfactory receptor neurons in the maxillary palps [23] is consistent with both the results of our binding assays and the lack of CquiOBP1 expression in peg sensilla on the maxillary palps (Figure 3G). This olfactory protein-based approach prompted us to test nonanal in the field and not to explore 1-octen-3-ol as an oviposition attractant. The latter compound has been demonstrated to attract other *Culex* species, but not *Cx. p. quinquefasciatus* [33], and furthermore it is unlikely to be an oviposition attractant. On the other hand, it has been demonstrated in laboratory bioassays that *Cx. p. quinquefasciatus* laid more eggs in water treated with candidate compound, nonanal, than in controls [17].

Field tests showed that gravid female traps baited with nonanal at all concentrations tested caught significantly more *Cx. p. quinquefasciatus* females than control traps, but the efficacy of the traps was dose-dependent. Catches in traps loaded with 0.15 ng/μl of nonanal (mean ± SEM, 19.1 ± 2.2 female/trap/night) were significantly higher ($N = 12$, Tukey HSD, $P < 0.01$) than captures in control water traps (2.5 ± 0.6 female/trap/night), and did not differ significantly from catches in infusion-baited traps (25.2 ± 1.2 female/trap/night). Captures decreased at higher and lower concentrations: 15 ng/μl (6.6 ± 1.4 female/trap/night), 1.5 ng/μl (7.5 ± 1.1 female/trap/night), and 0.015 ng/μl (7.7 ± 1.9 female/trap/night).

GC-EAD-based Identification of Attractants from Rabbit Chow Fermentation

We further prospected for test compounds by a conventional chemical ecology approach, that is, gas chromatography coupled with an electroantennographic detector (GC-EAD). In a project concurrent with this work, we have evaluated infusion-based gravid traps in California, and observed that proportionally more WNV-infected *Culex* mosquitoes were collected in gravid traps than in CO_2-baited traps, although total capture of mosquitoes in the latter was significantly higher than catches in traps baited with Bermuda grass or rabbit chow infusions [34]. Of particular note, rabbit chow-baited traps collected the largest numbers of urban *Cx. p. quinquefasciatus* mosquitoes in Los Angeles, even outperforming CO_2 traps. We, therefore, aimed at identifying electrophysiologically-active compounds from rabbit chow fermentations using GC-EAD, a technique previously utilized to identify attractants from Bermuda grass infusions [17]. With solid-phase micro extraction (SPME), three EAD-active peaks (Figure 8) were detected, including a compound with short retention time that probably would have been masked by the solvent peak in a conventional solvent extraction. Compounds 1, 2, and 3 were first identified by GC-MS and then confirmed with authentic standards to be TMA, nonanal, and 3-methylindole (skatole), respectively.

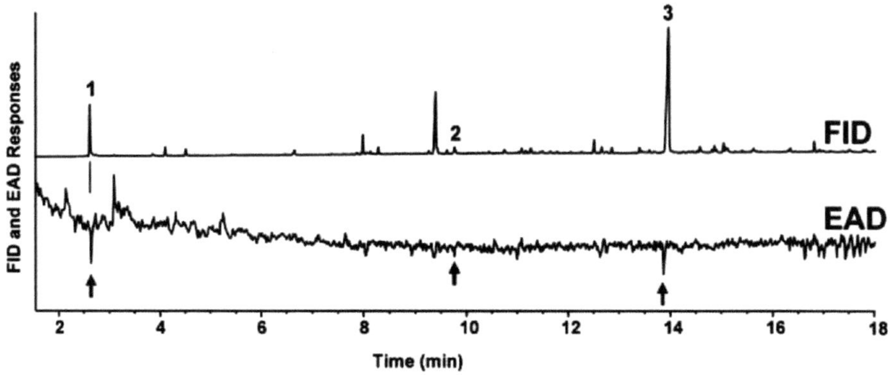

Figure 8. The GC-EAD analysis of rabbit chow fermentation products. The three EAD-active peaks (arrows) were identified as (1) trimethylamine, (2) nonanal, and (3) skatole.

In-depth Field Evaluation of *Cx. p. Quinquefasciatus* Attractants

Long-term field experiments to evaluate TMA and skatole individually as well as in binary and tertiary mixtures, including combinations with nonanal were conducted in Recife, Brazil. As opposed to field experiments with agricultural pests, evaluation of mosquito oviposition attractants is a time-consuming task due to heterogeneity and daily fluctuations of field populations, and, consequently, the need for high number of replicates to generate statistically reliable data. With the molecular and GC-EAD approaches we were able to concentrate our efforts on three potential attractants. First, we tested if individual compounds were attractants and then aimed at determining the optimal doses. The highest trap catches with TMA-baited gravid traps were obtained with 0.9 μg of TMA per liter (7.5 ± 1.3 females/trap/night), which was significantly higher than captures in control (water) traps (1.3 ± 0.5 females/trap/night; N = 5, Tukey HSD, $P < 0.01$). Captures decreased dramatically at lower concentrations of 90 ng/l (4.1 ± 1.1 females/trap/night), 9 ng/l (3.1 ± 1.2 females/trap/night), and 0.9 ng/l (2.3 ± 0.9 females/trap/night). Catches in traps baited with a higher than optimal concentration, 9 μg/l (1.7 ± 0.5 females/trap/night), did not differ significantly from those in control traps ($P > 0.05$). Previous laboratory and field assays showed that *Aedes albopictus* did not exhibit attraction, greater oviposition, or an electrophysiological response to TMA [35]. We also observed concentration-dependent performance of skatole-baited traps, in agreement with previously field studies [18]. Captures in traps baited with skatole in decadic solutions (20 pg to 0.2 μg of skatole per liter) were all significantly greater than catches in control traps (2.5 ± 0.8 female/trap/night), with the greatest captures being obtained with 2×10^{-3} μg/l (9.8 ± 2.1 females/trap/night) and 20×10^{-3} μg/l (7.4 ± 1.2 females/trap/night). Because these two dosages were not significantly different (Tukey HSD, $P > 0.05$), the higher concentration (20×10^{-3} μg/l) was adopted to compensate for evaporation in week-long experiments. The dose-dependence observed in our field tests is consistent with previous studies [18], but the optimal doses differ, probably because of differences in trap designs and, more importantly, differences in end-point measurements. While we evaluated performance on the basis of female capture, Mboera and collaborators focused on egg laying by counting the number of egg rafts.

Tests with binary mixtures showed that a combination of skatole and TMA (5.3 ± 1.1 female/trap/night; N = 8) did not improve captures compared to traps baited with skatole (4.3 ± 0.9 female/trap/night) or TMA alone (6.5 ± 1.3 female/trap/night). By contrast, captures in traps baited with a combination of nonanal and skatole (13.9 ± 1.9 female/trap/night; N = 17) were higher than those in traps with individual compounds (nonanal, 5.1 ± 0.9; skatole, 10.1 ± 2.7 female/trap/night). A synergistic effect was observed with a combination of nonanal and TMA (10.1 ± 2.5 female/trap/night; N = 10) compared to TMA alone (2.7 ± 0.6 female/trap/night) or nonanal alone (3.7 ± 1.2 female/trap/night). Due to the difficulty of simultaneously testing a large number of lures (three compounds at 3–5 dosages), these tests with binary versus individual compounds were conducted at different times of the year with different population levels, but later the optimal binary mixtures were compared at the same time with a tertiary mixture and positive control infusion. In these competitive field tests, captures in traps baited with binary or tertiary synthetic mixtures were significantly higher than those in negative control traps (Figure 9). Catches in traps baited either with nonanal plus TMA or skatole plus TMA were not significantly different (Tukey HSD, N = 19; $P > 0.05$) from captures in traps loaded with infusion (positive control). In the doses tested, the combination of the three compounds did not perform better than the binary mixtures. We adopted the mixture of nonanal and TMA for subsequent studies, because at the optimal concentrations this lure is odorless, performs comparably to infusion, and is suitable for use in human dwellings.

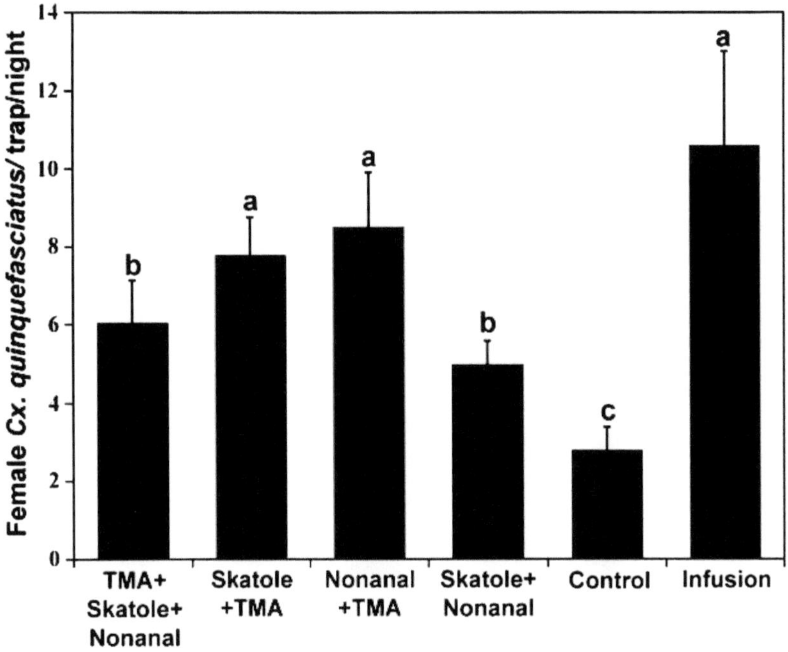

Figure 9. Field data comparing captures of female *Cx. quinquefasciatus* in synthetic mixtures- and infusion-baited traps. Catches in traps baited with nonanal and TMA, skatole and TMA, and infusion were not significantly different (Tukey HSD, $P > 0.05$), but the nonanal plus TMA lure is odorless and thus suitable for surveillance and use in monitoring population in human dwellings.

We then explored the possibility of synergism between this attractant mixture and MOP. Preliminary experiments with a previously tested dose (20 mg/trap) showed no significant difference between catches in traps baited with nonanal plus TMA compared to those with this binary mixture plus MOP. Follow-up experiments at lower doses demonstrated the same trend (Figure 10). The apparent discrepancy between our data and those previously reported [18, 19, 36] might be related to differences in behavioral measurements. We evaluated MOP for attraction (i.e., numbers captured) of gravid females whereas previous reports measured the effect of MOP and other attractants on egg laying. While the former is more important for surveillance, the latter seems to be a more tangible measurement for controlling mosquito populations. The differences in these results suggest that MOP might be an arrestant rather than an attractant thus leading to increased egg laying but not necessarily gravid female capture.

In summary, this work is the first report of translational research combining molecular basis of olfaction and chemical ecology to generate deliverable material for medical entomology. We have employed an OBP as a molecular target in binding assays in combination with a conventional chemical ecology approach to identify three compounds, namely, TMA, nonanal, and skatole, which were tested in extensive field studies. With this novel approach we developed a synthetic oviposition attractant mixture of nonanal and TMA, which is odorless and comparable in attraction to cumbersome infusions currently employed as standard lures. The newly developed user-friendly, synthetic mixture of readily available compounds holds considerable promise for future surveillance and management programs for *Cx. p. quinquefasciatus* and possibly other closely related *Culex* species with similar breeding requirements, which are major vectors of pathogens causing human diseases throughout the world.

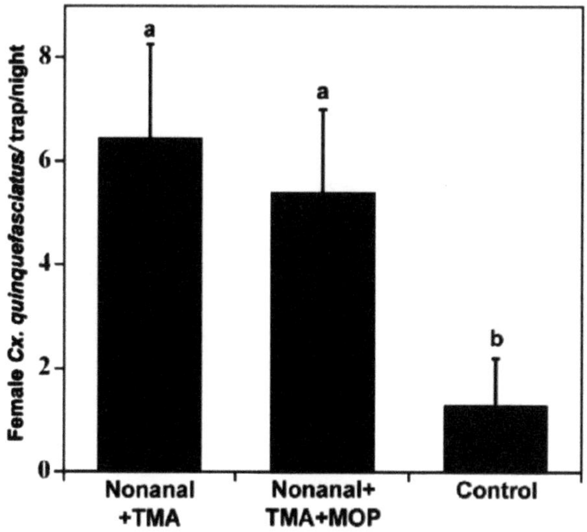

Figure 10. Catches in traps baited with a synthetic attractant alone or in combination with MOP. Captures in traps loaded with pheromone were not significantly different (Tukey HSD, N = 15, P > 0.05) from those in trap baited only with nonanal plus TMA.

MATERIALS AND METHODS

Protein Expression, Purification, and Antibody Production

One microgram of pET-22b(+) vector (EMD Chemicals, Gibbstown, NJ) was digested with 6 U of *Msc* I (New England Biolabs, Ipswich, MA) at 37°C for 3 hr. After purification of DNA by QIAquick PCR Purification Kit (Qiagen, Valencia, CA) the vector was digested with 7 U of *Bam* HI (New England Biolabs) at 37°C for 3 hr and subsequently gel-purified by QIAquck Gel Extraction Kit (Qiagen). The following primers were used for amplification of insert DNA: 5CquiOBP-1-KpnI, *5′-GGGGTAC/CC-GACGTTACACCgCGTCGtGA-3′* and 3CquiOBP-1-BamHI, *5′-CGCG/GATTCCT-TAAACCAGGAAATAATGCT-3′*. Slashes indicate cutting sites for *Kpn* I and *Bam* HI restriction enzymes, and lower cases in 5CquiOBP1-KpnI primer indicate bases replaced to overcome codon bias of *E. coli* and thus enhance protein expression. After amplification, and confirmation by sequencing, 16 μg of DNA was initially digested with 40 U of *Kpn* I (New England Biolabs) at 37°C for 3 hr, purified by QIAquick PCR Purification Kit, blunted by T4 DNA polymerase (New England Biolabs) with dNTP, and purified again by QIAquick PCR Purification Kit. Then, the DNA was digested with 20 U of *Bam* HI at 37°C for 3 hr and, gel-purified by QIAquick Gel

Extraction Kit, and ligated into prepared pET vector by T4 DNA ligase (New England Biolabs). CquiOBP1 was expressed in LB medium with transformed BL21 (DE3) cells, according to a protocol for perisplasmic expression of OBPs [7]. Proteins in the periplasmic fraction were extracted with 10 mM Tris·HCl, pH 8 by three cycles of freeze-and-thaw [37] and centrifuging at 16,000 × g to remove debris. The supernatant was loaded on a Hiprep™ DEAE 16/10 column (GE Healthcare Bio-Sciences, Piscataway, NJ). All separations by ion-exchange chromatography were done with a linear gradient of 0–500 mM NaCl in 10 mM Tris·HCl, pH 8. Fractions containing the target protein were further purified on a 20 ml Q-Sepharose Hiprep™ 16/10 column (GE Healthcare) and, subsequently, on a Mono-Q HR 10/10 column (GE Healthcare). The OBP fractions were concentrated by using Centriprep-10 (Millipore, Billerica, MA) and loaded on a Superdex-75 26/60 gel-filtration column (GE Healthcare) pre-equilibrated with 150 mM NaCl and 20 mM Tris HCl, pH 8. Highly purified protein fractions were concentrated by Centricon-10, desalted on four 5 ml HiTrap desalting columns (GE Healthcare) in tandem with water as mobile phase, and analyzed by LC-ESI/MS (see below), lyophilized, and stored at −80°C until use. The concentrations of the recombinant proteins were measured by UV radiation at 280 nm in 20 mM sodium phosphate, pH 6.5 and 6 M guanidine HCl by using the theoretical extinction coefficients calculated with EXPASY software (http://us.expasy.org/tools/protparam.html). An aliquot of highly purified CquiOBP1 was provided to Invitrogen (Camarillo, CA) for preparation of affinity purified CquiOBP1-specific rabbit antibody.

Immunohistochemistry (IHC)

Immunofluorescence was performed with modification of previously published protocols [38, 39]. Heads of 5- to 7-day-old female *Cx. quinquefasciatus* were dissected from adult mosquitoes anesthetized on ice and fixed overnight with 4% paraformaldehyde in 1 × PBS at 4°C. To dehydrate, the preparations were first rinsed in 1 × PBS and then incubated overnight in 24% sucrose in 1 × PBS at 4°C. After rinsing in 1 × PBS, heads were embedded in Tissue Tec® optimal cutting temperature medium (Sakura Finetek, Torrance, CA) and frozen at −22°C on the object holder. Sections (14 μm) were prepared at −24°C (Leica CM1850 Cryostat, Bannockburn, IL) and after thawing they were mounted on Superfrost Plus slides (Fisher, Pittsburgh, PA) and air dried for at least 30 min. Preparations were incubated at 4°C initially in 4% paraformaldehyde in 1 × PBS for 30 min, then in 1 × PBS for 10 min, and finally in PBST for 30 min. After that, sections were washed twice for 5 min in 1 × PBS and then the slides were incubated in PBST with 1% blocking reagent (Roche) for 30 min. The IHC was performed using affinity purified CquiOBP1-specific rabbit antibody diluted 1:1,000 in blocking solution. After washing five times with PBST, the sections were incubated with secondary antibody, Cy™3 linked goat anti-rabbit IgG (GE Healthcare, Piscataway, NJ) 1:500 in blocking solution, for 1 hr at room temperature in a humid box. Subsequently, slides were washed three times for 5 min each with PBS and sections were embedded in Vectashield mounting medium (Vector Laboratories, Burlingame, CA), covered and sealed with nail polish around the cover glass (Corning Labware and Equipment, Corning, NY). Confocal images were captured with a FV1000 Olympus Confocal Microscope system (Olympus America, Center Valley, PA).

WESTERN BLOT

Culex p. quinquefasciatus used in this study were from a laboratory colony originating from adult mosquitoes collected in Merced, CA in the 1950s and maintained under lab conditions at the Kearney Agricultural Center, University of California, as previously described [23]. Three to 5-day-old adult mosquitoes were anesthetized on ice. Antennae and maxillary palps were dissected and homogenized in ice-cold glass homogenizers with 10 mM Tris HCl, pH 8. Homogenized samples were centrifuged twice at 14,000 × g for 10 min at 4°C. After concentration, the supernatants were analyzed by native and SDS polyacrylamide gel electrophoresis (15% PAGE), and proteins were electroblotted onto polyvinyl difluoride (PVDF) membranes (Bio-Rad Laboratories, Hercules, CA). After treatment with 1% blocking reagent (Roche, Indianapolis, IN) in 1 × PBS (140 mM NaCl, 2.7 mM KCl, 1.8 mM KH_2PO_4, 10 mM Na_2HPO_4, pH 7.4) for 1 hr at room temperature, the membrane was incubated for 1 hr with affinity purified CquiOBP1-specific rabbit antibody diluted 1:2,000 with 1% blocking reagent. After washing four times with PBST (1 × PBS containing 0.1% Triton X-100), the membrane was incubated for 1 hr with anti-rabbit IgG, horseradish peroxidase (HRP) conjugate (dilution 1:5,000) (Millipore, Temecula, CA). Immunoreacting bands were detected by treatment with SuperSignal West Femto Maximum Sensitivity Substrate (Pierce, Rockford, IL).

Single Sensillum Recording

Recording from *Cx. p. quinquefasciatus* female antennae were performed as previously described for maxillary palps [23].

Binding Assays

Binding of CquiOBP1 to MOP and other test compounds was measured by incubating protein sample and test ligand, separating unbound and bound protein, extracting the bound ligand with hexane and quantifying by gas chromatography, according to a previously reported protocol [26]. The pH-Dependent binding of CquiOBP1 to MOP was confirmed by an independent competitive binding assays using NPN as a fluorescent reporter [27].

CHEMICALS

A random sample of the four isomers of MOP was prepared according to a previously reported method [20] and used only to determine the retention times of the four isomers after separation on a chiral column. Samples of racemic MOP, (5R,6S)-6-acetoxy-5-hexadecanolide, and (5S,6R)-6-acetoxy-5-hexadecanolide were gifts from Bedoukian Research Incorporated (BRI). The ^{1}H and ^{13}C NMR spectra of the racemic pheromone were consistent with published characterization data [40]. The ^{1}H NMR (400 MHz) 0.83 (t, 3H), 1.16–1.36 (br m, 16H), 1.53–1.66 (m, 3H), 1.72–1.98 (m, 3H), 2.04 (s, 3H), 2.35–2.46 (m, 1H), 2.51–2.60 (m, 1H), 4.28–4.34 (m, 1H), 4.91–4.97 (m, 1H); ^{13}C (100 MHz) 14.12, 18.26, 21.03, 22.68, 23.48, 25.26, 29.31, 29.40, 29.45, 29.52, 29.54, 29.56, 29.63, 31.89, 74.26, 80.51, 170.45, 170.87. The enantiomeric purity of the two stereoisomers was determined by gas chromatography equipped with a chiral column, Chiraldex GTA (25 m × 0. 25 mm; 0.125 μm; Astec,

Whippany, NJ), which was operated at constant temperature, 175°C. The enantiomeric excess of (5R,6S)- and (5S,6R)-MOP were estimated to be 99.5 and 99%, respectively. For binding assays, samples of the stereoisomers of MOP were first prepared by weighting the amounts of the compounds and then their concentrations were adjusted to have the same amounts of the two stereoisomers by gas chromatography. Eicosyl acetate was a gift from Fuji Flavor Co., Tokyo, Japan and racemic 1-octen-3-ol was a gift from BRI. Nonanal and TMA were purchased from Fluka (Buchs, Switzerland), NPN and skatole were from Aldrich Chemical Co. (St. Louis, MI).

Chemical Analysis

The GC-EAD was done with a gas chromatograph (HP 5890, Agilent Technologies, Palo Alto, CA) equipped with transfer line and temperature control units (Syntech, Kirchzarten, Germany). The effluent from the capillary column was split into EAD and flame ionization detector (FID) in a 3:1 ratio. Antennae from blood-fed female mosquitoes were placed in EAG probes of an AM-01 amplifier (Syntech) and held in place with Spectra 360 electrode gel (Parker Laboratories, Orange, NJ). The analog signal was fed into an A/D 35900E interface (Agilent Technologies) and acquired simultaneously with FID signal on an Agilent Chemstation. The GC-MS was obtained on a 5973 Network Mass Selective Detector (Agilent Technologies). Both GC-EAD and GC-MS were equipped with the same type of capillary column (HP-5MS, 30 m × 0.25 mm; 0.25 µm; Agilent Technologies). The temperature program started at 50°C for 1 min, increased at a rate of 10°C/min to 250°C, and held at this final temperature for 10 min. Both GCs were operated under splitless mode with the injection port at 230°C and purge time 2 min.

Other Analytical Procedures

Fluorescence measurements were done on a spectrofluorophotometer (RF-5301, Shimadzu, Kyoto, Japan) at 25 ± 1°C. Samples in 2 ml cell equipped with magnetic stir bar were excited at 337 nm, and the emission spectra were recorded from 350 to 500 nm, with emission and excitation slit widths of 1.5 and 10 nm, respectively. The spectra were obtained with the protein sample (CquiOBP1, 10 µg/ml in either 20 mM ammonium acetate, pH 7 or in 20 mM sodium acetate, pH 5) and after adding NPN (3.2 mM, 2 µl; final concentration, 3.2 µM) and MOP (3.2 mM, 1–3 µl; final concentrations, 1.6–4.8 µM). Circular dichroism (CD) spectra were recorded with a Jasco J-810 spectropolarimeter (Easton, MD) with CquiOBP1 (25 µg/ml) either in 20 mM ammonium acetate, pH 6.5 or in 20 mM sodium acetate, pH 5. The LC-ESI/MS was performed with a LCMS-2010 (Shimadzu, MD). High pressure liquid chromatography (HPLC) separations were carried out on a ZorbaxCB C8 column (150 × 2.1 mm; 5 µm; Agilent Technologies, Santa Clara, CA) with a gradient of water and acetonitrile plus 2% acetic acid as a modifier. The detector was operated with the nebulizer gas flow at 1.0 l/min and the curved desolvation line and heat block at 250°C.

Infusions

Rabbit chow infusions were prepared as previously reported [34]. For volatile collections, aliquots (20 ml) of fresh batches were transferred to 100 ml beakers. Volatile compounds were trapped from the headspace of each beaker with 3–4 SPME syringes introduced through a cover of parafilm (American National Can, Neenah, WI), with one of the syringes being used for GC-MS analysis. We employed SPME blue fibers (StableFlex™, 65 µm, polydimethylsiloxane/divinylbenzene partially crosslinked; Supleco, St. Louis, MI). Grass infusions for field tests were prepared by adding 30 g of fresh Indian goosegrass, *Eleusine indica* (Cyperales, Poaceae), to 2 l of water and incubating at 27 ± 2°C for 7 days.

Field Tests

Preliminary field tests in Davis and Sacramento, CA were discontinued after aerial sprays in the area in the summer of 2005 to mitigate the levels of West Nile virus-infected mosquitoes. Follow-up field tests (January, 2006–July, 2008) were conducted in Recife, Brazil, a city endemic for lymphatic filariasis, with abundant populations of *Cx. quinquefasciatus* that breed throughout the year [41]. Tests were conducted in the backyards of six residences with gravid mosquito traps (Bioquip, Rancho Dominguez, CA). To minimize inconsistencies observed in preliminary experiments due to variations in battery power and fan speed, traps were modified to run on AC power. Traps filled with 5 l of tap water were placed on the ground with an intertrap distance of at least 3 m. Traps were inspected and rotated every morning for a week, collecting chambers were replaced, and the trapped mosquitoes were counted after the collecting chambers were placed in a freezer for 10 min. All samples were identified morphologically and then confirmed as *Cx. p. quinquefasciatus* by PCR [42]. Test compounds were diluted by transferring each sample with a separate, disposable glass capillary to water in each trap's tub. Throughout the chapter, concentrations refer to the final concentrations of the attractants after dilution in 5 l of water. When MOP was tested in combination with other attractants, the former was released from microscope cover slides (ca. 2 × 2 cm, Corning Labware and Equipment), which were allowed to float on each trap's tub water. Preliminary data were obtained with 20 mg of pheromone, a dosage previously tested in the field [18]. Each glass cover was prepared by transferring small aliquots of a hexane solution of either racemic- or (5R,6S)-MOP and letting the solvent evaporate until the desired amount (100 µl, 200 mg/ml) was loaded on the glass. Additional experiments were conducted with a lower dosage (2 mg; 20 µl, 100 mg/ml or racemic-MOP). Data were transformed to log (x + 1) and analyzed by ANOVA and Tukey HSD (honestly significant differences).

KEYWORDS

- **CquiOBP1**
- ***Culex***
- **Gas chromatography-electroantennographic detection**
- **Immunohistochemistry**
- **Liquid chromatography-electrospray ionization mass spectrometry**
- **Odorant-binding protein**
- **Sensilla**

AUTHORS' CONTRIBUTIONS

Conceived and designed the experiments: Walter S. Leal. Performed the experiments: Walter S. Leal, Rosângela M. R. Barbosa, Wei Xu, Yuko Ishida, Zainulabeuddin Syed, Nicolas Latte, Angela M. Chen, Tania I. Morgan, Anthony J. Cornel, and André Furtado. Analyzed the data: Walter S. Leal, Yuko Ishida, Zainulabeuddin Syed, Anthony J. Cornel, and André Furtado. Wrote the chapter: Walter S. Leal.

ACKNOWLEDGMENTS

We thank Doug Pesak from Bedoukian Research for providing samples of racemic and enatiomerically pure MOP, Fuji Flavor Co. for a sample of eicosyl acetate, David Brown, Joao L. do Nascimento, Jose R. de Lima, Pedro R. Oliveira, Silvio R. V. Antunes, Edmar G. da Silva, Roberto A. Vasconcelos, for assistance with field tests, and Jim Ames, Julien Pelletier, and Chris Barker for their critique of an earlier version of the manuscript.

Chapter 11

Morphogenesis of the Rat Pre-implantation Embryo and Environmental Toxicant

Karla J. Hutt, Zhanquan Shi, David F. Albertini, and Brian K. Petroff

INTRODUCTION

Environmental toxicants, whose actions are often mediated through the aryl hydrocarbon receptor (AhR) pathway, pose risks to the health and well-being of exposed species, including humans. Of particular concern are exposures during the earliest stages of development that while failing to abrogate embryogenesis, may have long term effects on newborns or adults. The purpose of this study was to evaluate the effect of maternal exposure to the AhR-specific ligand 2,3,7,8-tetrachlorodibenzo-p-dioxin (TCDD) on the development of rat pre-implantation embryos with respect to nuclear and cytoskeletal architecture and cell lineage allocation.

We performed a systematic 3-dimensional (3D) confocal microscopy analysis of rat pre-implantation embryos following maternal exposure to environmentally relevant doses of TCDD. Both chronic (50 ng/kg/wk for 3 months) and acute (50 ng/kg and 1 μg/kg at proestrus) maternal TCDD exposure disrupted morphogenesis at the compaction stage (8–16 cell), with defects including monopolar spindle formation, f-actin capping and fragmentation due to aberrant cytokinesis. Additionally, the size, shape, and position of nuclei were modified in compaction stage pre-implantation embryos collected from treated animals. Notably, maternal TCDD exposure did not compromise survival to blastocyst, which with the exception of nuclear shape, were morphologically similar to control blastocysts.

We have identified the compaction stage of pre-implantation embryogenesis as critically sensitive to the effects of TCDD, while survival to the blastocyst stage is not compromised. To the best of our knowledge this is the first *in vivo* study to demonstrate a critical window of pre-implantation mammalian development that is vulnerable to disruption by an AhR ligand at environmentally relevant doses.

The AhR pathway is a widely expressed orphan receptor pathway activated by many environmental toxicants and carcinogens. The AhR ligands, including dioxins and polychlorinated biphenyls, induce a spectrum of developmental and toxic responses by modifying gene expression, altering hormonal profiles and disrupting cell proliferation and differentiation [1]. Epidemiological studies in adult human populations have linked dioxin exposure to defects in immune, neurological, and reproductive function, as well as cancer [2-4]. There is now a growing concern that tissue growth and differentiation during fetal development may be especially sensitive to dioxins and dioxin-like compounds. For example, accidental exposure of human mothers to some AhR ligands has been correlated with delayed growth and development, a number of physical abnormalities, as well as intellectual and behavioral deficits in their children [5]. Moreover, animal studies show embryonic lethality, teratogenesis, cleft palate, hydronephrosis, and growth retardation among the many adverse effects observed following gestational exposure to AhR ligands [6-8]. While, it is accepted that maternal exposure to AhR ligands during gestation is detrimental to the health of offspring, neither the mechanism of toxicity nor the exact stages of development affected by dioxins have been fully elucidated.

Past studies examining maternal dioxin exposure and subsequent fetal health have primarily focused on post-implantation embryogenesis, while the impact of environmental contaminants on the periconceptional and pre-implantation period remained largely unexplored. The maternal environment during this earliest window of development has been hypothesized as critical to the long term health of offspring [9]. For example, poor maternal nutrition around the time of conception and during pre-implantation development reduces birth weight in humans [10] and animals [11] and predisposes offspring to hypertension later in life [11]. Similarly, periconceptional exposure of mice to environmentally relevant doses of the environmental estrogen bisphenol A induces errors in meiotic chromosome segregation, yielding embryos that survive gestation but give rise to offspring with frank genetic deficits [12]. Therefore, toxic exposure during pre-implantation embryonic development could potentially induce long term effects on fetal and offspring health.

Compacting morulae may be particularly vulnerable to the effects of environmental toxicants. Compaction is a morphogenetic process during which mammalian embryos undergo major cytoplasmic, nuclear, and cytoskeletal remodeling events that lead to the establishment of apical-basal polarity [13-17]. Polarization permits the differentiative divisions that lead to the allocation of trophectoderm (TE) and the inner cell mass (ICM) for the placental and embryonic lineages, respectively [18]. Therefore, disruption of pre-implantation embryo morphogenesis by AhR ligands could conceivably have consequences for lineage allocation. This concept is supported by the finding that *in vitro* exposure of 2 cell mouse embryos to TCDD results in increased cavitation rates, a functional measure of TE differentiation [19]. Similarly, Tsutsumi et al. showed that *in vitro* exposure of 2 cell mouse pre-implantation embryos to very low levels of TCDD reduced the number of pre-implantation embryos that developed to 8 cells relative to controls, whereas blastocyst formation of the surviving 8 cell pre-implantation embryos was accelerated [20]. These studies suggest differential and stage specific effects of TCDD during pre-implantation development.

The extent to which TCDD, at environmentally relevant doses, perturbs pre-implantation mammalian development in intact reproductively fit animals has yet to be fully evaluated. Thus, in a well-established rat model, we studied the effect of maternal TCDD exposure on early embryogenesis with respect to blastomere nuclear and cytoarchitecture. We show specific nuclear and cytoskeletal modifications revealed from a systematic 3D confocal microscopy analysis of rat pre-implantation embryos following maternal exposure to TCDD. Both chronic and acute maternal TCDD exposure disrupted morphogenesis at the compaction stage (8–16 cell), with defects including monopolar spindle formation, f-actin capping, aberrant cytokinesis, and distortion of nuclear shape and position. Notably, maternal TCDD exposure did not compromise survival to blastocyst, which with the exception of nuclear shape, were morphologically similar to control blastocysts. These studies raise further concerns regarding the consequences of early embryo exposures to prevalent environmental toxicants like TCDD.

Chronic Maternal TCDD Exposure Disrupts 8–16 Cell Pre-implantation Embryo Morphogenesis

We first asked whether chronic maternal exposure to TCDD affected the number of pre-implantation embryos relative to controls. As shown in Table 1, exposure to TCDD had no effect on pre-implantation embryo number, suggesting that ovulation, fertilization efficiency, and embryonic survival were not influenced at these doses. These findings are consistent with previous studies demonstrating embryotoxicity only at considerably higher doses of TCDD [21]. Approximately 80% of pre-implantation embryos from control rats were morphologically normal 8–16 cell stage concepti, with regular shaped and sized blastomeres and no nuclear or cellular fragmentation (Table 1, Figure 1A–H). The apical and basolateral surfaces of each blastomere were smooth and the

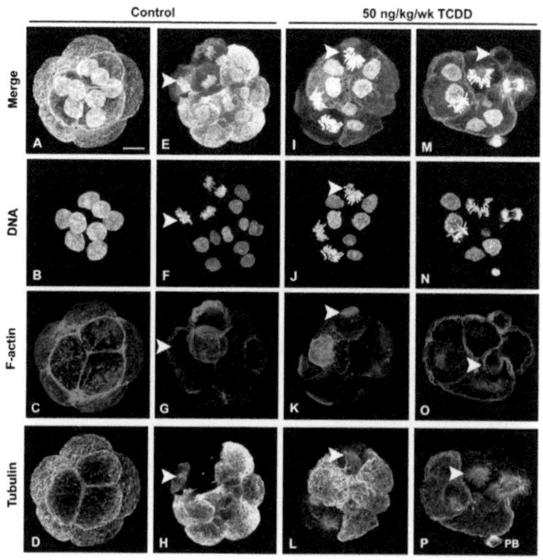

Figure 1. Chronic maternal TCDD exposure induces nuclear and cytoskeletal defects in compaction stage pre-implantation embryos. Compaction stage pre-implantation embryos from control and chronically exposed (50 ng/kg/wk TCDD) female rats were processed for visualization of microtubules, f-actin and DNA by confocal microscopy. (AD) Control 8-cell pre-implantation embryo with blastomeres of similar size and shape, basally positioned interphase nuclei and cytoplasmic microtubule arrays. F-actin is distributed at the cell cortex. (EH) Control 12-cell pre-implantation embryo with a normal bipolar mitotic spindle (E and H, arrows), metaphase chromosome configuration (F, arrow) and cortical f-actin localization (G, arrow). (IL) 50 ng/kg/wk TCDD exposed 9-cell pre-implantation embryo with abnormal mitotic spindles (L, arrow) and metaphase chromosome configurations (J, arrow), and enhanced f-actin cortical localization (K, arrow) in multiple blastomeres. (MP) 50 ng/kg/wk TCDD exposed 8-cell pre-implantation embryo with an anucleate fragment (M, arrow) and abnormal cytokinesis (O, arrows). Monopolar spindle (P, arrow). PB, polar body. Scale bar: 15 μm.

overall structure of controls was highly organized with respect to the relative position of blastomeres. Interphase blastomeres from controls had cytoplasmic microtubule networks, basally positioned nuclei and apically polarized f-actin localization. Centrally positioned bipolar spindles were consistently observed in mitotic blastomeres of control pre-implantation embryos (Figure 1E–H, arrow).

Table 1. Numbers and health status of pre-implantation embryos collected following chronic and acute maternal TCDD exposure.

Treatment (#animals)	Average# embryos/ animal (range)	Average# blastomere/ embryo (range)	# Normal embryos/ total embryos (%)
Chronic exposure (compaction)			
Control (3)	13.0 (12-14)	12.5 (7-18)	31/39 (79.5)
50 ng/kg/wk TCDD (3)	13.7 (13-IS)	11.3 (4-17)	1 5/41 (36.6)*
Acute exposure (compaction)			
Control (3)	13.3 (13-14)	9.5 (3-16)	32/40 (80.0)
50 ng/kg TCDD (S)	12.0 (11-13)	9.5 (3-16)	31/60 (5 1.7)*
1 μg/kg TCDD (6)	9.2 (3-13)	10.1 (6-16)	21/46 (45.7)*
Acute exposure (blastocyst)			
Control (6)	10.5 (9-12)	31.9 (13-52)	48/63 (76.2)
50 ng/kg TCDD (3)	9.7 (9-10)	32.1 (8-S8)	17/29 (58.6)
1 μg/kg TCDD (S)	9.0 (9-12)	37.5 (8-63)	20/45 (45.5)*

*Significantly different from control

In contrast, significantly lower proportions (~37%) of pre-implantation embryos from chronically exposed dams were morphologically normal (Table 1). Embryos from chronically exposed dams exhibited a range of defects, including irregularly sized and shaped blastomeres (Figure 1I–P). Moreover, f-actin staining at the apical and basolateral boundary of compacted blastomeres was highly irregular (Figure 1O). Analysis of interphase blastomere nuclei also revealed alterations in shape not evident in controls (Figure 2). In many cases these nuclei had an irregular boundary with one or more prominent projections from the nuclear surface. In addition to a more central nuclear position, chronic TCDD exposure impeded chromosome segregation or cytokinesis resulting in binucleate cells, anucleate fragments, and cells containing large nuclei or micronuclei (Figure 1M–P). In fact, pre-implantation embryos exposed to TCDD frequently exhibited monopolar spindles (13/18 spindles were monopolar). The solitary spindle pole was usually oriented toward the apical surface of blastomeres, with poorly aligned chromosomes located at the distal ends of microtubules (Figure 1I–P). In mitotic blastomeres, aberrant f-actin caps were adjacent to the spindle pole (Figure 1I–P).

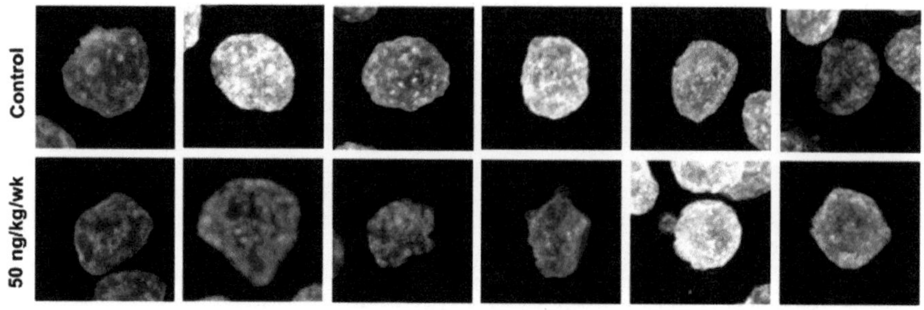

Figure 2. Nuclear profiles of compaction stage pre-implantation embryos from control and chronically exposed animals. The Z-series datasets for the DNA channel were compressed into a single plane and six randomly selected nuclei (each from a different pre-implantation embryo) for each group (Control and 50 ng/kg/wk TCDD) were compared.

Acute Periconceptional TCDD Exposure Disrupts 8–16 Cell Pre-implantation Embryo Morphogenesis

These initial findings prompted further analysis of cytoskeletal and nuclear characteristics in pre-implantation embryos to determine if limiting TCDD exposure to the time between oocyte maturation, ovulation, and implantation (i.e., the periconceptional period) would similarly modify pre-implantation embryo organization. Mature naturally cycling female rats were exposed to a single dose of TCDD (50 ng/kg or 1 μg/kg) or vehicle on the evening of proestrus and compaction (8–16 cell) and early blastocyst stage (32 cell or more) pre-implantation embryos were collected and analyzed.

Acute maternal TCDD exposure at the lower dose did not alter the number of compaction stage pre-implantation embryos relative to controls, but only ~52% of pre-implantation embryos from treated animals were normal (Table 1). Exposure to the higher dose of TCDD decreased the number of pre-implantation embryos, and an even lower proportion (~46%) of these pre-implantation embryos were normal (Table 1). A range of defects in nuclear and cytoskeletal integrity were observed (Figure 3D–L), including a dose dependent loss of microtubule and f-actin staining in some blastomeres (Figure 3J–L). Additionally, f-actin localization changed from a plasma membrane concentrated (Figure 3C) to a more diffuse pattern of stain (Figure 3I and L). In pre-implantation embryos from the 1 μg/kg treatment group, blastomeres often exhibited micronuclei (Figure 3K, arrow) and nuclei of different sizes. Further analyses of nuclear shape revealed a range of profiles deviating from being smooth surfaced in controls to more irregular contours detected in blastomere nuclei from either treatment group (Figure 4). Again, monopolar spindles with intense f-actin caps were evident at both low and high doses of TCDD (12/17 and 7/11 spindles, respectively, were monopolar) (Figure 3D–F). This was further confirmed by using pixel intensity line scans to monitor the topography of cortical f-actin staining in mitotic cells (Figure 3M and N). These analyses revealed equivalent intensity of f-actin around central spindles in control cells. However, as much as a 5-fold increase at the apical f-actin caps was detected in cells with monopolar spindles relative to the opposite side of the blastomere (Figure 3M and N). Frequently, astral-like microtubule fibers were detected between the spindle pole and cortex, accentuating the asymmetric displacement of chromosomes seen in pre-implantation embryos from treated animals (Figure 3N).

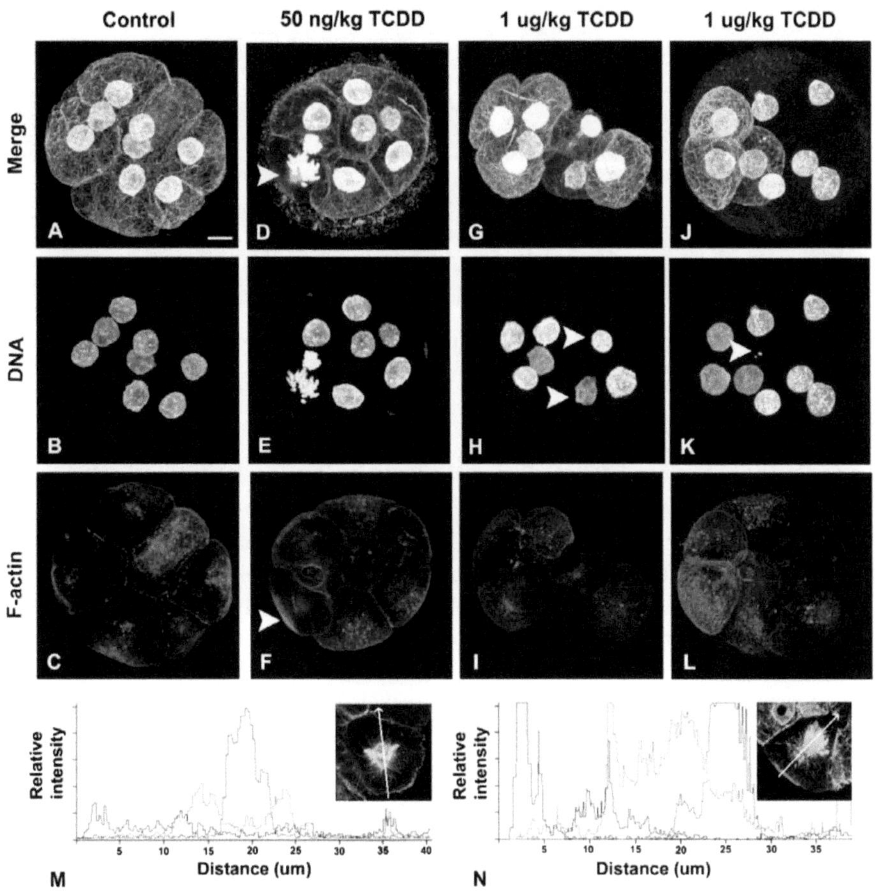

Figure 3. Acute maternal periconceptional TCDD exposure induces nuclear and cytoskeletal defects in compaction stage pre-implantation embryos. Compaction stage pre-implantation embryos collected from control and acutely exposed (50 ng/kg and 1 μg/kg TCDD) female rats and processed for visualization of microtubules, f-actin and DNA by confocal microscopy. (AC) Control 8-cell pre-implantation embryo. (DF) 50 ng/kg TCDD exposed 8-cell pre-implantation embryo with monopolar spindle (D, arrow) and f-actin cortical localization (F, arrow). (GI) 1 μg/kg TCDD exposed 8-cell pre-implantation embryo with distorted overall shape and centrally localized nuclei (H, arrows). (JL) 1 μg/kg TCDD exposed 8-cell pre-implantation embryo with global defects in the microtubule and f-actin networks; note micronuclei (K, arrow). Relative fluorescent intensity profiles for a normal bipolar spindle from a control pre-implantation embryo (M) and a monopolar spindle from a 50 ng/kg TCDD exposed pre-implantation embryo (N) are shown. Note the absence of tubulin fluorescence on the basal side of the aligned chromosomes and the intense f-actin signal at the apical cortex compared to the basal cortex (N). This is a different embryo to that shown in DF. Scale bar: 15 μm.

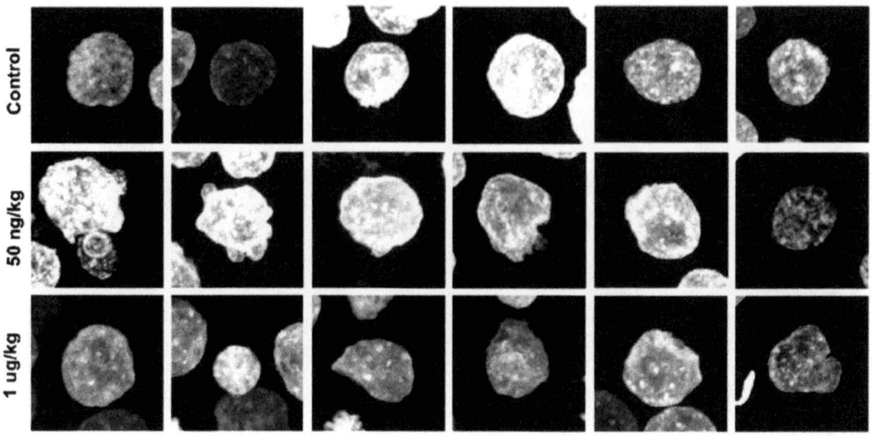

Figure 4. Nuclear profiles of compaction stage pre-implantation embryos from control and acutely exposed animals. The Z-series datasets for the DNA channel were compressed into a single plane and six randomly selected nuclei (each from a different pre-implantation embryo) for each group (Control and 50 ng/kg and 1 μg/kg TCDD) were compared.

Acute Periconceptional TCDD Exposure Permits Pre-implantation Embryo Survival to Blastocyst

We then asked whether the striking modifications evident at compaction were propagated through to the early blastocyst stage. In particular, it seemed likely that the occurrence of monopolar spindles would abrogate efficient cell cycle progression and negatively impact further pre-implantation embryo development. However, exposure to TCDD affected neither the number of pre-implantation embryos surviving to blastocyst nor the average number of cells within each blastocyst (Table 1), suggesting a normal rate of cell cycle progression in treated pre-implantation embryos.

The TCDD exposure at the higher dose did decrease the number of blastocysts exhibiting normal morphology (~46%) (Table 1), though TCDD associated abnormalities at the blastocyst stage were considerably less severe than those observed at the 8–16 cell stage. Blastocysts obtained from control animals typically contained 32 cells or more, formed blastocoel cavities and were comprised of cells with the distinct morphology of both TE and ICM (Figure 5A and B). The majority of blastocysts from treated animals also attained this general morphology. However, approximately half of the blastocysts exhibited discernable defects in one or more outer cells following TCDD exposure. Amongst these defects, cells failed to assume the squamous morphology typical of TE, exhibited a loss of interphase microtubule and f-actin networks and were binucleate or contained multipolar spindles (Figure 5C–J). However, such cytoskeletal and nuclear defects were usually observed within individual blastomeres of otherwise normal appearing blastocysts. Interestingly monopolar spindles were only rarely observed in blastocysts from treated animals (Figure 5C). Control and treated blastomeres exhibited a low level of pyknotic nuclei, which was restricted to the ICM. The number of blastomeres with pyknotic nuclei was similar in control and treated blastocysts.

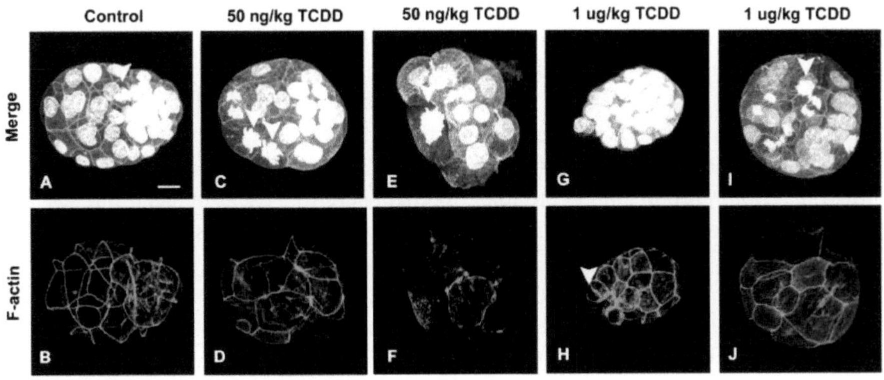

Figure 5. Acute maternal periconceptional TCDD exposure induces nuclear and cytoskeletal defects in early blastocysts. Blastocysts collected from control and acutely exposed (50 ng/kg or 1 μg/kg TCDD) female rats were processed for visualization of microtubules, f-actin and DNA by fluorescence confocal microscopy. (A, B) Control blastocyst with blastocoel and distinct ICM and TE cell populations; note normal bipolar spindle (arrow). (C, D) The 50 ng/kg TCDD exposed blastocyst with an abnormal metaphase blastomere (arrow) and slightly distorted cell shapes. (E, F) Developmentally delayed 50 ng/kg TCDD exposed blastocyst with an abnormally large mitotic spindle (arrow). (G, H) The 1 μg/kg TCDD exposed blastocyst with a compacted morphology. The F-actin and tubulin disorganization is apparent at one pole (arrow). (I, J) The 1 μg/kg TCDD exposed blastocyst with multipolar spindle (arrow) and irregular sized cells. Scale bar: 15 μm.

In comparing nuclear structure among blastocysts, it was noted that while TE cells contained nuclei resembling controls in the 50 ng/kg TCDD exposed group, at higher concentrations (1 μg/kg TCDD) nuclei were conspicuously smaller and irregular in shape (Figure 6). Despite this effect, TE cells from all groups exhibited prominent f-actin boundaries and given the development of the blastocoel, we concluded that this subpopulation of cells had differentiated into highly polarized epithelium able to support fluid transport. The consequences of maternal TCDD exposure on the ICM nuclear structure remains to be established.

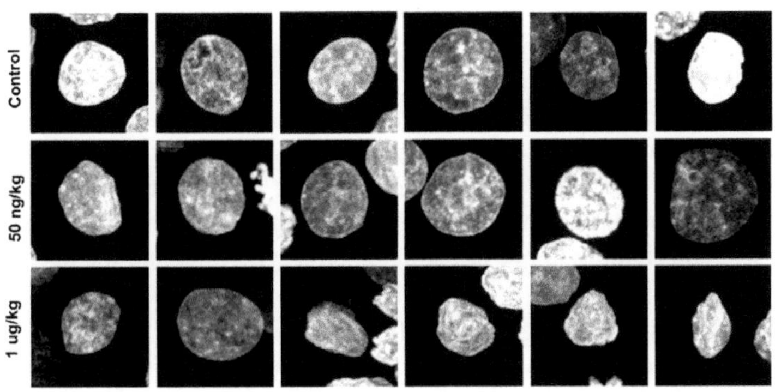

Figure 6. Blastocyst nuclear profiles from control and acutely exposed animals. The Z-series datasets for the DNA channel were compressed into a single plane and six randomly selected nuclei (each from a different blastocyst) for each group (Control, 50 ng/wk and 1 μg/kg TCDD) were compared. Only mural TE were analyzed.

DISCUSSION

Exposure to environmental toxicants, such as TCDD, prior to and during the earliest stages of pregnancy has been linked to developmental disabilities after birth in both human and animal studies [5-8]. We conducted a systematic high resolution confocal microscopy analysis of rat pre-implantation embryos that has revealed previously unappreciated morphogenetic defects following maternal TCDD exposure. We have shown that both chronic and acute maternal exposure to TCDD induced nuclear and cytoskeletal defects in pre-implantation embryo morphogenesis. Specifically, our studies revealed effects of TCDD on (1) mitotic spindle integrity, (2) chromosome alignment, (3) nuclear and cellular size and shape, and (4) cytokinesis efficiency. Furthermore, we have identified the compaction stage of pre-implantation embryogenesis as critically sensitive to the effects of TCDD, while survival and development to the blastocyst stage is not compromised. To the best of our knowledge, this is the first *in vivo* study to demonstrate a critical window of early mammalian development that is vulnerable to disruption by an AhR ligand at environmentally relevant doses.

The comparison of acute verses chronic maternal exposure to TCDD on subsequent embryo quality in the current study encompasses examination of both dose effects and transgenerational toxicant actions. Dams in the acute exposure paradigm were exposed to low and high doses of TCDD limited to a single administration immediately preceding ovulation. The half life of TCDD in the rat is approximately 3 weeks, implying significant exposure until collection of the embryo. In the chronic exposure model, continuous exposure to TCDD in the dam began in utero and continued until breeding sacrifice at 3 months of age. In other words, the acute exposures were comprised of exposures to both the mother and the embryo, while the chronic exposure entails exposure of grandmother, mother, and offspring. The results presented here show that both acute and chronic TCDD treatment protocols significantly compromised embryo quality.

It is likely that pre-implantation embryos are a direct target for TCDD, given that AhR is expressed throughout pre-impanation development [22] and that *in vitro* exposure of mouse pre-implantation embryos to TCDD accelerates differentiation of the blastocyst [19, 20]. However, due to the *in vivo* experimental design employed in this study, we can not rule out effects of TCDD on the oocyte nor on the mother's physiology, as factors contributing to the outcomes realized during pre-implantation embryo development. The AhR ligands are known endocrine disruptors [23] and have also been shown to compromise oocyte quality by inducing apoptosis in cumulus cells [24]. Additionally, recent studies have suggested that TCDD may actually accumulate in the follicular fluid and in the uterus, emphasizing the importance of the mother's physiology in contributing to pre-implantation embryo health and the necessity for *in vivo* studies [20, 25].

One of the most striking effects of TCDD, at both doses, was the induction of aberrant mitotic spindles and a failure in chromosome alignment in compaction stage pre-implantation embryos. The AhR ligands, such as TCDD, may be involved in generating meiotic spindle aberrations by causing local increases in the concentration of 2-methoxyestradiol (2-ME) [26]. The 2-ME binds tubulin and influences microtubule polymerization and function. It is an endogenous metabolite of 17β-estradiol normally present within granulosa cells and follicular fluid of the ovary. Exposure to elevated levels of 2-ME causes spindle abnormalities, chromosome congression failure and nondisjunction in mouse oocytes and it is conceivable that such abnormalities would be further propagated within the developing embryo. Interestingly, exposure of bovine pre-implantation embryos to 2-ME does not inhibit passage from morula to blastocyst, and the cell cycle proceeds despite aberrations in spindle morphology [27].

Similarly, exposure to TCDD did not inhibit development to blastocyst and the data presented here clearly demonstrate that the overall structure and morphology of treated blastocysts were similar to control blastocysts. However, the widespread prevalence of defects uncovered in compaction stage pre-implantation embryos of treated animals, together with the relative paucity of cells exhibiting defects at the blastocyst stage, is disconcerting. Thus, a central question raised by this work is "What is the fate of aberrant compaction stage blastomeres"? We suggest three scenarios that may contribute to the survivability of TCDD exposed pre-implantation embryos. First, defective cells may be eliminated by apoptosis during the transition from compacted pre-implantation embryo to blastocyst. Brison and Shultz showed apoptosis only occurs after compaction and is predominantly located in the ICM of the mouse [28]. However, TCDD exposure did not increase the incidence of pyknosis in pre-implantation embryos in this study and thus elimination of TCDD-induced defects by selective apoptosis seems unlikely. This is consistent with an earlier study in which *in vitro* exposure of 2, 4, or 8 cell mouse pre-implantation embryos to TCDD did not significantly increase the number of TUNEL positive cells, alter the Bax/Bcl-2 expression ratio, or change cell number at the blastocyst stage [22].

Alternatively, defective cells may initiate repair mechanisms to rectify errors. Surveillance mechanisms that alert cells to impending errors in chromosome segregation exist in many normal somatic cells exhibiting a stable euploid condition. Such cell cycle checkpoints are engaged in response to structural aberrations as wide-ranging as chromosome misalignment to centrosome number [29]. Notably, monopolar spindles with poorly aligned chromosomes are defects likely to activate cell cycle checkpoints. However, knockout and transgenic studies suggest that checkpoint controls may not be operational until the time of implantation in mouse embryos [30]. Furthermore, lack of checkpoint activity in human embryos is suggested by the extreme degree of aneuploidy and mixoploidy seen in association with defects in mitotic spindle organization [31, 32]. If cell cycle checkpoints were activated in rat pre-implantation embryos, one might expect either an increase in the mitotic index or decrease in total cell number to be evident in blastocysts derived from TCDD exposed animals compared to controls. While there was no obvious increase in the presence of mitotic figures or cell number, further studies will be required to fully investigate this possibility.

It is possible that our detection of monopolar spindles was biased by a prolonged prometaphase delay that yielded euploid daughter cells after a correction that we were unable to detect in fixed samples. While, this seems unlikely given the similar distribution of M-phase stages observed in control and treated pre-implantation embryos, on-going live cell recording experiments should resolve this dilemma. Moreover, cytogenetic assays will be needed to establish the incidence of aneuploidy, known to be elevated in human embryos exhibiting similar spindle defects, to better understand the status of checkpoint controls during this critical juncture during mammalian embryogenesis.

A final possibility is retention of defective blastomeres and their contribution to either or both of the lineages established in the blastocyst. While, it is interesting to speculate that the TE lineage naturally undergoes ploidy variations indicative of less stringent cell cycle checkpoint surveillance [33], our data to date cannot rigorously assign fates to the aberrant blastomeres present during compaction. Given the changes in nuclear shape and the cytoskeleton resulting from exposure to TCDD during oocyte maturation and pre-implantation development, the emergence of cell polarity and lineage assignment during the compaction process may be impacted in subtle ways and yet still have major consequences for the later stages of embryogenesis. For example, the disruption of nuclear architecture is the central factor underlying a physiologically diverse group of inherited diseases known as laminopathies [34]. Moreover, the epigenetic assignment of blastomeres to TE or ICM immediately precede the time we report here to be most sensitive to the effects of TCDD [35]. In this regard, Skinner and colleagues have recently demonstrated altered epigenetic marks following exposure to environmental estrogens leading to disease phenotypes in adult male offspring that were passed on to subsequent generations [36]. It was also recently shown that *in vitro* exposure of mouse pre-implantation embryos to TCDD increased methyl transferase activity, altered the methylation status of imprinted genes H19 and Igf2 and retarded subsequent fetal growth [37]. Thus, disruption of epigenetic programming made during compaction provides a plausible mechanism by which pre-implantation exposure to TCDD could affect later development. In this light, the persistence of nuclear defects in shape and position may be subtle indicators of disruption of chromatin remodeling and epigenetic reprogramming during oocyte maturation and development up to compaction.

MATERIALS AND METHODS

Animals

Female Sprague-Dawley rats (Charles River Laboratories) were housed under a 12L:12D photoperiod at an ambient temperature of 23 ± 2°C, with food and water ad libitum. All procedures were approved by the University of Kansas Medical Center Institutional Animal Care and Use Committee. In all experiments, pre-implantation embryos were obtained from naturally mated rats. Estrus cycles were monitored by vaginal cytology, with normal estrous cycle duration of 4–5 days [39]. Source and purity of TCDD: CAS 1746-01-6; MW, 321.9; purity, > 99%.

Experimental Design

In experiment 1, rats were exposed chronically to doses of TCDD that mimic exposure of high risk populations in humans [40, 41]. Weekly oral dosing was used in a previously validated regime [21, 42]. Initially, pregnant dams (n = 3 for each experimental group) received an oral dosing of TCDD (50 ng/kg) or corn oil vehicle (4 ml/kg) on day 14 and 21 of gestation and then on days 7 and 14 postpartum to provide in utero and postnatal lactational exposure, respectively. On postnatal day 21, female pups (n = 3 per experimental group) were weaned and orally dosed with TCDD (50 ng/kg) or corn oil vehicle (4 ml/kg), with dosing continued at weekly intervals thereafter. At 3 months of age, proven males were introduced on the evening of proestrus and mating was confirmed by the presence of sperm on vaginal cytology the following morning. Pre-implantation embryos were collected in FHM (Chemicon) media pre-warmed to 37°C by flushing oviducts and uteri on day 4.5 post coitum.

In experiment 2, female Sprague-Dawley rats (n = 3–6 for each experimental group) received a single oral dose (50 ng/kg or 1 μg/kg) of TCDD or corn oil vehicle (4 ml/kg) on the evening of proestrus and were housed with males of proven fertility. At the time of dosing rats were 50 days of age. Again, mating was confirmed by the presence of sperm on vaginal cytology the following morning. Pre-implantation embryos were collected in FHM (Chemicon) media pre-warmed to 37°C by flushing oviducts and uteri on day 4.5 or 5.5 post mating.

Immunofluorescence

Pre-implantation embryos were processed for microtubule, DNA and f-actin immunofluorescence as previously described [43]. Immediately following their collection, pre-implantation embryos were fixed for 30 min in 4% PFA at 37°C and stored at 4°C in wash solution comprising PBS supplemented with 2% BSA, 2% skim milk powder, 2% normal goat serum, 100 mM glycine, 0.01% Triton-X-100, and 0.2% sodium azide until processing for immunofluorescence. Pre-implantation embryos were extracted for 30 min at room temperature in 0.1% Triton-X-100 and incubated overnight at 4°C in wash solution. For immunostaining of microtubules, embryos were first incubated with mouse monoclonal anti-αβ tubulin (Sigma) diluted 1:100 in wash solution for 1 hr at 37°C, followed by Alexa 488 labeled goat anti-mouse IgG (Molecular Probes) diluted 1:1000 in wash solution for 1 hr at 37°C. The DNA was stained with Hoechst 33258 (1 μg/ml in wash solution) for 30 min and f-actin integrity was analyzed by staining with rhodamine labeled phalloidin (1 μg/ml in wash solution; Molecular Probes) for 30 min. Pre-implantation embryos were mounted under cover slips without compression in medium containing 50% glycerol and 25 mg/ml sodium azide.

Pre-implantation embryos were analyzed on a Zeiss LSM Pascal confocal imaging system mounted on a Zeiss Axioscope II using UV (405 nm), HeNe (543 nm), and Argon (488 nm) laser excitation. For every embryo, a complete Z-axis data set was collected at 0.8 μm intervals (~50 sections/embryo) using a x63 oil objective (na = 1.4). Laser power, gain, and offset settings were not changed during acquisition. Line scans and spatial restoration and 3D projections for each Z-series data set were computed and analyzed using Zeiss LSM 5 Image Browser.

Classification of Pre-implantation Embryos

Pre-implantation embryos were classified as abnormal if they contained blastomeres exhibiting one or more of the following: irregular size, irregular shape, weak or undetectable f-actin or tubulin, cellular fragmentation or micronuclei. Additionally, blastomeres containing metaphase-like chromosomes with mitotic spindles absent, or deviating from a focused bipolar microtubule array, were considered abnormal.

Statistical Analysis

Chi-square was used to analyze the proportion of normal and abnormal pre-implantation embryos. P-values of less than 0.05 were considered significant.

CONCLUSION

We have observed that compaction stage pre-implantation embryogenesis as critically sensitive to the effects of TCDD, while survival to the blastocyst stage is not compromised. The present work assumes particular relevance when considered together with recent evidence suggesting long term impacts on the health and well-being of offspring following environmental perturbations during the periconceptional and pre-implantation period [11, 38]. While, the pre-implantation embryo may exhibit resiliency and plasticity in obtaining a state of implantation competence, perturbations during this critical window of development may be propagated with further cellular expansion and could abrogate developmental processes that will only surface after birth. Identifying the specific targets of TCDD in the pre-implantation embryo, especially those that link cell cycle control with cytoskeletal and nuclear remodeling, represents an important avenue for future investigation.

KEYWORDS

- **Blastocysts**
- **Immunofluorescence**
- **Periconceptional**
- **Pre-implantation embryo**

AUTHORS' CONTRIBUTIONS

Karla J. Hutt carried out the embryo collection, analysis and interpretation of the data and drafted the manuscript. Zhanquan Shi dosed the animals and collected the embryos. David F. Albertini analyzed and interpreted the data and helped to draft the manuscript. Brian K. Petroff conceived of the study and participated in its design and coordination and helped to draft the manuscript. All authors read and approved the final manuscript.

ACKNOWLEDGMENTS

We wish to thanks our colleagues Drs. Renata Ciereszko and Kelli Valdez for their critical reading of this manuscript. This research was supported by NIH/NIEHS-012916 (Brian K. Petroff), ESHE Fund (David F. Albertini), Hall Family Foundation (David F. Albertini and Karla J. Hutt) and Biomedical Research Training Grant KUMC (Karla J. Hutt).

Chapter 12

Endosulfan in Farming Areas of the Western Cape, South Africa

Mohamed A. Dalvie, Eugene Cairncross, Abdullah Solomon, and Leslie London

INTRODUCTION

In South Africa there is little data on environmental pollution of rural water sources by agrochemicals.

This study investigated pesticide contamination of ground and surface water in three intensive agricultural areas in the Western Cape: the Hex River Valley, Grabouw, and Piketberg. Monitoring for endosulfan and chlorpyrifos at low levels was conducted as well as screening for other pesticides.

The quantification limit for endosulfan was 0.1 μg/l. Endosulfan was found to be widespread in ground water, surface water, and drinking water. The contamination was mostly at low levels, but regularly exceeded the European Drinking Water Standard of 0.1 μg/l. The two most contaminated sites were a sub-surface drain in the Hex River Valley and a dam in Grabouw, with 0.83 ± 1.0 μg/l (n = 21) and 3.16 ± 3.5 μg/l (n = 13) average endosulfan levels respectively. Other pesticides including chlorpyrifos, azinphos-methyl, fenarimol, iprodione, deltamethrin, penconazole, and prothiofos were detected. Endosulfan was most frequently detected in Grabouw (69%) followed by Hex River (46%) and Piketberg (39%). Detections were more frequent in surface water (47%) than in groundwater (32%) and coincided with irrigation, and to a lesser extent, to spraying and trigger rains. Total dietary endosulfan intake calculated from levels found in drinking water did not exceed the Joint WHO/FAO Meeting on Pesticide Residues (JMPR) criteria.

The study has shown the need for monitoring of pesticide contamination in surface and groundwater, and the development of drinking water quality standards for specific pesticides in South Africa.

As water pollution by pesticides can affect many biological systems, the widespread use of potentially harmful pesticides has recently come under scrutiny in South Africa [1, 2]. Once contaminated, the groundwater may take a long time to clear [3] and there is always the danger of bioaccumulation.

Expenditure on agrochemicals has increased markedly over the past decade [4] and a far greater variety of chemicals are used locally compared to other developing countries [5, 6]. There is, however, little environmental monitoring of pesticides [7].

Pesticide exposures are associated with a growing number of chronic health effects [8, 9], with local farm workers being at particular risk due to unsafe application methods [10] and adverse living and working conditions [5]. While, concern for water pollution by pesticides has mobilized considerable resources in other countries, particularly in the developed world, little research has been undertaken in South Africa [11]. Available literature [12-17] reports the presence of a number of pesticides in rivers and dams. In the Western Cape, Davies et al. [15] detected six pesticides in Elgin dams and three in Caledon dams, with endosulfan present in 26 of 27 Elgin dams at concentrations as high as 626 μg/l. Recently, Schultz et al. [13] found increased endosulfan presence in the Lourens River after washout during the first rains. However, no contamination was found in a study undertaken in the Hex River Valley, [17] probably due to the use of monitoring equipment with high detection limits. (Personal communication, Dr. John Weaver, Watertek, September, 1995). The aim of this study was to investigate pesticide pollution of water supplies in agricultural areas of the Western Cape, South Africa. The objectives were to identify rural water sources in the Western Cape at high risk of agrochemical contamination, to identify and quantify the presence of agrochemicals at these selected sites and to explore the implications for human health.

For assessment of the possible chronic health and environmental effects of long-term exposure to pesticides, extended monitoring of ground, surface and drinking water, as well as analytical techniques with sufficiently low levels of detection are essential.

MATERIALS AND METHODS

Identification of Study Areas ("Areas of Concern") and Sampling Sites

Identification of areas with the potential for water contamination by pesticides was conducted through review of secondary data, interviews with rural health care providers, farmers, environmental officers, and other agricultural personnel and field observation. Access to sites was negotiated with local agricultural organizations and assistance was sought from geohydrologists to identify areas and sites most vulnerable to pesticide contamination.

Three intensive agricultural districts, Piketberg, Grabouw, and the Hex River Valley were selected as study areas (Figure 1). All three areas have a Mediterranean climate with winter rainfall. The Hex River and Piketberg districts are semi-arid (receiving <300 mm rain per year), while Grabouw is in a high rainfall area (>400 mm/year).

Grape farming is practiced in the Hex River district (Figure 2). The most important source of water for drinking and irrigation is a mountain dam. Soil conditions are conducive to pesticides reaching the water table and contaminating groundwater (water table < 1 m, unconfined aquifer, coarse soils with low clay content) [18].

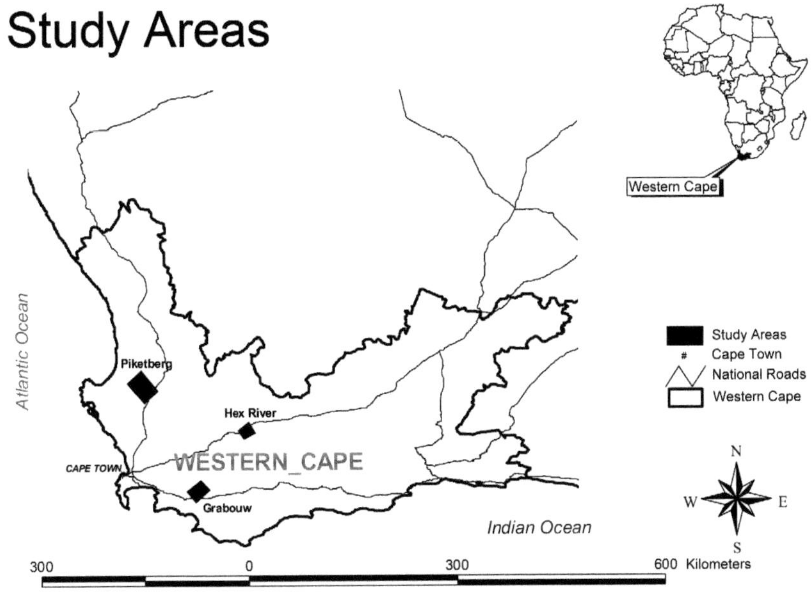

Figure 1. Location of study areas for pesticide sampling in the Western Cape, South Africa.

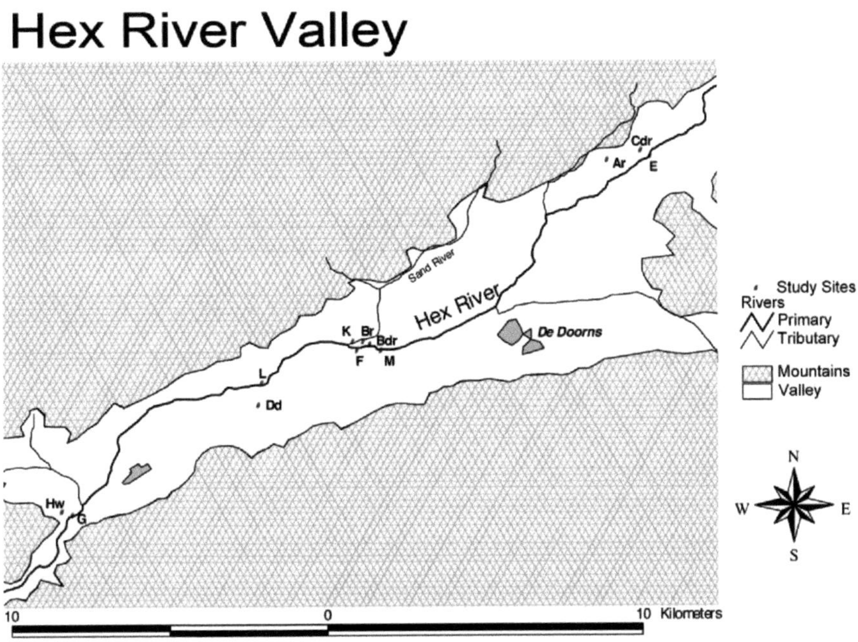

Figure 2. Location of sampling sites for pesticides in the Hex River Valley.

The Grabouw district (Figure 3) includes two pome fruit farming areas (Grabouw and Vyeboom). In both areas, the soil is complex but generally promotes run-off to surface water sites [18]. The high annual rainfall tends to encourage run-off.

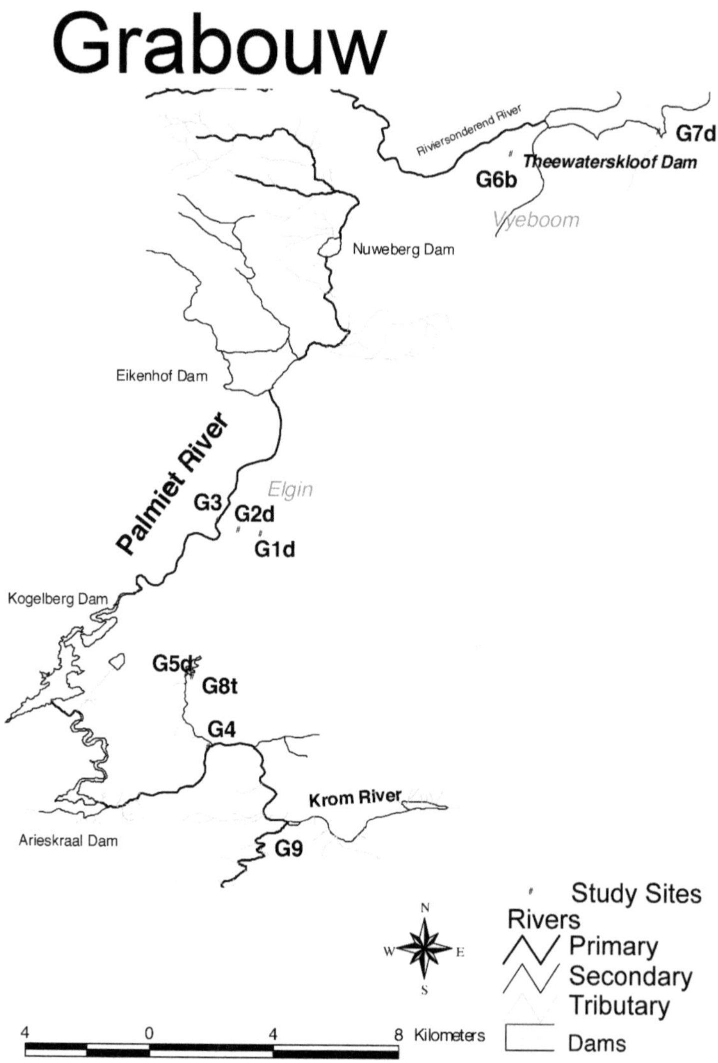

Figure 3. Location of sampling sites for pesticides in the Grabouw/Vyeboom Area.

Piketberg (Figure 4) is an important farming region covering a much larger area than the two other study districts. Fruit farming is practiced on the Piketberg mountains and wheat farming in the valley. The soil in both areas is multi-textured, but generally leachable and prone to run-off [18]. The water table is moderately shallow

(<5 m). There are substantially more wells in this area than in the Hex River and Grabouw. The Berg is the major river running through the area and water is purified for domestic consumption at a number of places along its course. The purification scheme at Wittewaters is a major source of drinking water in the rural Western Cape and is fed by the Misverstand Dam, situated amidst extensive wheat farms, where aerial spraying of pesticides is commonly practiced. The Berg River, flowing through fruit farming areas also flows into the Misverstand Dam.

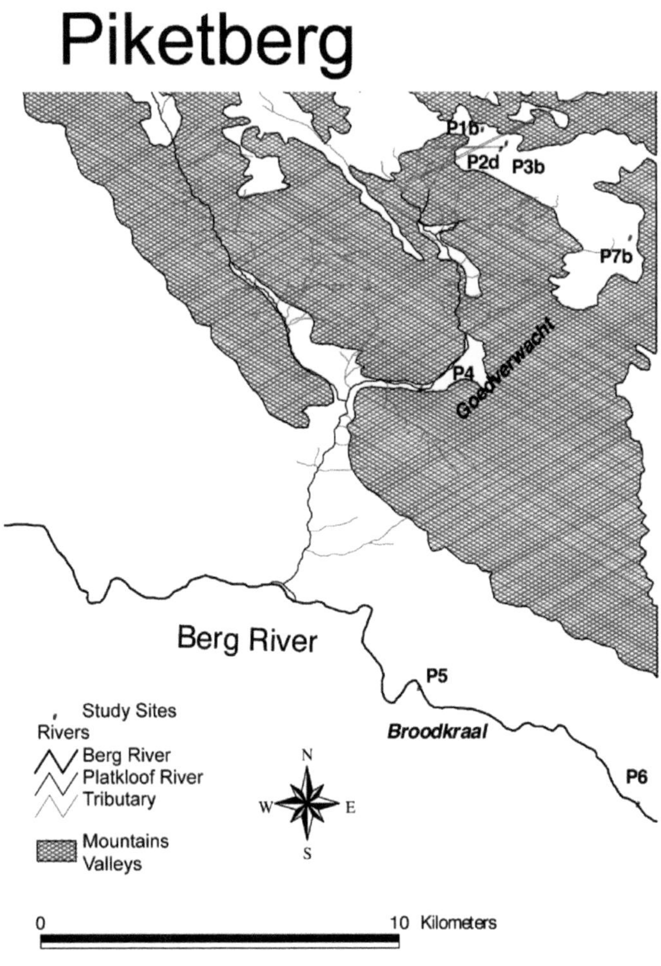

Figure 4. Location of sampling sites for pesticides in the Pikerberg Area.

Sampling sites in the three areas, summarized in Table 1 and shown in Figures 2, 3, and 4, were chosen to provide a spread of ground, sub-surface, and surface water. Some sampling points were added in the course of the study to enable a better understanding of contamination patterns at the different sites.

Table 1. Sampling points in the three study areas.

Hex River Valley	E*	River point high up the valley, towards the top of the production area
	F	River point in the middle of the river's course through the valley, at densely agricultural area
	G	River point at lowest end of the valley, after confluence with a fresh river from pristine area
	L*	River point between F and G
	M*	River point- between E and F, before discharge of Bdr
	Ar	Farm reservoir near vineyards containing spring and mountain water
	Br	Farm reservoir containing mountain water near vineyard
	Cdr	Open surface drain (I m); drains superficial vineyard run-off
	Bdr	Closed surface drain (I m); drains vineyard run-off from farm and from neighbouring farms.
	Dd	Open farm dam receives water from the Hex River
	H	Shallow well (5 m deep), containing groundwater used for domestic consumption
	J	Tap at Irrigation Board offices: representing potable water supply to the valley from distant mountain dam close to Ceres
	K*	Point on another river near F
Grabouw I Vyeboom	Gld	Dam receives irrigation drainage, flows into Palmiet river
	G2d	Farm dam, water pumped from Palmiet.
	G3	River point (Palmiet) in midst of intensive agriculture
	G4	River point lower in the course of the Palmiet
	G5d	Dam receiving water from Palmiet river, other dams and irrigation run-off; purified for domestic use.
	G6b	Well (in Vyeboom) used by farmers for domestic use, 30 m in depth
	G7*	A stream flowing from agricultural area in Vyeboom into the Theewaterskloof Dam (supplies just over 50% of Cape Town's drinking water). The stream is part of the dam when the dam is full.
	G8t	Tap water using output of G5d
	G9	River point on Palmiet after joining Krom
Piketberg	Plb	Well (depth= 100m) in intensive farming area; used for domestic water supply
	P2d	Dam receiving well and surface water but near the top of the mountain
	P3b	Well (depth = 70 m) in intensive farming area; used for domestic water supply
	P4r	Stream running down the mountain from P2d through Moravian Mission and into the Berg River. Used for domestic consumption.
	P5r	Site on Berg River mid-way further on from P4r
	P6r	Site on Berg River at pumping station providing municipal water
	P7b	Well (depth= I 00 m) on wheat farm on the plain below the mountai
	P8r	Tap at water purification scheme at Wittewaters (Berg River)

* River sites that could not always be sampled at a depth of I m

Sampling sites in the Hex River region were along the Hex River. The two sub-surface drains (Bdr and Cdr) eventually feed into the Hex River.

Grabouw/Vyeboom sites were selected on farms belonging to one of two major apple-packing co-operatives in the region.

Unlike the other two areas, in Piketberg sampling access to farms was arranged by the local environment officer, who is the municipal official responsible for public health functions in rural areas.

Field Sampling

Grab (manual) samples were collected by the project co-ordinator (AD) commencing February, 1998 in the Hex River, April, 1998 in Grabouw, and May, 1998 in Piketberg, and completed for all three areas in May, 1999. Samples were collected once monthly in each area on a rotating cycle, and twice in the week after the first rainfall trigger (>10 mm over 24 hr or > 15 mm over 48 hr), using a standardized procedure (the same method each time). Although an attempt was made to take dam and river samples as far away from the bank as possible and submerging sampling bottles about 1 m deep, this was not always possible due to problems with access and shallow water levels.

Samples were collected directly in clean, dry, 2.5 and 1 l amber glass bottles fitted with a screw cap lined with clean aluminum foil. Samples were kept at ambient temperature in a holding box for transport to the laboratory where they were stored in a refrigerator until extraction. Sample pH, water temperature, subjective assessment of water level (low, medium, or high) and the occurrence of spraying within 1 km of the sampling point were recorded (These results are not shown because they did not add to the interpretation of the findings (see London et. al. [18]).

Choice of Pesticides for Analysis

Not all pesticides could be monitored due to the prohibitive costs of multi-residue screening methods. Instead, a comprehensive list of pesticides used in the three areas was shortened [18] to 31 pesticides for analyses, conditioned by the availability, and existence of local methods for analysis.

Analyses were conducted jointly by the Analytical Chemistry laboratories of the Peninsula Technikon (PENTECH), which was the project laboratory, and the State Forensic (SF) Laboratory, both of which are in Cape Town. The SF undertook analyses conducted as a battery for all 31 pesticides (quantification limit, 0.1 µg/l) in line with their statutory function of providing a screening service for monitoring of pesticide residues in food. Based on their preliminary results and on anticipated findings, PENTECH developed methods to analyze five pesticides, including endosulfan (isomers I and II and endosulfan sulfate), BHC. Dichlorodiphenyltrichloroethane (DDT), dichlorvos, and chlorpyrifos.

Iprodione, azinphos-methyl, prothiofos, deltamethrin, and fenarimol were detected on six occasions at low levels at 10 different sites [18], whereas chlorpyrifos and endosulfan were detected on screening by either the State laboratory, or the Agricultural Research Council Laboratory (ARC) on a number of occasions. Consequently, PENTECH focused on investigating analytical methods for chlorpyrifs and endosulfan. The results for endosulfan, a commonly recognized endocrine disruptor [19, 20], are presented in this chapter. The results for the other pesticides did not change the overall findings of the study.

Analyses

Sample Extraction

The PENTECH used solid phase extraction following Environmental Protection Agency (EPA) methods [21, 22]. Samples, which were vacuum pre-filtered through

S&S filter paper (ref. No. 334508) were extracted within 7 days of collection (more than 80% were extracted within 3 days) using Bond Elute Extraction Cartridges (C18, 10 ml LRC, 500 mg sorbent mass). The column was conditioned with 2 volumes (2–10 ml) of ethyl acetate, and 1 volume each of methanol and deionized water.

High-pressure chromatography grade solvents were used. A 250 ml of filtered sample was column aspirated at 20–25 ml/min under vacuum. The column was then washed with 1 volume (10 ml) deionized water and thoroughly dried for 15 min under vacuum. Pesticides were eluted into a borosilicate glass vial with 2 × 10 ml ethyl acetate which was then left to evaporate at room temperature. A 1 ml hexane was added to dissolve the residue, for gas chromatography (GC) analysis.

Analytical Methodology

Standards were prepared from analytical standards (>98% purity) [21, 22]. The GC was used for identification and quantification of extracted samples using a Varian 3300 GC with an electron capture detector (ECD). A 2 μl sample was injected onto a capillary column with a BPX 5 stationary phase. The temperature was increased from 170°C at a rate of 7°C/min to 290°C and held there for 5 min. Injector and detector temperatures were 250°C and 300°C, respectively.

Quality Control and Quality Assurance

Quality Control at PENTECH included:

- Duplicate sampling and analysis of one site at least once per sample run. Duplicate samples were run after eight samples had been injected.
- A reagent blank and a laboratory control sample (LCS) run with each set of samples. Both were subjected to the same analytical procedure as those used on the study samples. The LCS was spiked with the target analytes at a concentration range expected for the samples in deionized water.
- Recoveries <70% or >130% for LCS prompted investigation and, if necessary reanalysis.
- Mixed standards injected prior to a sample run and at the end. Peak shape, resolution, and response evaluation by comparison with previous chromatograms was done to ensure optimal performance of the entire analytical system.

Quality assurance (QA) was with the GLP-accredited ARC and SF laboratories. Eleven (four Piketberg, two Grabouw, and five Hex River) samples sets were forwarded to the SF laboratory and two (Hex River) to the ARC, including one (Hex River) set to both laboratories. One set each to both laboratories included a duplicate sample of Bdr, with one falsely labeled (I). Additionally, all three laboratories analyzed a set of seven samples from Bdr (Hex River).

Samples for the SF laboratory were stored at 5°C and sent within 24 hr, while that for the ARC were couriered in polystyrene containers.

The SF laboratory used solid phase and the ARC liquid-to-liquid extraction. The ARC laboratory used a 2 m 3% OV-17 column and the SF laboratory, a DB1 column. All three laboratories use GC methods with ECDs but with different columns and temperature programs. This served to confirm pesticide identification. No confirmation with another detector was possible at PENTECH because of the lack of a second detector.

The results of the QA analyses [18], suggested that the laboratory analytical procedures followed in this study were able to achieve adequate precision and inter-laboratory agreement, consistent with normative practice for such strategies.

The quantification limit (empirically-derived quantification limit = 2 x Std Deviation of seven samples of low concentration of respective standard for endosulfan analyses) at PENTECH was 0.1 μg/l (Table 2).

Table 2. Quantification limits for endosulfan.

Isomer	Concentration of Standard Used (μg/L)	Empirical Mean Value (μg/L)	Empirically Derived Limit (μg/L)	EPA Limit (μg/L) [21]
Alpha-endosulfan	0. 171	0. 103	0. 11	0.030
Beta-endosulfan	0. 182	0.206	0.13	0.030
Endosulfan sulphate	0.266	0.290	0. 13	0.030

* Empirically-derived quantification limit= 2 × Std Deviation of 7 samples of low concentration of respective standard

Endosulfan data are quoted as the sum of isomers I and II plus endosulfan sulfate, unless otherwise specified. Endosulfan concentrations are expressed as μg/l, because of the different molecular weights of the isomers and endosulfan sulfate.

Field Results

Hex River

Table 3 lists and summarizes endosulfan levels and the number of detections in the Hex River region.

Table 3. Endosulfan levels detected in Hex River Valley.

Date	Sites and Concentration in μg/L												
	E	F	G	Cdr	Bdr	Ar	Br	Dd	H	J	K	L	M
11/2/98	nd	0.24	nd	nd	0.19	0.44	0.16	nd	ns	ns	ns	ns	ns
18/2/98	nd	0.32	nd	nd	0.37	0.11	nd	nd	ns	ns	ns	ns	ns
25/2/98	nd	0.24	nd	nd	0.18	(0.08)	nd	nd	ns	ns	ns	ns	ns
4/3/98	nd	0.29	nd	nd	2.22	0.28	0.204	nd	ns	ns	ns	ns	ns
11/3/98	nd	0.16	nd	(0.07)	1.53	0.16	nd	nd	ns	ns	ns	ns	ns
18/3/98	nd	0.22	nd	nd	1.81	0.14	nd	nd	ns	ns	ns	ns	ns
25/3/98	nd	0.20	nd	nd	1.10	(0.08)	nd	nd	ns	ns	ns	ns	ns
22/4/98	nd	0.26	nd	nd	0.43	nd	nd	nd	nd	ns	ns	ns	ns
12/5/98	nd	nd	nd	nd	(0.04)	nd	nd	nd	nd	ns	ns	ns	ns
19/5/98	nd	(0.06)	(0.03)	nd	0.23	0.06	ns	nd	nd	ns	ns	ns	ns
12/8/98	(0.03)	(0.04)	nd	nd	(0.03)	(0.02)	0.20	nd	nd	ns	ns	ns	ns
23/9/98	(0.03)	1.56	nd	ns	(0.01)	(0.02)	ns	0.2	nd	(0.03)	nd	nd	ns

Table 3. *(Continued)*

Date	Sites and Concentration in μg/L												
	E	F	G	Cdr	Bdr	Ar	Br	Dd	H	J	K	L	M
21/10/98	nd	(0.04)	0.264	ns	nd	0.19	(0.09)	(0.05)	0.23	nd	nd	ns	ns
12/11/98	nd	nd	nd	ns	0.13	nd	ns	nd	nd	nd	ns	nd	ns
18/11/98	nd	nd	nd	ns	0.58	nd	(0.06)	0.58	nd	nd	0.4	nd	ns
13/1/99	nd	nd	nd	ns	ns	nd	nd	0.25	0.89	ns	ns	ns	ns
24/2/99	nd	1.02	0.45	ns	1.84	nd	0.51	0.47	nd	0.15	ns	0.37	ns
17/3/99	ns	1.25	0.19	ns	3.86	1.02	nd	0.35	nd	0.62	ns	ns	ns
07/4/99	ns	0.54	nd	ns	0.79	nd	nd	(0.09)	nd	nd	ns	ns	ns
20/4/99	0.35	0.29	(0.05)	ns	1.48	ns	0.79	(0.08)	nd	ns	ns	ns	ns
26/4/99	0.47	0.27	nd	ns	0.59	nd	ns	0.2	ns	ns	ns	ns	ns
13/5/99	(0.02)	(0.03)	ns	ns	(0.03)	nd	ns	nd	ns	0.43	ns	ns	nd
Mean (SD)	0.05 (0.13)	0.32 (0.42)	0.05 (0.12)	0.006 (0.021)	0.830 (0.988)	0.124 (0.235)	0.112 (0.214)	0.103 (0.17)	0.086 (0.250)	0.154 (0.24)	0.133 (0.23)	0.0925 (0.185)	0 (0)
N	20	22	21	11	21	21	18	22	13	8	3	4	1
%positive samples*	25	82	24	9	95	57	39	41	15	50	33	25	0

ns = Not sampled; nd = not detected (no discernable peak, less than 0.01 μg/L () = less than quantification limit; * positive samples = samples in which endosulfan was detected (including those below the quantification limit)

The Bdr, the drain that receives sub-surface run-off from a number of different farms, consistently produced the highest detections. There were virtually no detections in the other drain Cdr, which dried up completely towards the end of the study.

The dam (Dd) had little contamination before September, 1998, but consistent detections thereafter. This might have been due to the decreased water level resulting from irrigation, thereby concentrating chemicals released from sediments, especially endosulfan, with a soil half-life of 120 days and sorption coefficient (Koc) of 17.52 l/g [23]. This explanation was, however, not supported by evidence of any significant pH changes in the dam water due to chemical release. The detections did correspond temporally with endosulfan spraying in the region.

Both reservoirs (Ar and Br) were erratically contaminated, generally at low levels.

River detections (points E, F, and G, Table 3) appeared to peak in mid-valley (F) and to be diluted in the lower valley (point G) after confluence with a tributary. However, point L, which lies between F and the confluence point, had similar levels of pesticide as G, suggesting that dilution occurs before L. Site L was, however, sampled only four times. There were no obvious point sources (e.g., pesticide mixing stands) identified along the course of the river (although a mixing stand was sited some 30 m from the river at L). No inference could be drawn on site K, lying on a river that joins the Hex River, because it was sampled only three times with endosulfan detected in one sample.

Detections during the spraying months, September to mid-October (endosulfan sprayed 1–2 times during this period), were low in all the sites. Subsequent irrigation

(September–May), especially during January–March (about 125 mm/month), appeared to be associated with enhanced detection because higher endosulfan levels were found during February–April, 1999 at most sites. Raised levels were also found in F and Bdr during the same period in 1998. Raised levels during January–April could also have been due to rainfall triggers on February 10, 16; March 7, 22; and April 21 in 1998 and January 9, 17, 25, 27, and March 9, 19 in 1999. There were detections in drinking water sources (H) and (J).

In summary, low-level endosulfan detections were widespread in the Hex River region. The Bdr and F (the mid-point of the river) were clearly "hot-spots" with regularly higher levels than other sites. Of the three mechanisms which could explain pesticide movement (rain washout, irrigation washout, and spray activities), irrigation and rainfall washout appear to be the most important although there is some temporal relationship to spraying. The effect is demonstrated in Figure 5 showing endosulfan detected in Bdr.

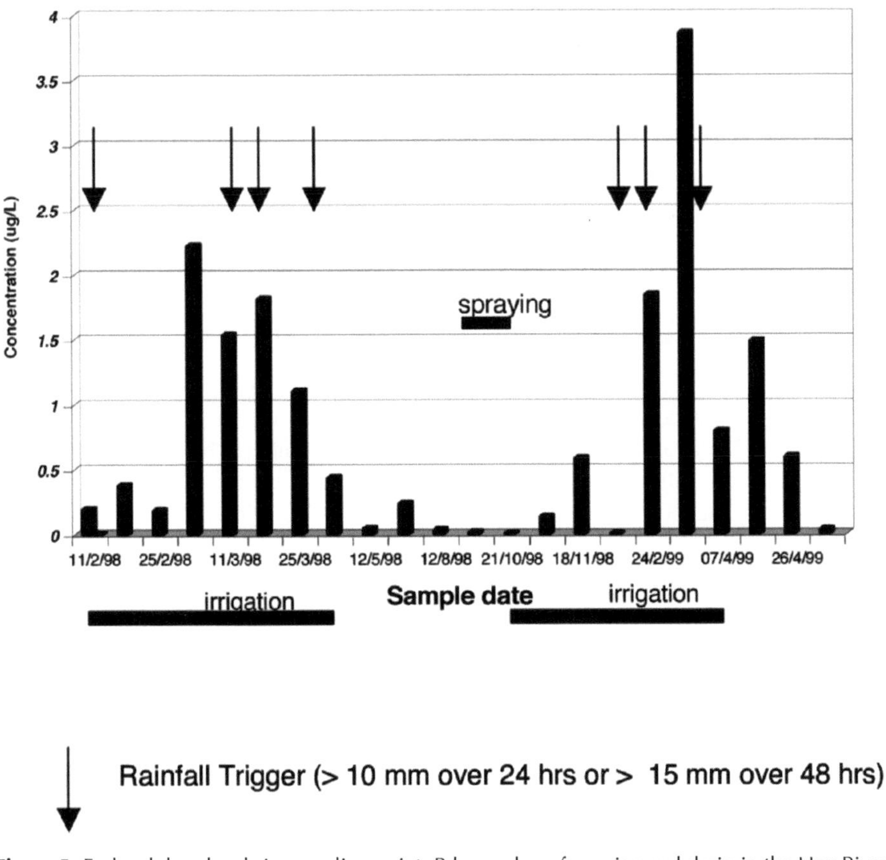

Figure 5. Endosulphan levels in sampling point, Bdr: a sub-surface vineyard drain in the Hex River valley Rainfall Trigger (>10 mm over 24 hr or >15 mm over 48 hr).

Piketberg

Table 4 presents total endosulfan levels for Piketberg, and also provides a summary of the detections.

Table 4. Endosulfan levels detected in Piketberg.

Date	Sites and Concentration in μg/L							
	P1b	P2d	P3b	P4r	P5r	P6r	P7b	P8t
13/5/98	0.13	nd	nd	(0.02)	nd	nd	ns	ns
20/5/98	ns	nd	ns	nd	nd	nd	nd	nd
1/7/98	nd	(0.09)	nd	nd	nd	nd	nd	nd
2/9/98	nd	0.12	nd	(0.01)	(0.04)	nd	(0.02)	ns
7/ 10/98	(0.05)	0.24	0.249	nd	nd	(0.07)	(0.01)	26.3
11/11 /98	0.13	0.20	(0.01)	0.20	0.07	0.25	1.15	0.06
25/ 11 /98	nd	nd	nd	nd	nd	nd	nd	nd
27/01 /99	nd	nd	nd	nd	1.05	ns	nd	1.123
17/02/99	0.47	0.67	nd	0.18	0.1	nd	0.21	(0.09)
10/03/99	0.44	0.13	nd	0.36	0.34	ns	0.59	0.16
31 /03/99	nd	nd	nd	0.24	nd	ns	nd	nd
22/04/99	nd	(0.08)	(0.08)	nd	nd	ns	0.27	nd
28/04/99	nd	nd	nd	nd	nd	nd	nd	ns
Mean (SD)	0.10 (0.17)	0.118 (0.185)	0.03 (0.07)	0.078 (0.123)	0.123 (0.294)	0.04 (0.088)	0.19 (0.35)	2.774 (8.277)
n	12	13	12	13	13	8	12	10
%positive samples*	42	54	25	46	38	25	50	50

ns = Not sampled; nd = not detected (no discernable peak, less than 0.01 μg/L () = less than quantification limit; * positive samples = samples in which endosulfan was detected (including those below the quantification limit)

There were detections in all sites, mostly during the irrigation period of February–March, 1999. Many of these sites are used for drinking water (P1b, P3b, P7b, P4, P8t), and include the purification scheme supplying a large area in the West Coast region (P8t). Rainfall triggers (May 6–11, 1998 and April 18–20, 1999) did not enhance contamination.

Endosulfan in P8t was substantially raised in October, 1998, coinciding with peak spraying (1–2 times) in surrounding fruit and grape growing areas. Endosulfan levels also peaked in two of the wells (P1b and P7b) shortly after the spraying period, suggesting movement through the soil after application.

Endosulfan in the two Berg River sites (P5 and P6) was lower than the Hex River, although one site, P6, was not sampled consistently due to inaccessibility. Higher levels in January, 1999 could partly reflect applications upstream in fruit and wine farming areas.

Detections in the dam (P2d) on occasion corresponded with those in the connecting stream (P4) lower down the water course.

Grabouw/Vyeboom

Table 5 presents total endosulfan levels detected in Grabouw, and also summarizes the results.

Table 5. Endosulfan levels detected in Grabouw.

DATE	SITES AND CONCENTRATION in μg/L								
	G3r	G4r	G7d	Gld	G5d	G2d	G8T	G6b	G9r
23/4/98	(0.01)	nd	(0.09)	(0.06)	(0.08)	1.08	ns	nd	ns
7/5/98	nd	nd	(0.09)	0.32	(0.09)	0.81	(0.09)	ns	ns
27/5/98	(0.06)	0.59	nd	0.24	nd	1.78	nd	ns	ns
29/7/98	nd	nd	nd	nd	nd	0.16	ns	ns	ns
6/9/98	nd	nd	nd	(0.03)	0.70	(0.05)	(0.07)	ns	ns
12/10/98	nd	0.10	0.20	(0.07)	nd	0.10	nd	ns	ns
17/11/98	0.18	0.98	0.90	1.09	1.61	4.41	0.49	ns	ns
2/12/98	0.62	nd	0.82	0.88	0.79	9. 11	0.54	ns	ns
18/1/99	0.50	1.09	1.14	0.34	1.2	5.84	0.59	0.26	ns
18/2/99	0.30	1.38	ns	0.96	0.50	ns	0.80	ns	ns
03/3/99	0.47	(0.03)	0.5	0.56	0.63	9.50	1.06	ns	ns
24/3/99	0.91	nd	0.91	nd	0.45	6.44	1.77	nd	ns
15/4/99	0.55	nd	0. 15	0.16	0.50	1.85	0.88	ns	0.29
23/4/99	nd	0.27	nd	nd	0.21	ns	0.17	ns	ns
05/5/99	nd	nd	nd	nd	0.10	nd	ns	nd	nd
Mean (SD)	0.10 (0.17)	0.296 (0.476)	0.35 (0.42)	0.31 (0.38)	0.46 (0.48)	3.16 (3 .5)	0.538 (0.532)	0.065 (0.13)	0.145 (0.205)
n	12	15	14	15	15	13	12	4	2
%positive samples*	42	58	64	73	80	92	83	25	50

ns = Not sampled; nd = not detected (no discernable peak, less than 0.01 μg/L () = less than quantification limit; * positive samples = samples in which endosulfan was detected (including those below the quantification limit)

Detection of endosulfan in all sites was consistent with the timing of spraying activities on farms in the area, with endosulfan levels being raised in November after the October spray and during January–March, 1999 after the December spray. The latter period also corresponds with maximum irrigation practices in the area. Regular trigger rains during May–December, 1998 and January, April–June, and October–December, 1999 also enhances endosulfan levels.

The time and level of endosulfan detected in the two sites on the Palmiet River was broadly similar.

Table 5 shows that the four dams sampled in the area, were consistently and relatively highly contaminated compared to other study sites. Raised levels in dams were measured well beyond the period of application. The pH data did not suggest any mobilization of sediments. Davies [15] also previously identified endosulfan as a common contaminant of dams in the Grabouw region.

Sites supplying drinking water yielded fairly consistent low levels of endosulfan. Particularly high levels for the dam (G2d) that supplies water to one household were noted. There was also intermittent presence of endosulfan in the stream (G7d) feeding the Theewaterskloof Dam supplying drinking water to Metropolitan Cape Town.

Overall Number of Samples with Endosulfan

Table 6 summarizes the number of samples in which endosulfan was detected above the water quality criterion (0.9 µg/l) of the Inland California Surface Water Plan (CAISWP, [24]), and those above and below the study quantification limit and exceeded the European Community (EEC) single pesticide limit (0.1 µg/L), in relation to study area and ground and surface water. Endosulfan was found most frequently in Grabouw, 72 (69%) out of 104 samples compared to Hex River, 85 (46%) out of 184 samples and Piketberg, 37 (39%) out of 94 samples (Table 6). Both ground and surface water sites regularly EEC [25] water standard (0.1 µg/l) used universally for all pesticides. Twenty-three percent of all samples (n = 194) exceeded the less stringent CAISWP [24] water quality criterion (0.9 µg/l).

Table 6. Number of samples in which endosulfan was detected in the three areas sampled, and in groundwater and surface water.

Endosulfan, area and number of samples in which endosulfan was detected (percentage)				
LEVEL	**ENDOSULFAN**			
	HEX RIVER	PIKETBERG	GRABOUW	TOTAL
ALL	85 (46)	37 (39)	72 (69)	194 (51)
> QL, EEC	60 (33)	24 (26)	59 (57)	143 (37)
> CAISWP	11 (6)	4 (4)	30 (29)	143 (37)
	Groundwater		Surface Water	
ALL	17 (32)		177 (47)	
>QL, EEC	12 (23)		131 (40)	
> EEC	7 (14)		38 (12)	

All: All samples in which endosulfan was detected >QL, EEC: Above study quantification and EEC single pesticide limit of 0.1 µg/L. EEC total pesticide limit= 0.5 µg/L; >CAISWP: Above 0.9 µg/L (30 day average)

The slightly higher frequency of endosulfan detected in Grabouw compared to Hex River and Piketberg might be explained by the more frequent rainfall and the higher levels of spraying with endosulfan during the irrigation period.

Although the results are based on relatively few groundwater sites and samples (only five sites in the three study areas, totaling 53 samples over the study), detections of endosulfan appear lower for groundwater (23%) compared to surface water (40%).

It is also worth noting that the SF laboratory sporadically detected a number of other pesticides commonly used in deciduous fruit farming in both the Hex River and Grabouw/Vyeboom areas. These detections (of azinphos-methyl, fenarimol, iprodione, deltamethrin, penconazole, and prothiofos) occurred at times more or less consistent with usage of these agents in the industry, and at relatively low levels (below 2 µg/l) although not as low as detections achieved at PENTECH. However, their presence in the samples adds consistency to the picture obtained and to the construct validity of the overall results.

DISCUSSION

This study shows evidence of consistent low-level endosulfan in rural water sources in the Western Cape and warrants greater attention to establishing mechanisms for pesticide surveillance of water sources in South Africa. That nineteen of the contaminated sites were drinking or domestic water sources is of particular concern. Comparison of the levels obtained to some human health guideline/standard would therefore be important. However, only two endosulfan drinking water standards (EEC, CAISWP) are available, with the EPA, WHO, and South Africa currently having no endosulfan standard [26-28]. With regard to aquatic safety the guideline is 0.003 μg/l in Australia [29] while in South Africa [28] the chronic effect value is set at 0.01 μg/l and the acute effect value, 0.02 μg/l.

Table 7 shows a modeling of daily intake of pesticides for study populations using selected sampling points for drinking water. The modeling assumes two scenarios: a worst case scenario where drinking water concentrations are characterized at the highest concentration detected at the site; and a scenario where the concentrations found at each site are averaged using a root mean square conversion. These are then used to estimate total daily intake of pesticide and compared to published acceptable daily intakes (ADIs) [30] to calculate a percentage of ADI derived through water consumption. Estimates which were determined assuming that the average person consumes 2 l of water per day and weighs 60 kg, were low when compared to WHO ADI. Drinking water intake is thought to pose a health risks if it exceeds 1–10% of ADI. Only the peak estimate for the site providing purified water to the West Coast exceeded 10%, while the average estimates of this site was also the only one that exceeded 1%. It is therefore reasonable to infer that these levels are not of immediate concern. However, it should be noted that the calculations in Table 7 do not take account of vulnerable groups such as children who have a higher consumption per kg body weight.

Table 7. Modeling of daily intake of endosulfan for study populations using selected sampling points for drinking water (μg/l).

AREA	Point	Peak concentration (μg/L)	Daily intake* based on peak		Root mean square concentration (μg/L)	Daily intake* based on root mean square concentration	
			(μg/kg)	%ADI		(μg/kg)	%ADI
ENDOSULFAN							
Grabouw	G6	0.26	0.009	0.14	0.13	0.004	0.07
	G7	1.14	0.038	0.63	0.53	0.017	0.30
	G8	1.77	0.059	0.98	0.74	0.025	0.41
Piketberg	P1	0.44	0.015	0.24	0. 15	0.005	0.09
	P3	0.25	0.009	0. 14	0.31	0.010	0. 17
	P4	0.36	0.012	0.20	0.14	0.005	0.09

Table 7. *(Continued)*

AREA	Point	Peak concentration (μg/L)	Daily intake* based on peak		Root mean square concentration (μg/L)	Daily intake* based on root mean square concentration	
			(μg/kg)	%ADI		(μg/kg)	%ADI
	P7	0.27	0.009	0.15	0.22	0.007	0.12
	P8	26.3	0.877	14.6	10.7	0.360	6.00
	J	0.62	0.021	0.34	0.27	0.009	0.15
	H	0.89	0.030	0.49	0.25	0.008	0.14

ENDOSULFAN: ADI < 0.006 mg/kg bw [30]; Note: Only results used where recoveries were > 70% and < 130 %; * Daily intake of water for adults assumed 2 L per day for an adult of 60 kg.

Nonetheless, thresholds for concern are being continually revised downward as more empirical evidence emerges. The presence of endosulfan, which has class two human toxicity, very high aquatic toxicity [29] and is a known endocrine disruptor with estrogenic effects comparable to estradiol [31] warrants attention.

A few studies have previously detected endosulfan in water sources [13, 15, 24, 32-34] and the levels found in this study are consistent with the range (0.1–100 μg/l) found in groundwater [24, 32] and surface water [33] in those studies. Pesticide detections in this study, however, appear to be more frequent than found in previous studies, probably a function of increased frequency of sampling [35]. Endosulfan spraying in Grabouw ranges from 0.5 to 1.5 kg active ingredient per hectare.

The findings in this study contrast with those found by Weaver [17] in the Hex River Valley in 1990, where no evidence was found for pesticides reaching ground water. However, that study analyzed a different set of pesticides, made use of less sensitive analytical techniques and focused primarily on groundwater. Detections in both surface and groundwater (including sampling point H, which was identical in the two studies) were in any case found to be low in this study. However, of importance is that detections are not confined to the Hex River but are ubiquitous in all three study areas. Out of 382 samples, there were 37% endosulfan detections above the EU limit of 0.1 μg/l.

Endosulfan has been reported as having a low pollution likelihood [36], but other factors such as soil characteristics, shallow water tables, and intensive spraying [24, 32, 37], could explain its relatively frequent detection in this study.

Endosulfan levels in all three areas were the highest and most frequent during January–March, corresponding mainly with irrigation practices, but also with rainfall events. Previously, Domagalski [35], also found irrigation to be an important trigger for both leaching and run-off events. Recently, Schultz et al. [13] found rainfall wash-out to increase endosulfan in the Lourens River located in the South-Western Cape (from 0.06 to 0.16 μg/l), but levels were substantially lower and detections less frequent, than that measured in this study and the effect of irrigation was not measured.

Correlation between rainfall and endosulfan detection in our study might therefore have been influenced by irrigation patterns.

The reliance on grab sampling was a limitation in the study. Intermittent monitoring may give false estimates of true exposures, or inadequate characterization of contamination patterns. For example, Domagalski [35] showed that thrice weekly sampling of surface water in the San Joaquin River Basin was more than twice as likely to identify concentrations exceeding state water standards than single weekly sampling. Efforts to develop methods that sample water sources on a continuous basis, to provide an integrated assessment of water contamination by pesticides, should be explored. Integrated sampling methods are, however, not practical at present.

Other limitations in the study include the use of manual grab samples, and the non-measurement of specific conductance and dissolved oxygen due to a lack of resources.

CONCLUSION

The results in the study indicate that monitoring of pesticide levels in South African water resources is warranted, preferably with cost-effective and practical methodologies. The findings also indicate that epidemiological studies investigating the health effects of endosulfan should be undertaken. Furthermore, policies aimed at reducing the potential contamination of water by pesticides need to be developed and implemented.

KEYWORDS

- **Endosulfan**
- **High-pressure chromatography**
- **Pesticides**
- **Quality assurance**
- **Sampling sites**
- **Water pollution**

AUTHORS' CONTRIBUTIONS

Mohamed A. Dalvie co-ordinated the study, assisted with the design, collected data, and drafted the manuscript. Eugene Cairncross was responsible for the design of the analytical methods, assisted with the design of the study and drafting of the manuscript. Leslie London was the principal investigator of this project, designed, and organized the study and assisted in drafting the manuscript. Abdullah Solomon was responsible for the laboratory analysis, assisted with data collection, and drafting of the manuscript. All authors read and approved the final manuscript.

ACKNOWLEDGMENTS

The authors wish to thank the Water Research Commission of the South African Department of Water Affairs and Forestry (DWAF) and the South African Medical

Research Council for financial support for this study. The assistance of Tom Robins of the University of Michigan School of Public Health and the Fogarty Centre for International Research Development; Jannie Walters (Wenkem); Chris Dain (Zeneca); Garth Hodges (Agrevo); John Levings (Two a day co-op); Mr. Watkins (Mechanical Engineering, UCT); WHO (donated WHO standards); Mr. S. van Niekerk and J. van Zyl (Elsenberg Agricultural College); Mr. A. Jacobs (Infrutec); Dr. O. Sisulu and Dr. G. Joubert (CSIR); John Weaver, Kevin Pieterse, and Gideon Tradouw (Watertek); Dr. Tawanda Masuka and Mr. Munro van der Merwe (ARC laboratories); Kevin Hearshaw (State Forensic laboratory); Hanlie van der Westhuizen; E. Truter and M. Loubscher (Department of Health, West Coast Region, Malmesbury); Kobus Hartman (UNIFRUCO); Alreta Louw (DWAF) and The Hex River Farmers association is also acknowledged.

COMPETING INTERESTS

The authors are not aware of any competing interests.

Chapter 13

Land Use Regression in Sarnia, "Chemical Valley," Ontario, Canada

Dominic Odwa Atari and Isaac N. Luginaah

INTRODUCTION

Land use regression (LUR) modeling is proposed as a promising approach to meet some of the challenges of assessing the intra-urban spatial variability of ambient air pollutants in urban and industrial settings. However, most of the LUR models to date have focused on nitrogen oxides and particulate matter. This study aimed at developing LUR models to predict BTEX (benzene, toluene, ethylbenzene, m/p-xylene, and o-xylene) concentrations in Sarnia, "Chemical Valley," Ontario, and model the intra-urban variability of BTEX compounds in the city for a community health study.

Using Organic Vapor Monitors, pollutants were monitored at 39 locations across the city of Sarnia for 2 weeks in October, 2005. The LUR models were developed to generate predictor variables that best estimate BTEX concentrations.

Industrial area, dwelling counts, and highways adequately explained most of the variability of BTEX concentrations (R^2: 0.78–0.81). Correlations between measured BTEX compounds were high (> 0.75). Although, most of the predictor variables (e.g., land use) were similar in all the models, their individual contributions to the models were different.

Yielding potentially different health effects than nitrogen oxides and particulate matter, modeling other air pollutants is essential for a better understanding of the link between air pollution and health. The LUR models developed in these analyses will be used for estimating outdoor exposure to BTEX for a larger community health study aimed at examining the determinants of health in Sarnia.

Volatile organic compounds (VOCs) are important outdoor air toxins suspected to increase chronic health problems in exposed populations [1, 2]. The BTEX are some of the common VOCs found in urban and industrial areas and are classified as "hazardous air pollutants" (HAPs) because of their potential health impacts [3]. Nonetheless, the evidence as to whether HAPs influence health effects remains equivocal. For example, while Leikauf [4] argued that there is insufficient evidence indicating that ambient HAPs exposure has the potential to exacerbate health problems such as asthma, the author acknowledged that once an individual with a health outcome (e.g., asthma) is sensitized to air pollution, they are more likely to respond to remarkably low concentrations of pollution. Furthermore, although low levels of VOCs might have no significant health impacts, the interaction between VOC species and other criteria pollutants might cause adverse health outcomes. Rumchev et al. [5] studied the linkages between domestic exposure to VOCs and asthma in young children in Perth, Western Australia, and found that exposure to VOCs increased the risk of childhood asthma.

Individual species within VOCs have also been examined for their health effects. For instance, the International Agency for Research on Cancer (IARC) [6] has classified benzene as a known human carcinogen based on evidence from epidemiologic studies and animal data. These studies have shown that exposure to benzene can cause acute nonlymphocytic leukemia and other blood disorders such as preleukemia and aplastic anemia [6, 7]. The US Department of Health and Human Services [8] also reported an association between occupational exposure to benzene and the occurrence of acute myelogenous leukemia. In Australia, Glass et al. [9] found an association between leukemia and cumulative benzene exposures that were considerably lower than the accepted level.

Besides benzene, other BTEX compounds are also suspected to adversely affect human health. The US Department of Health and Human Services [10] suggested that exposure to high dosages of toluene may cause headaches, sleepiness, kidney damage, and could impair an individual's ability to think clearly. Additionally, Chang et al. [11] reported that toluene exposure could exacerbate hearing loss in a noisy environment in Taiwan. While studying the association between several sites of cancer and occupational exposure to toluene in Montreal, Quebec, Gerin et al. [12] observed a doubling risk of esophageal cancer in subjects exposed to medium to high levels of toluene. Conversely, other studies that examined toluene as a possible risk factor for cancer did not find any significant association between exposure to toluene and cancer. For example, Antilla et al. [13] found no increase in overall cancer risk for cancers at specific tissues associated with exposure to toluene, except for a non-significant increase in the incidence of lung cancer in Finnish workers who were exposed to toluene for more than 10 years.

The evidence on the health effects of Ethylbenzene remains uncertain. Ethylbenzene has been linked to dizziness, throat, nose, and eye irritations and recent laboratory assessments have shown that long-term exposure to ethylbenzene may cause cancer [14, 15]. While reviewing the literature on the effects of low-level exposure to ethylbenzene on the auditory system, Vyskocil et al. [16] reported no evidence of ethylbenzene induced hearing loss after combined exposure to ethylbenzene and noise of workers in Quebec. In addition, acute exposure to xylenes could cause respiratory and neurological health problems in humans, while chronic exposure could affect the central nervous system [17]. On the other hand, work by the US Department of Health and Human Services [18] provided insufficient evidence showing that xylenes are potential human carcinogens.

Although, there is an understanding of the biological plausibility linking hazardous pollutants in the ambient environment to health effects, the evidence from toxicological, occupational and epidemiological studies are still frequently in discordance. This is partly due to different methodological issues. For instance, the threshold concentrations used in animal studies are frequently above those used in epidemiologic studies [4]. Also, researchers have documented that ambient (outdoor) air pollution concentrations used in epidemiologic studies may underestimate personal exposure because people spend most of their time indoors [19-21]. Despite this recognition, the argument is that the consistent pattern of outdoor air pollution when compared to indoor air pollution [20, 21] means that outdoor exposure estimates may still be useful for health studies where indoor air pollution data are unavailable. That is, outdoor air pollution estimates can be used as estimates of overall pollution pattern especially in highly polluted areas such as Sarnia where the correlation between indoor and outdoor air pollution may be high as a result of traffic and industry-related air pollution [22]. Hence, in the absence of indoor air pollution estimates, outdoor exposure patterns are sufficient for health studies [23].

The equivocal nature of the relationship between ambient air pollution and associated health effects [4, 24, 25] may be attributed to the challenges in the assessments of ambient air pollution for epidemiologic studies [26, 27]. Recently, different approaches have been proposed and utilized in addressing the challenges of estimating personal exposure to air pollution. For instance, kriging has been used both at the national and regional scale [26], but has been criticized for its inability to capture air pollution at very short distances [28]. Other studies have used proximity analysis and community average of pollution concentrations as proxies for exposure [29-31], however these approaches have also been criticized because of their high potential for exposure misclassification [32]. Microenvironment monitoring aims to address some of the exposure assessment challenges [33], but its suitability has been hampered by high costs related to data collection especially when dealing with a large cohort [34]. Traditionally, dispersion models are also used to estimate individual level exposure because they incorporate both spatial and temporal variations without the need for additional air pollution monitoring. The biggest challenge with dispersion models lies in their expensive data demands and lack of precision in the requisite meteorological or emissions data required for making accurate predictions [35, 36]. Since exposure estimation can have significant impacts on explaining relationships between exposure and health outcomes [37-39], there is a growing demand for improved and affordable ways of exposure estimation that can potentially capture the variability of air pollution for health studies in high polluted environments like Sarnia [32, 40].

The LUR modeling is proposed as a promising alternative approach to meet some of the challenges of assessing the intra-urban spatial variability of ambient air pollutants in urban and industrial settings because it can capture localized variation in air pollution more effectively and economically than some of the conventional approaches previously discussed [32, 35, 37, 40, 41]. The LUR modeling predicts outdoor ambient air pollution concentrations at given sites based on the surrounding land use, traffic, population, and dwelling counts, and physical characteristics such as elevation [35]. Several researchers [26, 27, 35] have provided critical reviews of LUR studies and emphasized the potential role of LUR models in estimating exposure to air pollution. However, most of the LUR models to date have focused on nitrogen oxides (NO_2 and NOx) and particulate matter (PM2.5, PM10). With potentially different health effects, modeling other air pollutants is essential for increasing our understanding of the link between air pollution and health. Consequently, the main objectives of this study were to: (1) develop LUR models to predict VOCs, specifically benzene, toluene, ethylbenzene, m/p-xylene, o-xylene, and total BTEX in Sarnia, and (2) determine the intra-urban variations of ambient benzene, toluene, ethylbenzene, m/p-xylene, o-xylene, and total BTEX to be used in a larger community health study.

MATERIALS AND METHODS

Study Area

The City of Sarnia (42° 58' N, 82° 22' W) is located in southwestern Ontario, Canada, on the border just east of Port Huron, Michigan, USA (Figure 1). Neighboring Canadian cities include London and Windsor. Sarnia has an approximate land area of 165 km^2 and a population of 71, 419 [42]. Both the city and surrounding communities are called "Chemical Valley" because more than 40% of Canadian chemicals are manufactured in this area [43]. Examples of the chemical industries in the area include Suncor, Bayer, Dow Canada, NOVA, and ESSO. Furthermore, one of the largest landfill sites in Canada known as Safety-Kleen is located in the region. These point sources in Sarnia are amongst the largest industrial polluters in Canada with the highest levels for some VOCs, such as 1–3 butadiene, compared to other polluters across the country [44]. Recently, the Canadian government designated the St. Clair region which includes Sarnia and 16 others as "Areas of Concern" based on a hypothesis that environmental pollution is negatively affecting the population in these areas [43, 45].

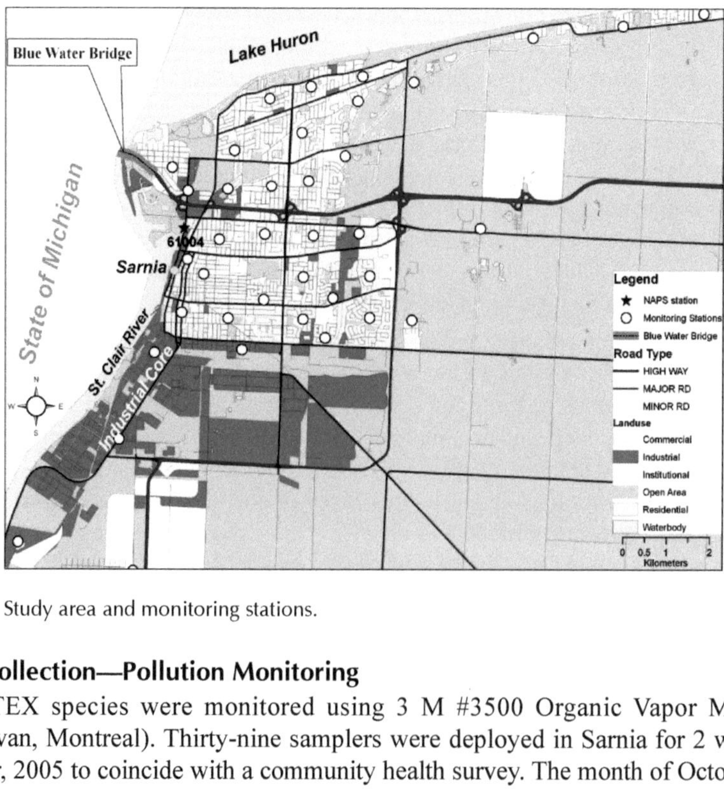

Figure 1. Study area and monitoring stations.

Data Collection—Pollution Monitoring

The BTEX species were monitored using 3 M #3500 Organic Vapor Monitors (Guillevan, Montreal). Thirty-nine samplers were deployed in Sarnia for 2 weeks in October, 2005 to coincide with a community health survey. The month of October best represents the average annual weather condition in Sarnia. Although, formal location-allocation techniques [46] were not used, the samplers were deployed based on a number of objective criteria to capture the spatial variability of BTEX compounds in areas of high population density. Samplers were located proportional to population size in each census tract. In addition, sites were selected to ensure sufficient variability in potential predictors (e.g., land use, road networks) (Figure 1). Hence, only two samplers were located within Vidal Street, the main traffic route through the industrial core, which served as the point of origin for the measures for this study to capture pollution near service areas. Vidal Street is called the industrial core because it is the major traffic feeder to industries in Sarnia (Figure 1). The rest of the sampling sites were at least 600 m away from the industrial core to ensure data accurately reflected diffused ambient pollution throughout the region rather than point sources. The samplers were installed at a height of 2.5 m on light poles after obtaining permission from the City of Sarnia and the Aamjiwnaang Indian Reserve. Global positioning systems were used to geocode the monitoring locations.

The exposed filters were sent to Air Monitoring and Analysis (Mississauga, Ontario) lab for analysis of all measured BTEX species. The samples were extracted with 2.0 ml of solvent and the compounds determined using gas chromatography—mass selective detector with a detection limit of 0.1 μg/l [47]. A multi-point calibration curve ($r^2 >= 0.99$) was used and the results were corrected with lab blank, deuterated internal standard and recovery. The 2-week BTEX measurements served as dependent variables in the developed LUR models.

Assessment of Spatial Trends

Sampling density was calculated as the number of samplers divided by the study area. Kriging was used as the spatial interpolation technique to examine how the different BTEX species were spatially distributed based on the sampling density. The spatial trends were examined using ArcMap 9.2.

Variable Generation

The predictors of BTEX species were extracted from several datasets including traffic counts, census data, street network, land use, and digital elevation models (DEMs). The traffic counts were annual average daily traffic (AADT) volumes collected in 2004 and compiled for major and minor roads by the City of Sarnia, the Administration and Engineering Department, and for highways by the Ontario Ministry of Transportation. Both the city and provincial traffic data were then combined in GIS to establish a comprehensive dataset for traffic counts based on road segments. Population and dwelling counts at the dissemination area (DA) level were generated from 2001 census data [42]. The street network and land use 2006 datasets were obtained from Desktop Mapping Technologies Inc (DMTI) via the Data Liberation System from the University of Western Ontario. The street network file had information on all three types of roads (minor, major, highway) segment-by-segment. Digital elevation data were used to generate the elevation for each sampled station at a 25 × 25 m grid resolution (DMTI).

The independent variables were generated within circular buffers that extended from the sampling locations at 50 m intervals out to 3000 m using ArcGIS. The predictor variables were conceptually grouped into four different broad categories: land use, road and traffic, population and dwellings, and physical geography. The land use category included areas (in hectares) of industrial, commercial, institutional, residential, open areas and water bodies that fall within the specified buffer radii with sampling sites as centers. The roads and traffic category included calculated lengths of minor and major roads and highways; and the total vehicle miles traveled (VMT) on the roads segments that fall within the buffer radii. The VMT was calculated as AADT counts multiplied by the road segment length within a specified buffer. Calculated VMT values were then summed as the total vehicle miles traveled for the monitored station within the specified buffer. The total population and dwelling counts were calculated as the ratio of each DA that fell within a specified buffer area and the total area of that DA multiplied by total population/dwelling counts of their respective DA. Meteorological data (e.g., wind direction) was not used in the analysis because there was only one functional meteorological station in the study area during the monitoring period. The physical geography category included the x, y coordinates, elevation, measured distances from monitoring stations to Vidal Street (industrial core), the Blue Water Bridge, minor and major roads and highways.

Model Selection

The natural logarithm of BTEX species were used in the LUR modeling because their distributions were skewed. The association between the geographic variables and the mean levels of measured air pollutants was analyzed using multiple linear regressions. Each of the buffers generated were individually screened through bivariate regression models using SPSS statistical software [48] to identify the variables that were highly correlated with measured BTEX species. Next, the most relevant univariate relationships were identified and then a stepwise multiple regression was conducted to find the most predictive models for benzene, toluene, ethylbenzene, m/p-xylene, o-xylene, and total BTEX (sum of all BTEX species). The final LUR models for BTEX and each species were identified as having a combination of variables with the highest coefficient of determination, R^2. Independent variables retained in the models had to have significant t-score ($p < 0.05$) and low collinearity with other variables (defined by a variance inflation factor < 2.0).

After the most predictive models were obtained, the standard regression diagnostics to identify outliers, leverage and influence values were performed. The individual influence of each measured concentration on the whole model was examined using the size-adjusted Cook's distance [49]. Points with calculated Cook's distance values greater than the cutoff (defined as 4/sample size) were removed because of their disproportionate influence on the most predictive models. The residuals were tested for Moran I (MI) spatial autocorrelation [50, 51]. Pearson correlations between significant independent variables in the most predictive models were also examined.

Two different cross-validation procedures to evaluate the precision of the optimized models were used. The first was a "leave-one out procedure" which involved removing one of the monitored sites and predicting the concentration at the omitted location [19, 52]. This procedure was repeated for all the sampling locations and the prediction error calculated as root mean squared error (RMSE)—the square root of the sum of the squared differences of the observed and the predicted concentration at removed locations [41]. A second cross-validation approach was performed in three random selections of 90, 80, and 50% of the samplers to predict BTEX concentrations at the remaining 10, 20, and 50% locations, respectively [52, 53]. The Chow test was used to determine whether the coefficients in the predictive regression models were similar to the coefficients of the three different validation trials in the second cross-validation [53, 54].

The surfaces of predicted BTEX concentrations were created by applying the coefficients of the predictive model equation and generating predicted surfaces with a 5 × 5 m resolution. The correlation between kriged and LUR modeled BTEX concentrations were calculated for each sampling site. All data management and statistical analyses were performed using SPSS statistical software [48]. Spatial autocorrelation and surface generations were performed using ArcGIS 9.2.

Two of the samplers were lost due to vandalism. The two samplers were 600 and 2,800 m away from the industrial core and 8,200 m apart. The calculated sampling density of 0.24 was higher than for other Canadian studies in Hamilton (0.08), Toronto (0.16), and Montreal (0.18) [32, 53, 55, 56]. With the general distribution and sampling density, the two lost samplers would likely have no significant effect on the different BTEX models. Table 1 presents the summary statistics of the BTEX compounds from the remaining 37 locations. Arithmetic means of the compounds were 0.93 ± 0.56 $\mu g/m^3$ for benzene, 2.58 ± 1.35 $\mu g/m^3$ for toluene, 0.46 ± 0.23 $\mu g/m^3$ for ethylbenzene, 1.21 ± 0.61 $\mu g/m^3$ for (m + p) xylene, and 0.49 ± 0.25 $\mu g/m^3$ for o-xylene. Toluene was the most abundant compound at all sampling sites followed by benzene.

Table 1. Distribution of BTEX concentrations at measured sites.

					Percentiles		
	Mean	**SD**	**Min**	**Max**	**25th**	**50th**	**75th**
Benzene	0.93	0.56	0.28	3.36	0.56	0.86	1.15
Toluene	2.58	1.35	0.85	6.88	1.67	2.20	3.42
Ethyl benzene	0.46	0.23	0.15	1.06	0.30	0.39	0.58
(M+P) xylene	1.21	0.61	0.40	2.81	0.78	1.08	1.56
0-xylene	0.49	0.25	0. 15	1. 19	0.32	0.43	0.63
Total BTEX	5.67	2.88	1.83	14.50	3.69	4.91	7.36

Table 2 compares monthly (there were only four measurements for the month of October, 2005: the 1st (Saturday), 7th (Friday), 19th (Wednesday), and 25th (Tuesday)) and 5-year (2001–2005) means of BTEX concentrations measured at the National Air Pollution Surveillance (NAPS) station (#61004). The average ambient concentrations of the 3 sampling points closest to the station (Figure 1) were chosen for comparison following Atari et al. [55] and Miller et al. [47]. In general, the 2-week average concentrations of benzene (1.07 μg/m3), toluene (3.35 μg/m3), ethylbenzene (0.56 μg/m^3), and total BTEX (7.19 μg/m^3) at the 3 sampling points closest to the station were slightly lower than the monthly and 5-year means measured at the NAPS station (Table 2). The 2-week average concentrations of (m + p) xylene (1.43 μg/m^3) and o-xylene (0.58 μg/m^3) measured at the 3 sampling points closest to the station were slightly higher than the monthly and 5-year means measured at the NAPS station. The differences could be attributed to the fact that (m + p) xylene and o-xylene are more photochemically reactive than their counter parts [57], and different measuring instruments were used. Environment Canada used 6 l Summa canisters at the NAPS stations [58] while 3 M samplers were used in this study.

Table 2. Comparison between NAPS and sampled BTEX data.

	NAPS Data[a]		Sampling Data[b]
	Monthly average[c]	**5 years average[d]**	**2-week average**
Benzene	1.93	1.40	1.07
Toluene	3.58	3.80	3.35
Ethyl benzene	0.64	0.57	0.56
(M+P) xylene	1.19	1.25	1.43
0-xylene	0.37	0.40	0.58
Total BTEX	7.71	7.25	7.19

[a] Measured data at the National Air Pollution Surveillance (NAPS);
[b] 2-week average data at 3 sampling points closest to the NAPS station;
[c] The monthly average is the mean of the only 4 records available for October 2005;
[d] 5 years average (2001–2005) concentration

The measured BTEX species are highly correlated to each other (Table 3). The kriged surfaces of measured BTEX concentrations showed similar patterns with high concentrations along the industrial core. Because of the high correlation between BTEX species and their similar patterns in the kriged surfaces, only two surfaces are shown (Figure 2). The benzene surface has a slightly more localized pattern when compared to the other BTEX species. Table 4 shows the Pearson correlation coefficients between measured, kriged, and LUR modeled concentrations at the sampling locations. The correlation between measured and kriged concentrations were low for ethylbenzene ($r = 0.38$), (m + p) xylene ($r = 0.16$) and o-xylene (0.14). Likewise, the correlation between kriged and LUR modeled concentrations at the sampling locations were low for ethylbenzene ($r = 0.46$), (m + p) xylene ($r = 0.31$) and o-xylene ($r = -0.19$). Kriged o-xylene concentrations were consistently lower than the LUR modeled concentrations at the sampling locations.

Table 3. Pearson correlation between measured ambient BTEX compounds.

	Toluene	Ethyl benzene	(M+P) xylene	0-Xylene	Total BTEX
Benzene	0.817**	0.845**	0.752**	0.755**	0.865**
Toluene		0.973**	0.965**	0.969**	0.991 **
Ethyl benzene			0.977**	0.973**	0.989**
(M+P) xylene				0.994**	0.966**
0-Xylene					0.970**
**** Significant at 95%**					

Table 4. Correlations between measured, kriged and LUR modeled concentrations at sampling locations in Sarnia.

	Measured	LUR	Measured	LUR
	Benzene		**Toluene**	
LUR	0.793**		0.719**	
Kriged	0.927**	0.828**	0.999**	0.736**
	Ethylbenzene		**(M+P) Xylene**	
LUR	0.384**		0.162	
Kriged	0.892**	0.458**	0.713**	0.307
	0-Xylene		**BTEX**	
LUR	0.135		0.614**	
Kriged	0.711 **	-0.186	0.988**	0.658**
**** Significant at 95%**				

The calculated Moran's indices for benzene (MI = –0.02), toluene (MI = 0.01), ethylbenzene (MI = –0.04), (m + p) xylene (MI = –0.03), o-xylene (MI = –0.03), and total BTEX (MI = –0.03) residuals of the most predictive models indicate no significant autocorrelation. Table 5 shows the final LUR models for predicting the concentrations of benzene, toluene, ethylbenzene, (m + p) xylene, o-xylene, and total BTEX. The model for benzene (R^2 = 0.78) included industrial land use within 1,600 m, dwelling counts within 1200 m, and length of highway within 800 m. The model for toluene had an R^2 of 0.81 including industrial land use within 2,800 m, open area within 600 m, and length of highway within 800 m as significant predictors. The model for ethylbenzene (R^2 = 0.81) included industrial land within 2,600 m, dwelling counts within 1,400 m, and length of highway within 800 m. The model for (m + p) xylene and o-xylene had similar R^2 of 0.80 including industrial land use within 1,600 m, dwelling counts within 1,200 m, and length of highway within 800 m. The total BTEX model had a coefficient of determination (R^2) of 0.81 including industrial land use within 2,500 m, dwelling counts within 1,400 m, and length of highway within 900 m showing significant contribution to the model. The positive regression coefficients indicate that concentrations of BTEX compounds increase as the values of the independent variables (e.g., industrial area) rise, while the negative coefficients indicate a decrease in concentrations as the values of the predictor variables (e.g., area of open space) increase. All variables in the six models are significant at the 95% level of confidence. None of the variables in the final models were significantly correlated with each other (Table 6).

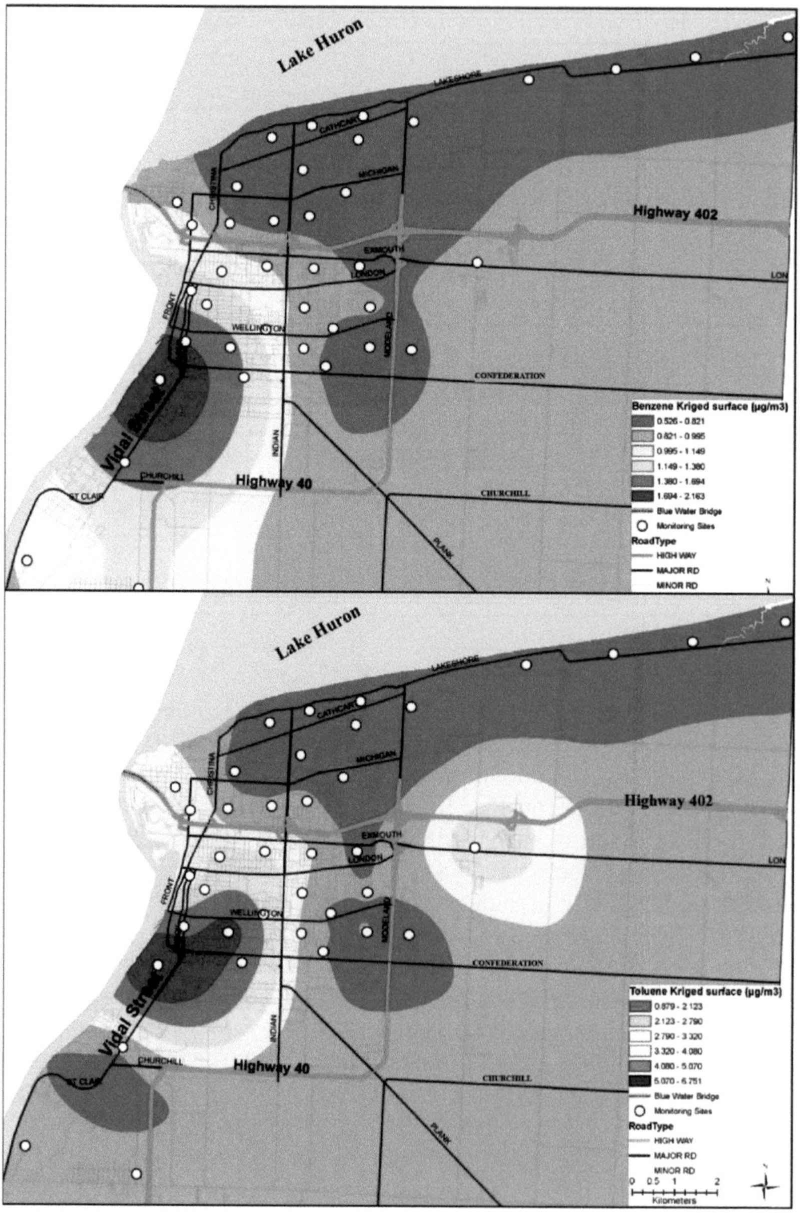

Figure 2. Kriged surfaces for measured benzene and toluene.

Table 5. Land use regression model results for BTEX compounds.

Variables	Benzene	Toluene	Ethyl benzene	(M+P) xylene	0- xylene	BTEX
Intercept	-1.086 ± 0.141	0.486 ± 0.092 5.258**	-1.879 ± 0.127,- 14.758**	-1.079± 0.127,- 8529**	-2.009 ± 0.144,- 13.996**	0.587 ± 0.119, 4.927**
Industry 1600 m	0.005 ± 0.001 (0.640), 9.409**	...	...	0.003 ± 0.000 (0.339), 6.887**	0.003 ± 0.00 1 (0.526), 5.929 **	...
Industry 2500 m	...	...	...	...	...	0.002 ± 0.00 1 (0.464), 9.095**
Industry 2600 m	...	...	0.002 ± 0.000 (0.570), 10.084 **	...	...	...
Industry 2800 m	...	0.002 ± 0.000 (0.558), 8.181**	...	...	...	...
Open 600 m	...	-0.007 ± 0.002 (0.119),-3.444**	...	...	...	...
Dwelling 1200 m	0.002 ± 0.00 1 (0.066), 3.153**	...	...	0.004 ± 0.00 1 (0.398),7.379 **	0.003 ± 0.00 1 (0.219), 5.724 **	...
Dwelling 1400 m	...	...	0.002 ± 0.00 1 (0.157), 4.081**	...	...	0.003 ± 0.000 (0.224), 4.997**
Highway 800 m	0.076 ± 0.024 (0.07 3), 3.162**	0.094 ± 0.021 (0.134), 4.543 **	0.079 ± 0.022 (0.081), 3.61 1**	0.062 ± 0.021 (0.060), 2.926**	0.062 ± 0.022 (0.056), 2.889 **	...
Highway 900 m	...	...	...	...	...	0.079 ± 0.0 18 (0. 126),4.418**
Model R^2	0.779(0.757)	0.81 I (0. 792)	0.808 (0.790)	0.797 (0.776)	0.800 (0.780)	0.813 (0.794)[b]
Average V/F[c]	1.01	1.13	1.07	1.01	1.13	1.05
Model validation	...	...	...	...	...	...
R^2	0.75-0.81	0.77-0.86	0.79-0.86	0.78-0.81	0.77-0.79	0.80-0.84
RMSE[d]	0.25 - 0.87 11gim³	0.16 - 0.55 11gim³	0.14 - 0.17 11gim³	0.27- 0.42 11gim³	0.07 - 0.21 11gim³	0.58 - 1.48 11gim³

Note: [a] C = coefficient, SE = Standard Error, Cont = R^2 contribution; [b] Numbers in brackets are adjusted R^2; [c] VIF = Variance Inflation Factor; ** Significant at 95%; [c] RMSE = Root mean square error

Figure 3 shows the relationship between the observed and predicted pollutants based on their natural logarithmic scales. The scatterplots reflect the strength of each of the developed models and demonstrate that the models fit the observations well with no significant outliers. The spatial pattern of the predicted BTEX species concentrations showed expected characteristics (Figure 4) compared to their kriged surfaces. The predicted surfaces reflected the significant variables with industrial area, dwelling counts and traffic showing significance. The numerous petrochemical industries along the industrial core and dwelling counts showed significant influences on the modeled surfaces. The predicted surfaces have more detailed variability compared to the kriged surfaces of measured concentrations.

Table 6. Pearson correlation between significant variables in the most predictive LUR models.

Benzene		
	Dwelling counts within 1200 m	Length of highway within 800 m
Industrial Land Use within 1600 m	0.028	-0.058
Dwelling counts within 1200 m		-0.056
Toluene		
	Open area within 600 m	Length of highway within 800 m
Industrial Land Use within 2800 m	-0.322	-0.164
Open area within 600 m		-0.107
Ethyl benzene		
	Dwelling counts within 1400 m	Length of highway within 800 m
Industrial Land Use within 2600 m	0.014	-0.205
Dwelling counts within 1400 m		0.221
(M+P) Xylene	Dwelling counts within 1200 m	Length of highway within 800 m
Industrial Land Use within 1600 m	0.048	-0.119
Dwelling counts within 1200 m		0.042
0-Xylene		
	Dwelling counts within 1200 m	Length of highway within 800 m
Industrial Land Use within 1600 m	0.042	-0.095
Dwelling counts within 1200 m		-0.034
BTEX	Industrial land use	Length of highway within 900 m
Dwelling counts within 1400 m	0.010	-0.179
Industrial Land Use within 2500 m		0.176

The results of the validation approaches are provided in Table 5. The BTEX root mean square error predicted in this study were somewhat lower than the average estimated error of 1.72–2.15 μg/m^3 for BTEX concentrations reported by Aquilera et al. [19] who used similar approaches for cross-validation. Overall, the predicted benzene, toluene, ethylbenzene, (m + p) xylene, o-xylene, and total BTEX concentrations correspond nicely with measured concentration suggesting that these models are capable of predicting reliable concentrations. The Chow test results were not significantly different between the predictive models and the three different tests suggesting that the benzene, toluene, ethylbenzene, (m + p) xylene, o-xylene and total BTEX models developed were quite stable.

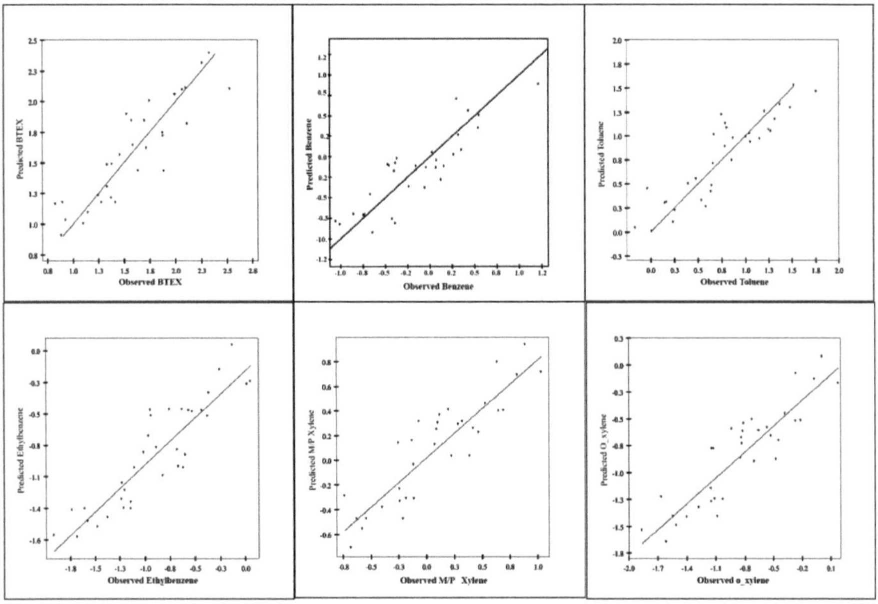

Figure 3. Observed versus predicted BTEX, benzene, toluene, ethylbenzene, m/p xylene and o-xylene (logarithmic scale) based on the best land use regression models.

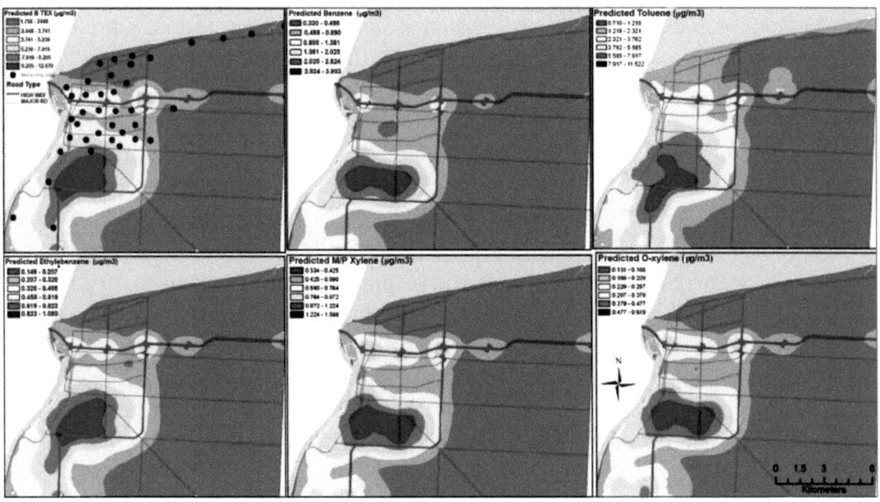

Figure 4. Land use regression predicted surfaces for BTEX compounds.

DISCUSSION

The aim of this study was to model the intra-urban variations of ambient VOCs including benzene, toluene, ethylbenzene, (m + p) xylene, o-xylene, and total BTEX for use in a large health study aimed at examining the determinants of health in sentinel high exposure environments. Although, most of the significant variables were similar in the six models, their individual contributions to the models were significantly different. For example, while industrial land use within 1,600 m was significant in both (m + p) xylenes and o-xylene models, the effect of industry (34% and 53%, respectively) differed in the two models (Table 5). These differential influences support the need for modeling the different air pollutants [55].

When compared to other LUR models developed in Munich [59], El Paso [60], Sabadell [19] and Windsor, Ontario [61], the significant variables in the present study showed considerably larger buffer radii. For example, Wheeler et al. [61] reported significant highway buffer radii of 50 m and 100 m for benzene and toluene models, respectively. In this study, we found significant highway buffer radii of 800 m for both benzene and toluene models (Table 5). The later result was also larger than the 300 m buffer radius reported by Beckerman et al. [62] when examining the variability of traffic-related pollutants around an expressway in Toronto, Ontario. The differences could be due to the unusually large number of petrochemical facilities in Chemical Valley, hence the broader distribution of ambient air pollutants in the area. The larger buffer radii found in this study potentially limits the generalizablility and transferability of the developed LUR models to areas of similar contextual and compositional characteristics [26].

When compared to other models developed in Sabadell [19], Munich [59], and Windsor, Ontario [61], the results of the various models of BTEX species are considerably different, further suggesting the need to model air pollutants in their various contexts rather than depending on proxies [37, 55]. The benzene model (R^2 = 0.78) showed comparable coefficient of determination when compared to a similar model developed in Munich, Germany (R^2 = 0.80) [59] but slightly higher than the R^2 of a model developed in Windsor, Ontario, Canada (R^2 = 0.73) [61]. The toluene model showed high coefficient of determination (R^2 = 0.81) compared to similar models developed in Windsor (R^2 = 0.46) [61] and Munich (R^2 = 0.76) [59], while the coefficient of ethylbenzene (R^2 = 0.81) was comparable to the coefficient reported in Munich (R^2 = 0.79) [59]. The BTEX model developed in this study showed high coefficient of determination (R^2 = 0.81) as compared to an R^2 of 0.74 reported by Aquilera et al. [19] in Sabadell, Spain. Differences in the R^2 could be due the contextual factors in the various cities. Although the industrial area exhibited varying influences in each of the models (Table 5), the results support the view that the numerous petrochemical industries are significantly affecting the VOC concentrations in Sarnia, Chemical Valley. If possible, it is important to model each air pollutant of interest to better analyze, determine, and understand personal exposures for health studies.

Besides industrial area, dwelling counts also emerged as a strong determinant of the intra-urban variation of BTEX concentration in Sarnia (Table 5). These results are consistent with other researchers [46] who found dwelling counts to influence the intra-urban variation of air pollution. The view is that high dwelling counts may influence heavy traffic and emissions [63]. The results also indicate that a combination of land use and dwelling counts could be used to estimate exposure to air pollution, especially BTEX compounds.

The correlations between BTEX species in this study showed slightly different coefficient ranges compared to other studies in Canada and the US [62, 64]. This research has slightly narrow coefficient ranges (0.76–0.99) (Table 3) compared to the coefficient ranges (0.53–0.89) reported in Toronto, Canada [62]. The difference could be due to the numerous petrochemical industries in the region. While examining the concentration and co-occurrence of VOCs in the US, Pankow et al. [64] reported comparable correlation ranges (0.78–0.99) between BTEX species. The high correlation coefficients in this study suggest that BTEX species are emitted by similar sources and it might be possible to monitor only one or two of BTEX species in Sarnia [47].

When compared to the measured concentrations (Table 4), kriging showed higher correlation coefficients (0.71–0.99) compared to the LUR modeled concentrations (0.14–0.79) for BTEX and all its individual components. The LUR models showed high correlations with measured concentrations for benzene ($r = 0.79$), toluene ($r = 0.72$), and BTEX ($r = 0.61$) but considerably lower correlation coefficients for ethylbenzene ($r = 0.38$), (m + p) xylene (0.16) and o-xylene ($r = 0.14$). When the kriged concentrations were compared to the LUR modeled concentrations at the monitoring sites, benzene ($r = 0.83$), toluene ($r = 0.73$), ethylbenzene ($r = 0.51$), and BTEX ($r = 0.66$) showed significantly higher correlations compared to (m + p) xylene ($r = 0.31$) and o-xylene ($r = -0.19$). The LUR models underestimated o-xylene concentration at the sampling locations compared to kriging. The correlation results suggest that LUR modeling could be an efficient interpolator for benzene, toluene, and ethylbenzene but not for xylenes in a highly polluted area like Sarnia. The effectiveness of kriging in Sarnia may be due to the uniqueness of the area. As mentioned, Sarnia is a relatively small region with about 40% of Canada's chemicals manufactured in the region [43].

Similar to other LUR studies, the benzene, toluene, ethylbenzene, (m + p) xylene, o-xylene, and total BTEX models were developed based on a two-week monitoring campaign. The high network deployment, monitoring, and chemical analysis cost did not permit an extensive monitoring campaign. In spite of the short-term monitoring, the models developed captured the intra-urban variability of total BTEX and its associated species in Chemical Valley. When compared, the 2-week measured concentrations at the 3 sampling locations closest to the NAPS station had comparable patterns with the monthly and 5-year average concentrations at the station suggesting that the measured ambient BTEX concentrations in this study were reliable. Hence, although seasonal variations may affect the temporal trend of modeled air pollution concentration, seasonality would have little influence on the spatial and geographic patterns of pollution because of the numerous petrochemical facilities in the region [53, 55, 63, 65]. Subsequently, seasonal variation may not greatly influence chronic health outcomes because, as observed in this research, the 2-week concentrations adequately represent mean annual concentration in Sarnia (see also Lebret et al. [65]).

CONCLUSION

Despite the potential limitations of this research, including the short-term monitoring campaign, the development of LUR models is a relatively affordable approach that clearly offers an advantage over traditional exposure estimation methods such as dispersion models [35]. From the models developed, it is evident that in addition to industrial emissions, traffic related VOC pollutions cannot be ignored in Chemical Valley and in similar industrial areas. Because of their prevalence and potential to cause adverse health outcomes, it is crucial to model VOCs such as BTEX for increasing the research communities understanding of the link between air pollution and health. The modeled ambient air pollution surfaces generated in this study suggest that some residents may be disproportionally exposed to high air pollutants. The results suggest the need for environmental policies that help reduce industrial pollution and assist residents to reduce and cope with daily industrial exposures. The LUR modeling of benzene, toluene, ethylbenzene, (m + p) xylene, o-xylene, and total BTEX models are used to estimate personal exposure for a large community health study aimed at examining the determinants of health in a government labeled area of concern.

KEYWORDS

- **Ethylbenzene**
- **Kriging**
- **Photochemically reactive**
- **Xylenes**

AUTHORS' CONTRIBUTIONS

Dominic Odwa Atari and Isaac N. Luginaah conceived the study and were involved in the preparation of the manuscript. Dominic Odwa Atari was involved in the analysis and interpretation of the results and preparation of the paper. All authors have read and approved the final manuscript.

ACKNOWLEDGMENTS

The authors would like to thank the Ontario Ministry of Transportation for providing assistance in accessing the highways traffic count data, the City of Sarnia Works Administration and Engineering Department for providing the local daily traffic counts, and the Aamjiwnaang Indian Reserve for their support. The authors would also like to thank Dr. Iris Xu and her students at the University of Windsor for assisting in air quality data collection. We thank all the reviewers for their constructive suggestions and comments. This project is funded by the Social Sciences and Humanities Research Council of Canada grant (# 410-2004-0159) and Canada Research Chair funding to Dr. Luginaah.

COMPETING INTERESTS

The authors declare that they have no competing interests.

Chapter 14

Chemical and Physical Properties of Saline Lakes in Alberta and Saskatchewan

Jeff S. Bowman and Julian P. Sachs

INTRODUCTION

The Northern Great Plains of Canada are home to numerous permanent and ephemeral athalassohaline lakes. These lakes display a wide range of ion compositions, salinities, stratification patterns, and ecosystems. Many of these lakes are ecologically and economically significant to the Great Plains Region. A survey of the physical characteristics and chemistry of 19 lakes was carried out to assess their suitability for testing new tools for determining past salinity from the sediment record.

Data on total dissolved solids (TDS), specific conductivity, temperature, dissolved oxygen (DO), and pH were measured in June, 2007. A comparison of these data with past measurements indicates that salinity is declining at Little Manitou and Big Quill Lakes in the province of Saskatchewan. However, salinity is rising at other lakes in the region, including Redberry and Manito Lakes.

The wide range of salinities found across a small geographic area makes the Canadian saline lakes region ideal for testing salinity proxies. A nonlinear increase in salinity at Redberry Lake is likely influenced by its morphometry. This acceleration has ecological implications for the migratory bird species found within the Redberry Important Bird Area.

Canada's Northern Great Plains contain a region of athalassohaline (of a different ionic composition than seawater) lakes extending from Manitoba, across the southern half of Saskatchewan, and into Alberta. This region includes both permanent and ephemeral bodies of water (playas) and displays a range of TDS from near 0 to 370 g l^{-1} [1]. These lakes and playas can be categorized by specific conductivity as fresh (>80 $\mu s\ cm^{-1}$), oligosaline (800–8,000 $\mu s\ cm^{-1}$), mesosaline (8,000–30,000 $\mu s\ cm^{-1}$), polysaline (30,000–45,000 $\mu s\ cm^{-1}$), eusaline (45,000–60,000 $\mu s\ cm^{-1}$), and hypersaline (<60,000 $\mu s\ cm^{-1}$) as described by Cowardin [2, 3] and LaBaugh [3]. A survey of some of the lakes in this system was completed by Hammer [1, 4] who compiled detailed information on ion composition and began a time series for several lakes that indicated a secular increase in salinity during the last century. Hammer also presented seasonal variations of TDS in five saline lakes, highlighting the strong connection between seasonal water budgets, water projects, and salinity in these basins. The geochemical nature of the lakes is well described in a review by Last and Ginn [5], who investigated the wide variation in sediment composition between lakes of the Western Great Plains of Canada and attributed those differences to underlying geological processes.

The importance of Canada's saline lakes from an environmental and economic standpoint cannot be understated. Many of the larger saline lakes are critical habitat for migratory birds [6-9]. Redberry Lake itself is designated as a UN Biosphere Reserve [10]. Numerous lakes in the system play a role in Saskatchewan's mineral industry including Big Quill (potassium sulfate production), Chaplin (sodium sulfate extraction), and Ingebright (past sodium sulfate extraction). Other lakes, such as Patience Lake, are greatly affected by mineral extraction [11]. Untapped sodium sulfate deposits exist under several lakes including Muskiki, Little Manitou, and Bitter Lakes [12]. Big Quill is also home to a fishery for *Diaptomus* and *Artemia*, both of which are sold as aquarium food.

A considerable amount of research has been done on phytoplankton and primary production in Canadian saline lakes [13-17]. However relatively little data is available on the diversity of prokaryotic life in these lakes. Although, extensive research has been conducted throughout the world on the microbiology of evaporative hypersaline environments, these environments are typically thalassohaline. Where, well studied lakes are athalassohaline they tend to be "soda lakes"; highly alkaline lakes containing a high concentration of carbonate [18, 19]. Recently Grasby and Londry [20] reported on the microbiology of spring-fed saline lake systems in Manitoba. Sørensen et al. [21], Sorokin et al. [22], and others have described the microbial community structure of hypersaline lakes with high concentrations of chloride and sulfate, a characteristic shared with many Canadian saline lakes. However, from an ionic and microbiological standpoint the lakes of Saskatchewan and Alberta represent a unique environment, with sulfate the dominant anion over chloride in most lakes.

Regardless of their ionic composition saline lakes support communities of phototrophic, chemoautotrophic, and heterotrophic microorganisms that in turn support a community of multicellular organisms [20, 23-25]. In the most saline lakes multicellular life is limited to the brine shrimp *Artemia* sp. [5, 26, 27]. The invertebrate communities of saline lakes play a critical role in the ecology of the Great Plains by supporting large populations of migratory birds nesting locally and *enroute* to and from the high Arctic and Asia [6, 8, 9, 28-30]. Redberry, Big Quill, and the Chaplin Lakes in particular are noted for their importance to migratory bird populations.

The wide range of salinities represented among the lakes of Canada's Great Plains makes them ideal for testing the biochemical responses of prokaryotic and eukaryotic microorganisms to changes in salinity. These responses can ultimately be exploited for the reconstruction of salinity in the geological record [31]. Calibrating such biochemical responses requires accurate data on salinity and water chemistry. The purpose of this study is to identify suitable, accessible lakes across a wide salinity range for comparing biochemical responses to salinity. In addition we will augment the extensive geological data compiled by Last and Ginn [5] with a snapshot of water column data including TDS, DO, pH, specific conductivity, and sediment characteristics. These parameters are known to change dramatically on a seasonal and annual basis. Some of these changes can be observed by using these data to extend the time series begun by Hammer for Little Manitou, Big Quill, Manito, and Redberry Lakes [32].

A total of 19 lakes were investigated across eastern Alberta and Saskatchewan from June 9 to 15, 2007. The location where data was collected, specific conductivity, TDS, and absolute salinity (in ppt) for each lake are shown in Table 1. Locations relative to the geography of the region are shown in Figure 1. Figure 2 shows satellite images of all lakes investigated and is printed with the permission of Google Earth.

Table 1. Sampling location, average specific conductivity (from all measured depths), and TDS for lakes investigated.

Lake	Sample Date	Latitude (degrees)	Longitude (degrees)	Average Specific Conductivity (mS cm·1 at 25oC)	TDS (g L· 1), depth (m)	Absolute Salinity (ppt)
Patience Chappice	6/10/2007	52. 1354	106.33	201 .35 110.90	161.16,0	139.5
West Chaplin West	6/15/2007	50.13328	110.3698	108.50	193.58, .1	163.6
Division I	6/13/2007	50.43772	106.6845		235.74, .1	183.7
Bitter	6/15/2007	50.13317	109.8707	106.20		
West Chaplin Center Division	6/13/2007	50.43772	106.6845	99.00	192.39, 0	154.8
Ingebrigh West	6/15/2007	50.36329	109.3169	97.36	251 .80, .1	160.0
Chaplin Lake SE Division	6/13/2007	50.41832	106.6735	89.08	144.82, .1	125.7
West Half Chaplin Lake NE Division	6/13/2007	50.43968	106.6458	84.75		
Muskiki	6/11/2007	52.35197	I 05.7771	57.93	121.47, . 1	91 .9
Freefight	6/14/2007	50.39685	109. 1149	57.71	I 00.98, 3	94.5
Little Manitou West	6/10/2007	51.71965	105.3944	56.40	85.14, 0	73 .7
East Half West Chaplin Lake NE Division	6/13/2007	50.43978	106.641	47.35	60. 14, 0	54.1
West Chaplin West Division 2	6/13/2007	50.41922	1 06.7065	46.62	161 .93, 0	137.4
Little Manitou East	6/10/2007	51.71957	105.3944	46.28		
Manito	6/9/2007	52.79052	109.7864	38.58	38.50, .5	38.0
Big Quill	6/11/2007	51.78784	104.3234	20.32	24.87, 0	22.9
Redberry	6/9/2007	52.70732	107.2082	18.72	44.52, 0	40.3
Hughes Bay	6/13/2007	50.40552	106.6555	15.6		
Midtskogen	6/13/2007	50.40552	106.6555	8.35		

Figure 1. Locations of lakes investigated. (A) Patience (B) Chappice (C) Chaplin (D) Bitter (E) Ingebright (F) Muskiki (G) Freefight (H) Little Manitou (I) Manito (J) Big Quill (K) Redberry. Printed with the permission of Google Earth.

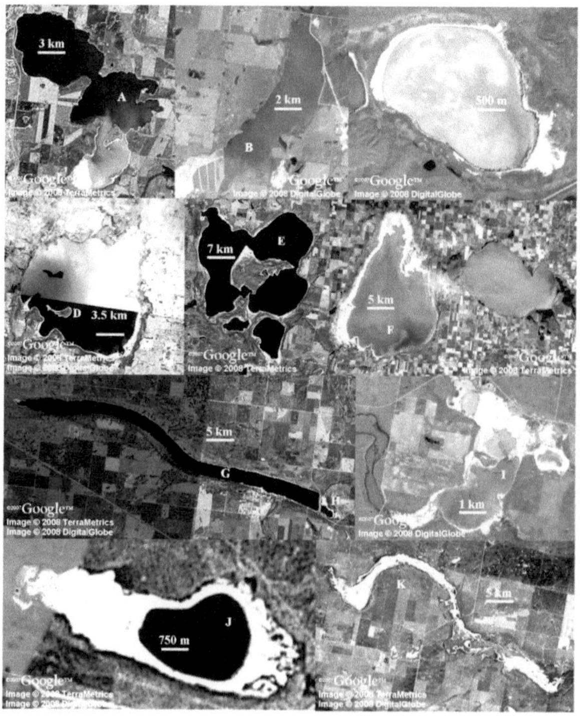

Figure 2. Satellite images of lakes investigated. (A) Muskiki (B) Patience (C) Chappice (D) Redberry (E) Manito (F) Big Quill G) Little Manitou West (H) Little Manitou East (I) Ingebright (J) Freefight (K) Bitter. Printed with the permission of Google Earth. All images are oriented with north at the top of the image.

Patience Lake

Patience Lake is a shallow, permanent, hypersaline lake east of Saskatoon in an area of extensive potash extraction. It is unusual as a NaCl system in a region dominated by Na_2SO_4 systems. This was first noted by Hammer [1, 11], who cited the dumping of potash mine tailings in the lake as the reason for this abnormality. Measurements were taken from a boat near the north end of the lake (specific location given in Table 1). The lake was well mixed and oxygenated throughout the water column. The DO ranged from 4.89 mg l^{-1} at the surface to 4.57 mg l^{-1} at 2 m, the maximum depth sounded. The pH also stayed constant with depth, from 8.75 at the surface to 8.76 at 2 m. Water at Patience Lake was turbid; the Secchi depth was measured at 3 m. Water samples taken here began to precipitate solids of unknown composition almost immediately upon collection. Sediment sampled from the lake bottom at 2 m was black in color, smelled strongly of hydrogen sulfide, and had a hydrocarbon sheen. A salt crust made deep penetration into the sediment difficult and recovered sediment contained a number of large salt crystals. Primary ions in Patience Lake were Na^+ and Cl^- at 71.6 and 89.1 milliequivalent percentage of the sum of cations or anions respectively in 1978 [1].

Freefight, Ingebright, Bitter, and Chappice Lakes

Freefight, Ingebright, and Bitter Lakes are located in the southwestern portion of Saskatchewan near the Alberta border while Chappice is just west of the border in Alberta (Figure 1). Although, located within a small geographic area the lakes are quite dissimilar. Freefight Lake is a deep water, permanently stratified lake [33]. Ingebright, Bitter, and Chappice Lakes are shallow and ephemeral.

Freefight Lake is eusaline within the mixolimnion. Water column chemical and physical properties to a depth of 10.5 m are shown in Figure 3 and Table 2. Although, sampling was limited to this depth, the lake is known to approach a depth of 25 m [33]. Figure 3 shows a shallow, weak pycnocline at 2 m and a chemocline at 6 m below which suboxic conditions exist. A strong region of primary productivity was identified at 4 m based on a supersaturation of DO. The Secchi depth was determined to be 2.5 m. Sediment recovered from 10.5 m smelled strongly of hydrogen sulfide suggesting that sulfate reduction is occurring. Access to the southern shore of Freefight Lake was found via fenced rangeland.

Chappice Lake is a shallow, hypersaline playa [34] located approximately 80 km west of Freefight Lake. Readings taken from just below the surface several meters offshore recorded a DO concentration of 4.82 mg l^{-1} and pH of 9.09. Access to Chappice was by rangeland to the south of the lake.

Hypersaline Ingebright Lake is the site of a mothballed sodium sulfate extraction plant owned by Saskatchewan Minerals and the largest Na_2SO_4 deposit in North America [5]. Measurements were made from a boom between the former evaporation pond and reservoir pond in 1.4 m of water. The current state of the plant allows for a free exchange of water between the evaporation pond and reservoir pond. Conductivity, pH, and DO were constant with depth. Specific conductivity ranged from 97.33 mS cm^{-1} at 25°C on the surface to 97.39 mS cm^{-1} at 25°C at 1 m. The DO ranged from 1.39 mg

l^{-1} on the surface to 1.30 mg l^{-1} at depth while pH remained constant at 8.43. The Secchi depth was determined to be 0.75 m. Access to Ingebright was across rangeland to the west of the lake to the point where the evaporation and reservoir ponds exchange.

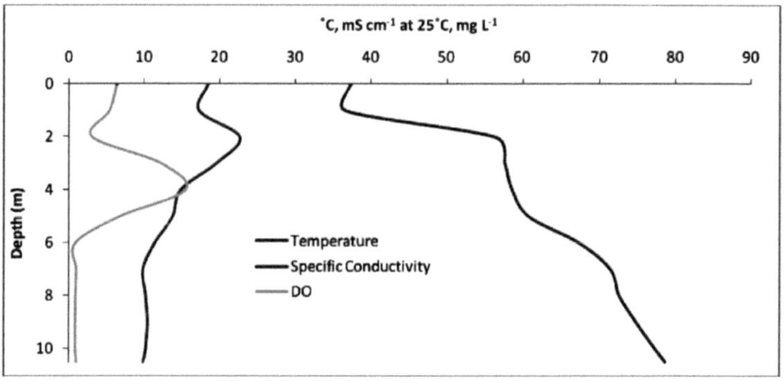

Figure 3. Freefight Lake. Freefight depth profiles for temperature (°C), specific conductivity (mS cm^{-1} at 25°C), and DO (mg l^{-1}). Some weak stratification can be seen along with a primary production maximum at 4 m.

Table 2. Chemical and physical properties for Manito, Redberry, Freefight, Little Manitou East, Little Manito West, and Big Quill Lakes.

Lake	Depth (m)	Temperature CC)	Specific Conductivity (mS cm^{-1} at 25' C)	DO (mg L·')	pH
Manito	0	15.78	38.23	7.83	9.53
	0.5	15.77	38.31	7.8	9.53
	1	15.75	38.31	7.69	9.53
	1.5	15.69	38.31	7.59	9.53
	2	15.67	38.31	7.53	9.54
	2.5	1 5.4	38.31	7.54	9.53
	3	15.67	38.31	7.53	9.53
	3.5	IS. I	38.31	7.43	9.53
	4	14.92	38.32	7.3	9.53
	4.5	14.48	38.37	7.09	9.52
	5	14.27	38.41	6.48	9.52
	S.S	13.97	38.53	6.31	9.5
	6	6.12	41.5	1.38	9.43
Red berry	0	15.19	17.68	9.77	8.82
	2	15.18	17.68	9.71	8.83
	3	15.17	17.68	9.74	8.83
	4	15.15	17.68	9.5	8.82
	5	15.08	17.69	9.8	8.83
	S.S	14.81	17.69	9.66	8.83
	6	14.44	17.69	9.78	8.83
	6.5	14.15	17.69	9.76	8.82
	7	14.04	17.76	9.86	8.8
	7.5	7.18	19.66	14.73	8.73
	8	4.13	20.11	15.07	8.78
	9	3.17	20.31	14.4	8.74
	9.5	2.46	20.43	13.75	8.72
	10	2.06	20.5	12.95	8.71
		2.02	20.51	12.93	8.7

Table 2. *(Continued)*

Lake	Depth (m)	Temperature CC)	Specific Conductivity (mS cm^{-1} at 25' C)	DO (mg L·')	pH
Little Manitou West	0	16.16	53.25	7.83	8.65
	1.5	16.05	53.25	7.36	8.65
	2.5	15.87	53.3	7.06	8.64
	3.5	15.42	53.53	6.3	8.62
	4	14.51	54.04	5.5	8.6
	4.3	13.23	63.12	0.31	8.43
		12.16	64.32	0.5	8.42
Little Manitou East	0	17.49	26.3	8.95	8.76
	0.5	17.58	26.7	8.91	8.76
	1	18.39	51.04	12.98	8.7
	1.5	18.55	53.25	10.18	8.65
	2	17.41	59.1	20.62	8.54
Freefight	0	18.55	37.36	6.43	8.87
	1	17.37	36.67	5.46	8.93
	2	22.66	56.49	3.18	8.77
	4	19.62	57.71	12.24	8.79
	5	14.95	58.68	15.56	8.7
	6	13.87	60.87	6.85	8.58
	7	11.44	67.48	1	8.25
	8	9.86	71.55	1.07	8.06
	9	10.23	72.76	0.99	7.99
	10	10.47	74.96	0.95	7.89
	10.5	10.24	77.44	0.95	7.81
		9.9	78.77	1.07	7.77
Big Quill	0	16.03	20.27	6.25	8.71
	1	16.03	20.28	6.25	8.71
	2	15.89	20.29	6.02	8.70
	3.3	15.62	20.44	5.39	8.70

Bitter Lake is an extensive network of large, hypersaline playas lying on the border between Alberta and Saskatchewan and is adjacent to the playas of Many Island Lake. The exact sampling location is given in Table 1. After the Chaplin Lakes system Bitter Lake hosted the largest abundance of microbial mats of the lakes investigated. Because of its shallow nature *in situ* measurements at Bitter Lake were not possible. Measurements made from a water sample indicated a specific conductivity of 106.2 mS cm^{-1} at 25°C and pH of 8.79. Sediment recovered in 10 cm of water was black, sandy, firmly packed, and smelled weakly of hydrogen sulfide. Historically the major cation in Bitter Lake was Na^+ comprising 23.8% of total cations, while the major anions are Cl^- and SO_4^{2-} which comprised 38.3% and 60.4% of total anions, respectively [1]. Access to the southern shore of Bitter Lake was made across fenced rangeland from Canada Highway 1.

Chaplin Lakes

The Chaplin Lakes are located near the town of Chaplin in south central Saskatchewan. Reconstructed from several lakes into a network of solar evaporation ponds by Saskatchewan Minerals for a sodium sulfate extraction facility, the Chaplin Lakes represent a wide range of salinities across a small geographic area. The configuration of the evaporation ponds in 2007 is shown in Figure 4. Water chemistry data for the

Chaplin Lakes is shown in Table 3. Ion concentrations for the current Chaplin Lakes are not known, however historic data for "Chaplin East" and "Chaplin West" are available [1]. These data indicate that SO_4^{2-} was the major anion in both lakes comprising 88% of the total anions in Chaplin West and 90.8% in Chaplin East. In the case of both lakes the major cation was Na+, making up 82.6% of the total cations in Chaplin West and 88.7% in Chaplin East [1].

Figure 4. Satellite image of the Chaplin Lakes. (A) Uren Bay (B) West Chaplin Lake West Division #2 (C) West Chaplin Lake West Division #1 (D) West Chaplin Lake Center Division (E) West Chaplin Lake North East Division West Side (F) West Chaplin Lake South East Division (G) Hughes Bay (H) Midtskogen Bay (I) East Chaplin Lake North Division North Side (J) West Chaplin Lake North East Division East Side (K) East Chaplin Lake North Division South Side (L) East Chaplin Lake South Division. Printed with the permission of Google Earth.

All of the Chaplin Lakes with a specific conductivity above 40 mS cm^{-1} contained abundant microbial mat communities similar to the one pictured in Figure 5. Sediment collected from West Chaplin West Division 1, West Chaplin North East Division, West Chaplin Center Division, and West Chaplin South East Division smelled strongly of hydrogen sulfide. Access to the Chaplin Lakes was from Provincial Highway 58 and, with permission, from private causeways controlled by Saskatchewan Minerals.

Table 3. Chemical and physical properties of the Chaplin Lakes.

Lake	Designation in Fig. 4	Temperature ("C)	Specific Conductivity (mS cm^{-1} at 25°C)	pH	TDS (g L^{-1})	Absolute Salinity (ppt)	DO (mg L^{-1})
West Chaplin West Division I	C	20.45	108.5	8.66	235.74	184	2.58
West Chaplin Center Division	D	19.97	99.00	8.99	192.39	155	6.00
West Chaplin SE Division	F	17. 11	89.08	8.8	144.82	126	-
West Chaplin NE Division West Side	E	13.31	84.75	8.94	-	-	.96
West Chaplin NE Division East Side	J	16.94	47.35	9.19	60. 14	54	5.68
West Chaplin West Division 2	B	18.6	46.62	-	152.31	137	-
Midtskogen	H	19.27	18.35	9.01		-	7.75
Hughes Bay	J	21 .30	15.60	9.27		-	-

Figure 5. Microbial mat at West Chaplin Lake North East Division. All of the Chaplin Lakes with a specific conductivity above 40 mS cm^{-1} at 25°C contained microbial mat communities similar to the one pictured here, from West Chaplin Lake West Division #1.

Muskiki Lake

Muskiki Lake is a shallow, hypersaline playa northeast of Saskatoon. Maximum depth sounded was 0.75 m. The water column was well mixed; specific conductivity ranged from 57.82 mS cm^{-1} at 25°C at the surface to 58.06 mS cm^{-1} at 25°C at depth. The DO ranged from 5.74 mg l^{-1} at the surface to 5.16 mg l^{-1} at depth, and pH remained constant with depth at 8.59. Sediment recovered from the bottom of the lake was black in color, contained a number of large salt crystals (intermixed within the sediment), and had a strong tar-like odor. Historically, Muskiki had the highest percentage of SO_4^{2-} relative to other anions of any lake investigated at 93.5% [1]. The primary cation present is Na^+ at 52.6% of total cations [1]. Access to Muskiki was found along the west shore where Provincial Highway 2 approaches the lake.

Little Manitou East and West

The Little Manitou system is comprised of two lakes southeast of Saskatoon. The two lakes are separated by a semi-permeable causeway formed by Provincial Highway 365. Stratified and eusaline (within the mixolimnion), Little Manitou West is considerably larger and deeper then Little Manitou East. Measurements were made to a depth of 4.3 m at the location given in Table 1, maximum lake depth is reported as 5.2 m [4]. Water chemistry data is presented in Table 2 and depth profiles of temperature, specific conductivity and DO are shown in Figure 6. A strong chemocline was observed between 3.5 and 4 m in which DO dropped from 5.5 to 0.3 mg l^{-1} while salinity and specific conductivity increased from 57.82 to 63.12 mS cm^{-1} at 25°C. A weak thermocline was observed below 2.5 m. Sediment recovered from a depth of 4.3 m consisted of mostly algal material. It was not possible to determine at what depth this algal material originated, however given the anoxic bottom water it was unlikely to have come from below 3.5 m.

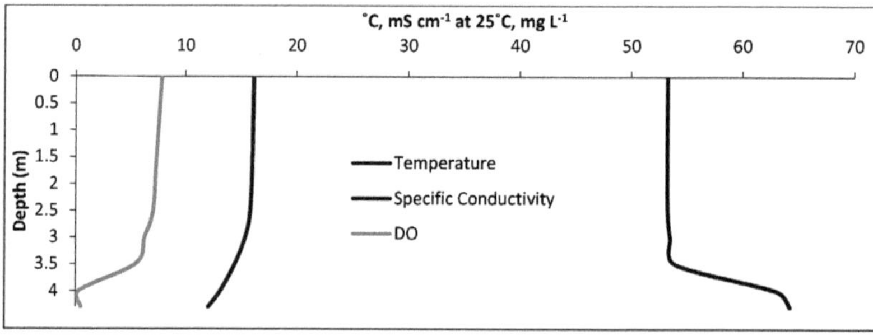

Figure 6. Little Manitou West. Depth profile for Little Manitou West with temperature (°C), specific conductivity (mS cm^{-1} at 25°C), and DO (mg l^{-1}). The strong chemocline between 3.5 m and 4 m can be clearly seen.

Little Manitou East is a shallow, eusaline lake with a Secchi depth exceeding the 2 m depth of the lake. The turbidity and salinity of Little Manitou East allows for abundant aquatic vegetation and a well oxygenated water column. Water chemistry and physical properties are listed in Table 2. Figure 7 shows temperature, specific conductivity, and DO profiles for the lake. Temperature remained relatively constant with depth throughout the water column. The oxygen and salinity maxima were on the bottom at 2 m. Sediment recovered from a depth of 2 m was black, with a strong hydrogen sulfide odor and contained some aquatic vegetation. Access to East and West Little Manitou was from the causeway separating the two lakes.

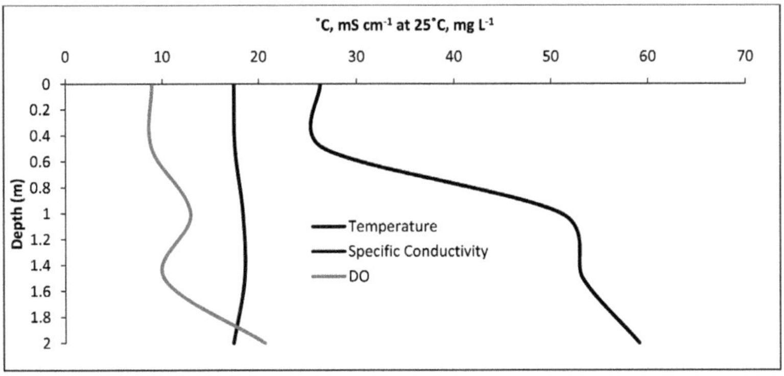

Figure 7. Little Manitou East. Depth profile for Little Manitou East with temperature (°C), specific conductivity (mS cm^{-1} at 25°C), and DO (mg l^{-1}). Some correlation between DO and specific conductivity can be seen throughout the water column.

Manito Lake

Manito Lake is a large, deep, stratified, polysaline lake located northwest of Saskatoon near the Alberta border. Data were collected to a depth of 6 m at the location listed in Table 1, maximum lake depth is given as 21.5 m [4]. The Secchi depth was determined to be 1.8 m. Sediment recovered from a depth of 6 m smelled strongly of hydrogen sulfide. Historically the dominant cation in Manito was Na^+ at 89.5% of total cations and the dominant anion was SO_4^{2-} at 67.1% of total anions [1]. Water chemistry and physical properties in Manito are presented in Table 2. Depth profiles for temperature, DO, and specific conductivity are shown in Figure 8. Manito Lake exhibited a well mixed surface layer from 0 to 5 m. At 5.5 m a strong pycnocline and chemocline were found. Access to Manito Lake was from Provincial Highway 40 at the northwest corner of the lake.

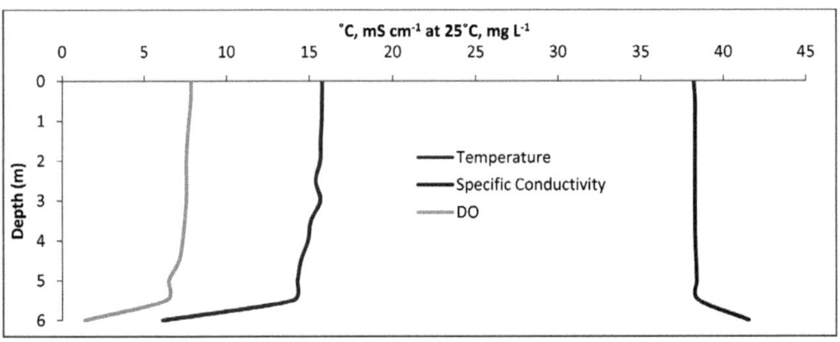

Figure 8. Manito. Depth profile for Manito with temperature (°C), specific conductivity (mS cm^{-1} at 25°C), and DO (mg l^{-1}). A strong pycnocline and chemocline can be seen at a depth of 5.5 m.

Big Quill Lake

Big Quill is the largest member of a three lake system east of Saskatoon that includes Little Quill Lake and Mud Lake. It is also home to a potassium sulfate production facility and a commercial fishery for *Diaptomus and Artemia.* Despite its large surface area the deepest point sounded at Big Quill was 3.3 m, although deeper spots exist. Interviews with commercial fisherman revealed that numerous deep holes can be found in the lake as a result of past use as a bombing range by the Canadian RAF. The Secchi depth was 1 m. Historically the major cation in Big Quill was Na^+ at 46.5% of total cations [1]. The predominant anion was SO_4^{2-} at 84.4% of total anions [1]. Water chemistry and physical properties are listed in Table 2. The lake was well mixed to the bottom with little variation in the measured parameters. Access to Big Quill was from the southern shore near the Big Quill Resources potassium sulfate plant.

Redberry Lake

Redberry Lake is a large mesosaline lake located north of Saskatoon at the Redberry Lake UN Biosphere Reserve. The reserve is a critical habitat for several species of migratory birds, most notably the American white pelican and piping plover [29]. Based on turbidity Redberry was the most oligotrophic lake investigated with a Secchi depth of 4 m. This comparison is consistent with observations made by Waiser and Roberts [35]. Measurements were made in 10 m of water, maximum lake depth is 17 m [35]. Water chemistry data for Redberry are presented in Table 2 and the depth profiles of temperature, specific conductivity, and DO with depth are shown in Figure 9. This figure shows a strong parallel between specific conductivity and DO down to the oxygen maximum at 7.5 m. Both a strong pycnocline and chemocline were present at this depth. Sediment from 10 m was gray with a green and cream colored covering and did not smell of hydrogen sulfide. Historic ion concentrations in Redberry Lake are 93.1% of total anions for SO_4^{2-} and 67.4% of total cations for Mg^{2+} [1]. Redberry was the only lake investigated where Mg^{2+} was the dominant cation. Access to Redberry Lake was through the park entrance on the west side of the lake.

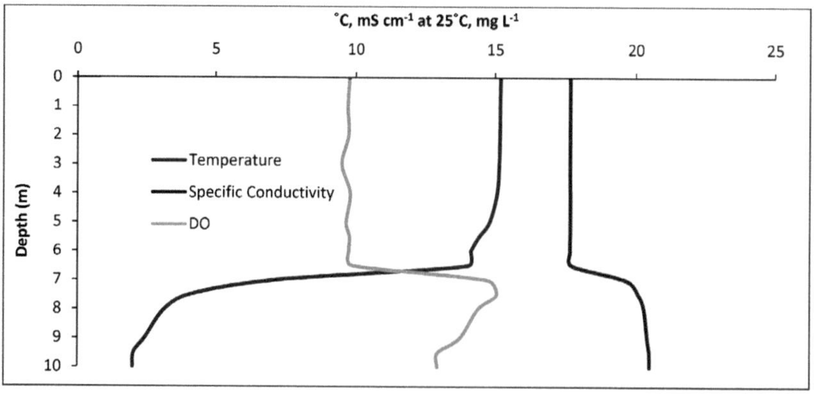

Figure 9. Redberry Lake. Depth profile for Redberry Lake with temperature (°C), specific conductivity (mS cm^{-1} at 25°C), and DO (mg l^{-1}). The DO is seen to parallel specific conductivity down to the oxygen maximum at 7 m.

DISCUSSION

Long Term Salinity Trends

Figure 10 shows the continuation of a time series compiled by Hammer [1] for Little Manitou West, Big Quill, Redberry, and Manito. The change in TDS for Redberry can be described by a second order polynomial with an R^2 value of 0.94. A projection of this trend to 100 ppt is displayed. The TDS (in ppt) decreased from 1971 to the present in Little Manitou and Big Quill, and increased in Redberry and Manito. In the case of Little Manitou the mechanism of this decrease is known to be the diversion of water into the lake from the South Saskatchewan River [1]. The start of this diversion coincides with the 1968 salinity maximum shown in Figure 10.

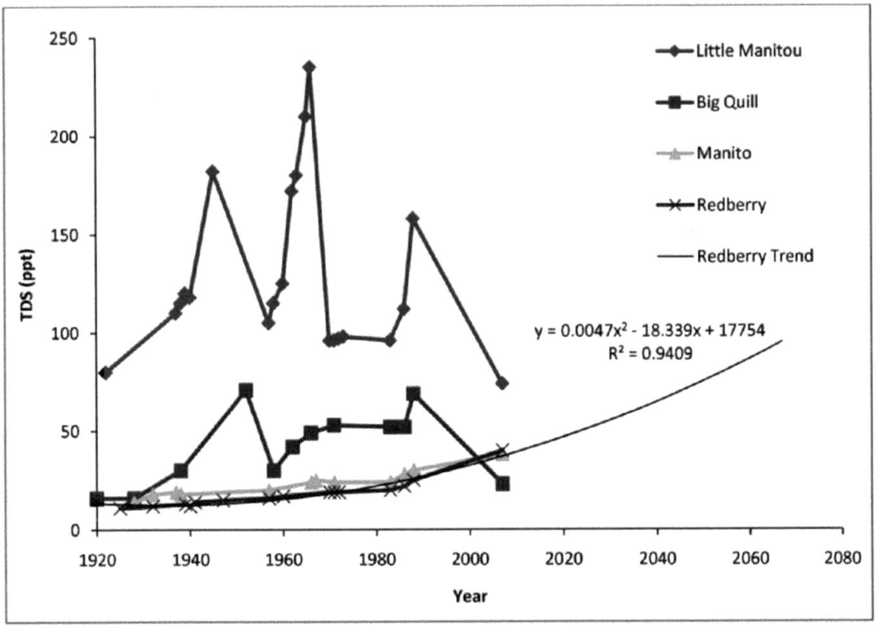

Figure 10. Extension of Hammer time series. Data for all lakes from 1920 to 1988 [32]. Some data for the period 1990–2007 is available but has not been used due to reporting in g l^{-1} or specific conductivity [7, 43]. The TDS has decreased slightly since 1971 for Little Manitou West. A sharper decrease can be seen for Big Quill Lake, although the lake is still above historic minimums. Redberry and Manito Lakes have both increased in TDS since 1971.

The slow decrease in salinity at Big Quill can also be attributed to water management. Historic lows for TDS in Big Quill are near 15 ppt [32]. Placement of a dam between Big Quill and Little Quill in 1936 initiates a period of widely varying salinity [32]. During times of high precipitation water is allowed to flow into Big Quill from the Little Quill Drainage, decreasing salinity. During times of low precipitation water is retained in Little Quill [19]. Both of these measures are augmented by the large surface area to volume ratio (307.4 km^2:449 × 10^6 m^3) of Big Quill that amplifies the response of salinity to both precipitation and evaporation [4]. The potential economic effects of decreasing salinity in Big Quill Lake cannot be overlooked. Both the commercial fishery and potassium sulfate production facility require a certain minimum concentration of salt. This makes understanding the mechanisms controlling salinity in Big Quill a priority and water managers should be aware of the minimum values necessary for commercial and ecological viability.

Water management projects do not involve the Redberry or Manito Lakes directly, and the salinity of these lakes can be expected to respond naturally to evaporation and precipitation with some effects from aquifer depletion or perforation [36]. Currently evaporation exceeds precipitation in most North American prairie regions, leading to the expectation of increasing salinities over time [16, 32, 36-38]. A closer look at any particular basin will require local long term precipitation records. Unfortunately, long term precipitation records are not available for Manito Lake; climate data for the nearby station in Lloydminster is only available from 1983 on [39]. However such records are available for a weather station 48 km south of Redberry Lake near the city of Saskatoon [39].

Figure 11 shows annual total precipitation (rain and snow) from 1925 to 2006 for the Saskatoon weather station. Data for the years 1930 and 1941 are not available. A linear trend line for these annual precipitation totals has a slope of –0.347 mm yr^{1}, indicating that precipitation totals have remained relatively constant since 1930. The Redberry Lake catchment is estimated to be 1,180 km^2 [29]. For a drainage this size a deficit of 0.347 mm of water in a given year results in 413 m^3 less water available to dilute the terminal basin every year, a small fraction of the estimated 1999 lake volume of 2.53×10^8 m^3 (1.63×10^{-6}% of lake volume). Figure 11 also shows annual estimated evapotranspiration determined by a modified Thornwaite equation [40]. This equation does not take into account the strong evaporative effects of wind direction and speed in a prairie environment. However diurnal and annual variations in wind direction and speed have remained constant since they were first recorded in 1957 [39]. The steady average rate of evapotranspiration suggests that the nonlinear increase in salinity shown in Figure 10 does not result from a decadal-scale shift in climate. Another factor, such as basin morphology may be responsible.

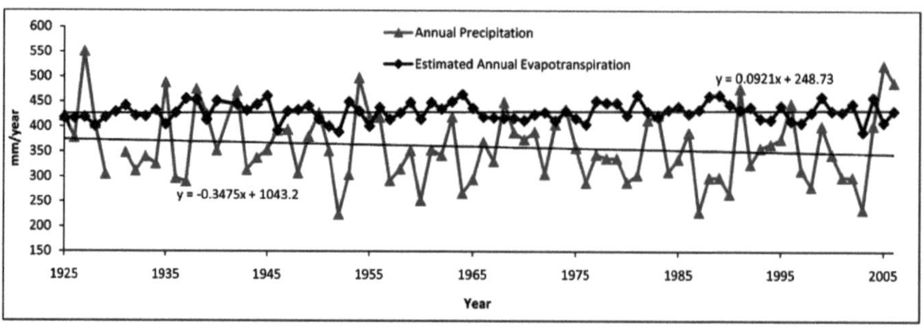

Figure 11. Annual precipitation and evapotranspiration for Redberry Lake. Relatively constant rates of precipitation and evapotranspiration (calculated with a modified Thornwaite equation [40]) are shown for Redberry Lake. This suggests a mechanism other than climate for the nonlinear increase in salinity at Redberry.

Redberry Lake possesses a roughly conical basin morphology, as illustrated by Hammer in Figure 12 [4]. As the surface area to volume ratio of Redberry Lake increases with decreasing lake depth, the fraction of the lake which evaporates each year will increase in a nonlinear manner. Figure 13 illustrates this effect. Annual TDS for Redberry Lake are determined as a function of time according to the equation shown in Figure 10. The close correlation between lake surface elevation above sea level [29] and TDS (which increases inverse to the decrease in volume) can be seen along with a nonlinear decrease in the ratio of lake elevation to TDS.

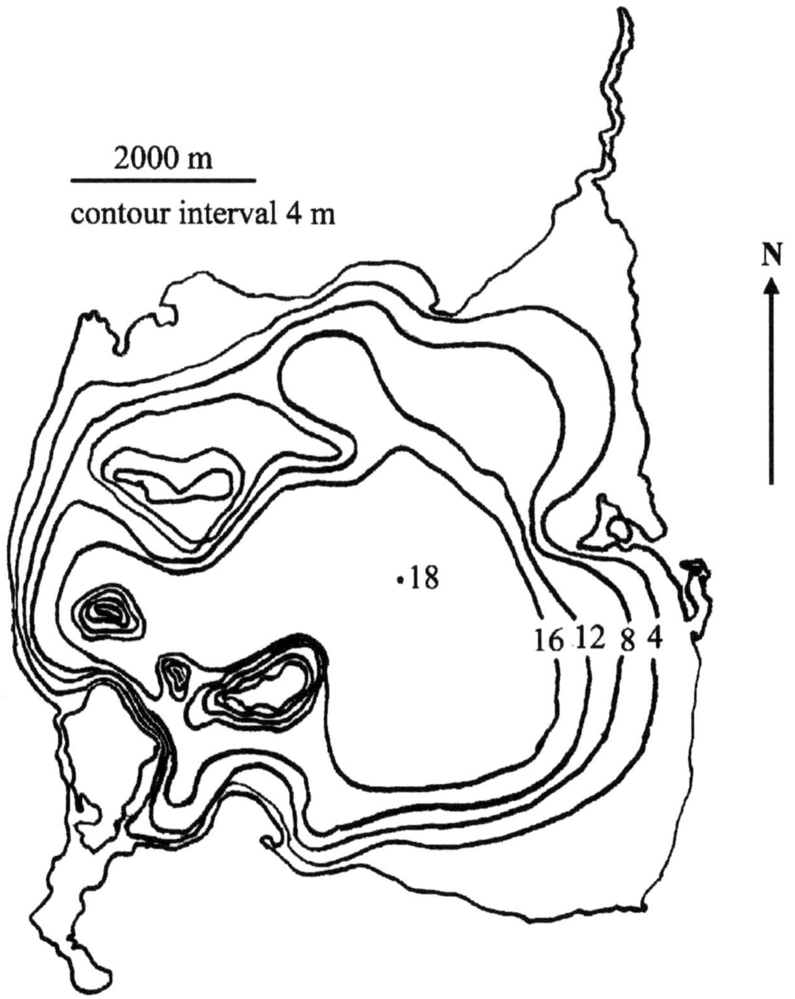

Figure 12. Redberry Lake bathymetry. This illustration of the bathymetry for Redberry Lake is redrawn from Hammer [4]. The roughly conical shape of the basin can be clearly seen. As the lake level drops, surface area increases relative to volume. This increases evaporation relative to volume, resulting in a nonlinear increase in salinity.

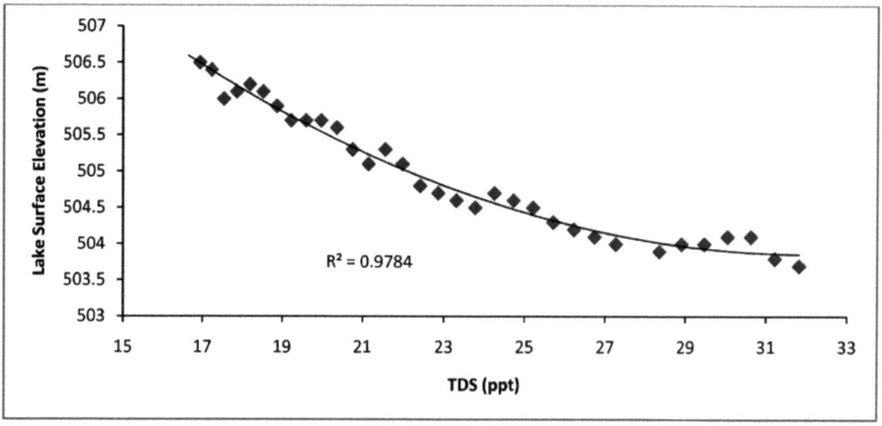

Figure 13. The TDS versus lake surface elevation at Redberry Lake. The TDS was derived by the equation displayed in Figure 10. Derived TDS are shown to correlate well with surface elevation [29] at Redberry Lake. This correlation reflects the increasing rate of salinity change at Redberry.

Due to the ecological significance of the lake increasing salinity at Redberry is cause for concern, just as the decreasing salinity at Big Quill is cause for economic concern. If the current trend for Redberry Lake continues, salinity may reach 100 ppt by 2,070 as shown in Figure 10. The effects of rising salinity have already been seen on several native and introduced species of fish at Redberry Lake. Reports from the 1920s indicate that northern pike inhabited Redberry Lake, a testament to the low salinity at that time [29]. Lake whitefish, walleye, and rainbow trout were all stocked in the mid-20th century, but no population has persisted. Several small species of marine fish remain at Redberry including the brook stickleback or *Culaea inconstance* [29]. Salinity at Redberry is already exceeding values typically found in marine environments and it is unlikely that any fish species will remain far into the future.

The effect of rising salinity at Redberry and other prairie lakes on invertebrate species is less clear. It has been demonstrated that high salinity does not necessarily correlate with low productivity [14, 16, 23], though there is some disagreement as to whether biomass declines as salinity increases [6, 23]. There is good evidence that phytoplankton and invertebrate species richness declines with rising salinity [6, 13, 23, 25, 27]. What effect this loss of prey diversity will have on migratory bird populations at Redberry Lake remains to be seen. Reduced food availability due to rising salinity is cited as a likely reason for the delayed nesting of white winged scoters at Redberry [30].

The Impact of Salinity on Other Parameters

Salinity does not appear to be a determining factor for other chemical characteristics. No correlation was observed between TDS and DO, pH, or specific conductivity. Although pH is controlled in part by the concentration of Ca^+ and CO_3^{2-}, these ions are thought to be present in low concentrations compared to the major ions Na^+, SO_4^{2-}, K^+, and Cl^- [1]. This differentiates these lakes from the "soda" lakes common in other hypersaline environments where Ca^+ and CO_3^{2-} make up a significant portion of the ions present. The pH is probably more strongly controlled by the production and utilization of CO_2 within the water column [5].

Finally, no correlation was observed between specific conductivity and TDS. This highlights a fundamental problem in investigating athalassohaline lakes; the widely used practice of determining salinity as a function of specific conductivity does not work when the anion charge to ion ratio is not 1:1 [41]. The dominant anion within the Great Plains of Western Canada's lakes is almost always SO_4^{2-}, but it is never the only anion present in high concentrations. This makes specific conductivity useful only as a rough guide to the salinity of a specific lake (high specific conductivity will always correlate with high TDS but not by any set conversion factor), or as an indicator of change within a single lake. Despite its limitations specific conductivity has been included here to provide a means of comparison with other sources.

MATERIALS AND METHODS

During June, 2007, 19 lakes were investigated throughout the Northern Great Plains region of Canada. The DO, specific conductivity, water temperature, and pH were measured with an YSI-556 multiprobe. Where shore conditions and lake depth permitted, readings were taken by boat from the point at which lake depth began to stabilize as determined by a Hawkeye Digital Sonar handheld sonar system. Where it was not possible or practical to deploy the YSI-556 multiprobe by boat, it was deployed several meters from shore by wading into the lake and/or throwing the probe into deeper water. In some cases it was necessary to collect water in a graduated cylinder before a reading could be made. In these cases DO values are not reported. At the highest salinity lakes serial dilutions were made to insure accurate readings of specific conductivity. In all cases serial dilutions verified the initial readings. When the boat was deployed the Secchi depth was determined by lowering a Secchi disk on a marked rope to the depth from which it could not be distinguished from the surrounding water without the aid of polarizing glasses.

Filtered and unfiltered water samples were collected in situ using a Masterflex peristaltic pump to draw water from depth. Water depths reported as 0 m indicated that water was drawn from just below the surface. Samples were filtered using 142 mm Pall–Gelman A/E glass fiber filters with a 1.0 μm pore diameter. Filtered and unfiltered water samples were collected in PET1 plastic bottles sealed with electrical tape to minimize evaporation. The high alkalinity present at many of the sampled lakes caused deterioration of the PET1 bottles necessitating a transfer of the samples to LDPE bottles within 3–5 days. Sediment samples were collected either with a stainless steel Van Veen sediment grab from the boat or with a trowel from shore.

The TDS were determined in the laboratory using a variation on the procedure for determining filterable residue described by the EPA's Office of Research and

Development [42]. Due to the high concentration of salts present a smaller aliquot was used then is recommended by this procedure. In addition a lower initial heat was used to minimize the "popping" and resulting loss of material caused by faster, hotter evaporations. Three 10 ml aliquots of filtered water were transferred by pipette into pre-weighed aluminum weighing pans. The specific gravity of each aliquot was determined by weighing the pan and aliquot on a precision scale, then subtracting the weight of the pan. Pans were evaporated at 60°C for 20 hr, then at 180°C for 2 hr at which time a stable weight was reached, indicating that all water had been driven from the hydrated salts. Residues were cooled in a desiccator prior to reweighing. Care was taken to insure minimum time outside of the desiccator for each sample. The TDS is reported as the average weight of the three measurements and reported in g l^{-1}. To eliminate small errors in measuring the initial 10 ml aliquot, salinity is also reported as absolute salinity following the guidelines presented in Anati [41]. This index provides the most accurate salinity measure for comparison with future work and across lakes with varying ionic compositions.

An effort was made to measure suspended particle concentrations in the lab from unfiltered water collected in the field. A 50 ml aliquot of water was measured in a volumetric flask then filtered through a pre-weighed 47 mm Pall membrane filter with a pore diameter of 0.45 μm. For samples of high salinity the filter was rinsed briefly with distilled water. Filters were then dried at 60°C for 2 hr at which time a constant weight was reached. Filters were cooled in a desiccator then weighed on a precision scale. The suspended particle concentration was determined as the weight of the filter and particles minus the weight of the filter converted to g l^{-1}. A second set of TDS values was determined from the filtrate produced in this procedure. The same technique was used as that already described with the exception that only one 50 ml aliquot was evaporated for each sample. Results from this effort suggest that it is not possible to separate particulate organic matter (POM) from inorganic precipitates with a high degree of accuracy or precision by membrane filtration.

CONCLUSION

The saline lakes of Canada's Great Plains provide a suitable testing ground for salinity proxies. Relatively easy access is available to lakes ranging in TDS from below 23 ppt to 184 ppt. In addition the high sulfate, low calcium carbonate nature of many of the lakes represents a unique environment.

Salinity is increasing at Redberry and Manito Lakes but is decreasing at Little Manitou and Big Quill Lakes. Long term variations in the salinity of other lakes in the region remain unknown. It is necessary to develop an understanding of the mechanisms behind these changes in order to predict future salinity trends. The economic and ecological importance of many of these lakes makes such an investigation a priority. It is known that the decreasing salinity in Little Manitou and Big Quill Lakes are caused by water management projects. The nonlinear increase in salinity at Redberry and Manito Lakes may result from basin morphometry combined with a negative water balance during the 20th century.

KEYWORDS

- **Chaplin Lakes**
- **Dissolved oxygen**
- **Great Plains**
- **Muskiki Lake**
- **Patience Lake**
- **Saline lakes**
- **Total dissolved solids**
- **Water chemistry**

AUTHORS' CONTRIBUTIONS

Jeff S. Bowman drafted the manuscript, participated in the design and execution of the field work, and conducted analyses of TDS and suspended particles. Julian P. Sachs devised the project, helped to draft and edit the manuscript, directed the field team, and oversaw all aspects of data collection. All authors read and approved the final manuscript.

ACKNOWLEDGMENTS

This work was supported by the National Science Foundation under Grant No. 0639640 (to Julian P. Sachs), by the American Chemical Society's Petroleum Research Fund under Grant No. 46937-AC2 (to Julian P. Sachs) and by the Gary Comer Science and Education Foundation (to Julian P. Sachs). We would like to thank Chase Stoudt for his invaluable assistance in the field, Dr. Orest Kawaka for his assistance preparing and outfitting the sampling gear, Saskatchewan Minerals, Big Quill Resources, and the Redberry Lake Biosphere Reserve for allowing access to sampling sites, Laurie Balistrieri and Jim Murray for the use of their pump and filtration equipment. We are grateful to the University of Washington School of Oceanography for material and financial support, and the Hutchings family for material support.

COMPETING INTERESTS

The authors declare that they have no competing interests.

Permissions

Chapter 1: Protein Pattern Changes in *Zea mays* Plants from Nitrate was originally published as "Evaluation of protein pattern changes in roots and leaves of *Zea mays* plants in response to nitrate availability by two-dimensional gel electrophoresis analysis" in *BioMed Central 8:23, 2009*. Reprinted with permission under the Creative Commons Attribution License or equivalent.

Chapter 2: Pesticide-related Illness Diagnosed in Primary Care was originally published as "Pesticide-related illness reported to and diagnosed in Primary Care: implications for surveillance of environmental causes of ill- health" in *BioMed Central 7:6, 2009*. Reprinted with permission under the Creative Commons Attribution License or equivalent.

Chapter 3: Effects of Increased Litterfall in Tropical Forests was originally published as "Increased Litterfall in Tropical Forests Boosts the Transfer of Soil CO_2 to the Atmosphere" in *PLoS ONE 12:12, 2007*. Reprinted with permission under the Creative Commons Attribution License or equivalent.

Chapter 4: Evolution of Warning Coloration was originally published as "Explaining the Evolution of Warning Coloration: Secreted Secondary Defence Chemicals May Facilitate the Evolution of Visual Aposematic Signals" in *PLoS ONE 6:3, 2009*. Reprinted with permission under the Creative Commons Attribution License or equivalent.

Chapter 5: Chemically Diverse Toxicants Effects on Precursor Cell Function was originally published as "Chemically diverse toxicants converge on Fyn and c-Cbl to disrupt precursor cell function" in *PLoS BIOLOGY 2:6, 2007*. Reprinted with permission under the Creative Commons Attribution License or equivalent.

Chapter 6: Estrogen-like Activity of Seafood from Chemical Contaminants was originally published as "Estrogen-like activity of seafood related to environmental chemical contaminants" in *BioMed Central 3:30, 2006*. Reprinted with permission under the Creative Commons Attribution License or equivalent.

Chapter 7: Microbial Contamination and Chemical Toxicity was originally published as "Microbial contamination and chemical toxicity of the Rio Grande" in *BioMed Central 4:22, 2004*. Reprinted with permission under the Creative Commons Attribution License or equivalent.

Chapter 8: Heavy Metal Tolerance in *Stenotrophomonas maltophilia* was originally published as "Heavy Metal Tolerance in *Stenotrophomonas maltophilia*" in *PLoS ONE 2:6, 2008*. Reprinted with permission under the Creative Commons Attribution License or equivalent.

Chapter 9: Chemical Defense in Marine Biofilm Bacteria was originally published as "Marine Biofilm Bacteria Evade Eukaryotic Predation by Targeted Chemical Defense" in in *PLoS ONE 7:23, 2008*. Reprinted with permission under the Creative Commons Attribution License or equivalent.

Chapter 10: Reverse and Conventional Chemical Ecology in *Culex* Mosquitoes was originally published as "Reverse and conventional chemical ecology approaches for the development of oviposition attractants for Culex mosquitoes" in *PLoS ONE 8:22, 2008*. Reprinted with permission under the Creative Commons Attribution License or equivalent.

Chapter 11: Morphogenesis of the Rat Pre-implantation Embryo and Environmental Toxicant was originally published as "The environmental toxicant 2,3,7,8-tetrachlorodibenzo-*p*-dioxin disrupts morphogenesis of the rat pre-implantation embryo" in *BioMed Central 1:2, 2008*. Reprinted with permission under the Creative Commons Attribution License or equivalent.

Chapter 12: Endosulfan in Farming Areas of the Western Cape, South Africa was originally published as ""'Contamination of rural surface and ground water by endosulfan in farming areas of the Western Cape, South Africa in *BioMed Central 3:10, 2003*. Reprinted with permission under the Creative Commons Attribution License or equivalent.

References

1

1. Marschner, H. (1995). *Mineral Nutrition of Higher Plants*. Academic Press Limited, London.
2. Barker, A. V. and Bryson, G. M. (2007). Nitrogen. In *Handbook of Plant Nutrition*. A. V. Barker and D. J. Pilbeam (Eds.). CRC Press, Boca Raton, pp. 21–50.
3. Stitt, M. (1999). Nitrate regulation of metabolism and growth. *Curr. Opin. Plant Biol.* **2**, 178–186.
4. Brouquisse, R., Masclaux, C., Feller, U., and Raymond, P. (2001). Protein hydrolysis and nitrogen remobilisation in plant life and senescence. In *Plant Nitrogen*. P. J. Lea and J. F. Morot-Gaudry (Eds.). Springer-Verlag Berlin Hidelberg, Hidelberg, pp. 275–293.
5. Stitt, M., Müller, C., Matt, P., Gibon, Y., Carillo, P., Morcuende, R., Sheible, W. R., and Krapp, A. (2002). Steps towards an integrated view of nitrogen metabolism. *J. Exp. Bot.* **53**, 959–970.
6. Miller, A. J. and Cramer, M. D. (2004). Root nitrogen acquisition and assimilation. *Plant Soil* **274**, 1–36.
7. Scheible, W. R., Morcuende, R., Czechowski, T., Fritz, C., Osuna, D., Palacios-Rojas, N., Schindelasch, D., Thimm, O., Udvardi, M. K., and Stitt, M. (2004). Genome-wide programming of primary and secondary metabolism, protein synthesis, cellular growth processes, and the regulatory infrastructure of *Arabidopsis* in response to nitrogen. *Plant Physiol.* **136**, 2483–2499.
8. Jackson, L. E., Burger, M., and Cavagnaro, T. R. (2008). Roots, nitrogen transformation and ecosystem services. *Annu. Rev. Plant Biol.* **59**, 341–363.
9. Lawlor, D. W., Lemaire, G., and Gastal, F. (2001). Nitrogen, plant growth and crop yield. In *Plant Nitrogen*. P. J. Lea and J. F. Morot-Gaudry (Eds.). Springer-Verlag Berlin Hidelberg, Hidelberg, 343–367.
10. Hirel, B., Bertin, P., Quillere, I., Bourdoncle, W., Attagnant, C., Dellay, C., Gouy, A., Cadiou, S., Retailliau, C., Falque, M., and Gallais, A. (2001). Towards a better understanding of the genetic and physiological basis for nitrogen use efficiency in maize. *Plant Physiol.* **125**, 1258–1270.
11. Hagedorn, F., Bucher, J. B., and Schleppi, P. (2001). Contrasting dynamics of dissolved inorganic and organic nitrogen in soil and surface waters of forested catchments with Gleysols. *Geoderma* **100**, 173–192.
12. Owen, A. G. and Jones, D. L. (2001). Competition for amino acids between wheat roots and rhizosphere microorganisms and the role of amino acids in plant N acquisition. *Soil Biol. Biochem.* **33**, 651–657.
13. Orsel, M., Filleur, S., Fraisier, V., and Daniel-Vedele, F. (2002). Nitrate transport in plants: Which gene and which control? *J. Exp. Bot.* **53**, 825–833.
14. Ullrich, C. I. and Novacky, A. J. (1990). Extra and intracellular pH and membrane potential changes by K^+, Cl^-, H_2PO_4 and NO_3 uptake and fusicoccin in root hairs of *Limnobium stoloniferum*. *Plant Physiol.* **94**, 1561–1567.
15. McClure, P. R., Kochian, L. V., Spanwick, R. M., and Shaff, J. E. (1990). Evidence for cotransport of nitrate and protons in maize roots. I. Effects of nitrate on the membrane potential. *Plant Physiol.* **93**, 281–289.
16. Meharg, A. A. and Blatt, M. R. (1995). NO_3^- transport across the plasma membrane of *Arabidopsis thaliana* root hairs: Kinetic control by pH and membrane voltage. *J. Membrane Biol.* **145**, 49–66.
17. Crawford, N. M. and Glass, A. D. M. (1998). Molecular and physiological aspects of nitrate uptake in plants. *Trends Plant Sci.* **3**, 389–395.
18. Huang, N. C., Liu, K. H., Lo, H. J., and Tsay, Y. F. (1999). Cloning and functional characterization of an *Arabidopsis* nitrate transporter gene that encodes a constitutive component of low-affinity uptake. *Plant Cell* **11**, 1381–1392.

19. Espen, L., Nocito, F. F., and Cocucci, M. (2004). Effect of NO_3^- transport and reduction on intracellular pH: An *in vivo* NMR study in maize roots. *J. Exp. Bot.* **55**, 2053–2061.
20. Palmgren, M. G. (2001). Plant plasma membrane H^+-ATPases: Powerhouses for nutrient uptake. *Annu. Rev. Plant Physiol. Plant Mol. Biol.* **52**, 817–845.
21. Santi, S., Locci, G., Monte, R., Pinton, R., and Varanini, Z. (2003). Induction of nitrate uptake in maize roots: Expression of putative high affinity nitrate transporter and plasma membrane H^+-ATPase isoforms. *J. Exp. Bot.* **54**, 1851–1864.
22. Sondergaard, T. E., Schulz, A., and Palmgren, M. G. (2004). Energization of transport processes in plants. Roles of the plasma membrane H1-ATPase. *Plant Physiol.* **136**, 2475–2482.
23. Oaks, A. and Hirel, B. (1985). Nitrogen metabolism in roots. *Annu. Rev. Plant Physiol.* **36**, 345–365.
24. Meyer, C. and Stitt, M. (2001). Nitrate reduction and signalling. In *Plant Nitrogen.* P. J. Lea and J. F. Morot-Gaudry (Eds.). Springer-Verlag Berlin Hidelberg, Hidelberg, pp. 37–59.
25. Hirel, B. and Lea, P. J. (2001). Ammonia assimilation. In *Plant Nitrogen*. P. J. Lea and J. F. Morot-Gaudry (Eds.). Springer-Verlag Berlin Hidelberg, Hidelberg, pp. 79–99.
26. Crawford, N. M. (1995). Nitrate:Nutrient and signal for plant growth. *Plant Cell* **7**, 859–868.
27. Paul, M. J. and Foyer, C. H (2001). Sink regulation of photosynthesis. *J. Exp. Bot.* **52**, 1383–1400.
28. Forde, B. G. (2002). Local and long-range signalling pathways regulating plant responses to nitrate. *Annu. Rev. Plant Biol.* **53**, 203–224.
29. Wang, R., Guegler, K., LaBrie, S. T., and Crawford, N. M. (2000) Genomic analysis of a nutrient response in *Arabidopsis* reveals diverse expression patterns and novel metabolic and potential regulatory genes induced by nitrate. *Plant Cell* **12**, 1491–1509.
30. Rossignol, M. (2001). Analysis of the plant proteome. *Curr. Opin. Biotech.* **12**, 131–134.
31. Roberts, J. K. M. (2002). Proteomics and future generation of plant molecular biologists. *Plant Mol. Biol.* **48**, 143–154.
32. Yarmush, M. L. and Jayaraman, A. (2002). Advances in proteomic technologies. *Annu. Rev. Biomed. Eng.* **4**, 349–373.
33. Patterson, S. D. and Aebersold, R. H. (2003). Proteomics: The first decade and beyond. *Nat. Genet. Suppl.* **33**(Suppl.), 311–323.
34. Jorrín-Novo, J. V., Maldonato, A. M., Echevarría Zomeòo, S., Valledor, L., Castllejo, M. A., Curto, M., Valero, J., Sghaier, B., Donoso, G., and Redonado, I. (2009). Plant Proteomics update (2007–2008): Second-generation proteomic techniques, an appropriate experimental design, and data analysis to fulfil MIAPE standards, increase plant proteome coverage and expand biological knowledge. *J. Proteomics* **72**, 285–314.
35. Lawrence, C. J., Dong, Q., Mary, L., Polacco, M. L., Seigfried, T. E., and Brendel, V. (2004). MaizeGDB, the community database for maize genetics and genomics. *Nucleic Acids Res.* **32**, D393–397.
36. Porubleva, L., Velden, K. V., Kothari, S., David, J., Oliver, D. J., Parag, R., and Chitnis, P. R. (2001). The proteome of maize leaves: Use of gene sequences and expressed sequence tag data for identification of proteins with peptide mass fingerprints. *Electrophoresis* **22**, 1724–1738.
37. Majeran, W., Cai, Y., Sun, Q., and van Wijk, K. J. (2005). Functional differentiation of bundle sheath and mesophyll maize chloroplasts determined by comparative proteomics. *Plant Cell* **17**, 3111–3140.
38. Dembinsky, D., Woll, K., Saleem, M., Liu, Y., Fu, Y., Borsuk, L. A., Lamkemeyer, T., Fladerer, C., Madlung, J., Barbazuk, B., Nordheim, A., Nettleton, D., Schnable, P. S., and Hochholdinger, F. (2007). Transcriptomic and proteomic analyses of pericycle cells of the maize primary root. *Plant Physiol.* **145**, 575–588.
39. Bahrman, N., Le Gouls, J., Negroni, L., Amilhat, L., Leroy, P., Lainé, A. L., and Jaminon, O. (2004). Differential protein expression assessed by two-dimensional gel electrophoresis for two wheat varieties

grown at four nitrogen levels. *Proteomics* **4**, 709–719.

40. Bahrman, N., Gouy, A., Devienne-Barret, F., Hirel, B., Vedele, F., and Le Gouis, J. (2005). Differential change in root protein pattern of two wheat varieties under high and low nitrogen nutrition levels. *Plant Sci.* **168**, 81–87.
41. Foyer, C. H., Ferrario-Méry, S., and Noctor, G. (2001). Interactions between carbon and nitrogen metabolism. In *Plant Nitrogen*. P. J. Lea and J. F. Morot-Gaudry (Eds.). Springer-Verlag Berlin Hidelberg, Hidelberg, pp. 237–254.
42. Sivasankar, S., Rothstein, S., and Oaks, A. (1997). Regulation of the accumulation and reduction of nitrate by nitrogen and carbon metabolites in maize seedlings. *Plant Physiol.* **114**, 583–589.
43. Klein, D., Morcuende, R., Stitt, M., and Krapp, A. (2000). Regulation of nitrate reductase expression in leaves by nitrate and nitrogen metabolism is completely overridden when sugars fall below a critical level. *Plant Cell Environ.* **23**, 863–871.
44. Huppe, H. C. and Turpin, D. H. (1996). Appearance of novel glucose-6-phosphate dehydrogenase isoforms in *Chlamydomonas reinhardtii* during growth on nitrate. *Plant Physiol.* **110**, 1431–1433.
45. Wang, Y. H., Garvin, D. F., and Kochian, L. V. (2001). Nitrate-induce genes in tomato roots. Array analysis reveals novel genes that may play a role in nitrogen nutrition. *Plant Physiol.* **127**, 345–359.
46. Li, M., Villemur, R., Hussey, P. J., Silflow, C. D., Gantt, J. S., and Snustad, D. P. (1993). Differential expression of six glutamine synthetase genes in *Zea mays*. *Plant Mol. Biol.* **23**, 401–407.
47. Sakakibara, H., Kawabata, S., Hase, T., and Sugiyama, T. (1992). Differential effects of nitrate and light on the expression of glutamine synthetases and ferredoxin-dependent glutamate synthase in maize. *Plant Cell Physiol.* **33**, 1193–1198.
48. Rockel, P., Strube, F., Rockel, A., Wildt, J., and Kaiser, W. M. (2002). Regulation of nitric oxide (NO) production by plant nitrate reductase *in vivo* and *in vitro*. *J. Exp. Bot.* **53**, 103–110.
49. Igamberdiev, A. U., Bycova, N. V., and Hill, R. D. (2006). Nitric oxide scavenging by barley hemoglobin is facilitated by a monodehydroascorbate reductase-mediated ascorbate reduction of methemoglobin. *Planta* **223**, 1033–1040.
50. Lamattina, L., Garcìa-Mata, C., and Pagnussat, G. (2003). Nitric oxide: The versatility of an extensive signal molecule. *Annu. Rev. Plant Biol.* **54**, 109–136.
51. Stöhr, C. and Stremlau, S. (2006). Formation and possible roles of nitric oxide in plant roots. *J. Exp. Bot.* **57**, 463–470.
52. Zhao, D. Y., Tian, Q. Y., Li, L. H., and Zhang, W. H. (2007). Nitric oxide is involved in nitrate-induced inhibition of root elongation in *Zea mays*. *Ann. Bot.* **100**, 497–503.
53. Peschke, V. M. and Sachs, M. M. (1994). Characterization and expression of transcripts induced by oxygen deprivation in maize (*Zea mays* L.). *Plant Physiol.* **104**, 387–394.
54. Igamberdiev, A. U. and Hill, R. D. (2004). Nitrate NO and haemoglobin in plant adaptation to hypoxia: An alternative to classic fermentation pathways. *J. Exp. Bot.* **408**, 2473–2482.
55. Fritz, C., Palacios-Rojas, N., Fell, R., and Stitt, M. (2006). Regulation of secondary metabolism by the carbon-nitrogen status in tobacco: Nitrate inhibits large sectors of phenylpropanoid metabolism. *Plant J.* **46**, 533–548.
56. Kingston-Smith, A. H., Bollard, A. L., and Minchin, F. R. (2005). Stress-induced changes in protease composition are determined by nitrogen supply in non-nodulating white clover. *J. Exp. Bot.* **56**, 745–753.
57. Simões, I. and Faro, C. (2004). Structure and function of plant aspartic proteinases. *Eur. J. Biochem.* **271**, 2067–2075.
58. Askura, T., Watanabe, H., Abe, K., and Arai, S. (1995). Rice aspartic proteinases, oryzasin, expressed during seed ripening and germination, has a gene organization distinct from those of animal and microbial aspartic proteinases. *Eur. J. Biochem.* **232**, 77–83.
59. Raven, J. A. (1986). Biochemical disposal of excess H^+ in growing plants? *New Phytol.* **104**, 175–206.

60. Sakano, K. (1998). Revision of biochemical pH-stat: Involvement of alternative pathway metabolisms. *Plant Cell Physiol.* **39**, 467–473.

61. Britto, D. T. and Kronzucker, H. J. (2005). Nitrogen acquisition, PEP carboxylase, and cellular pH homeostasis: New views on old paradigms. *Plant Cell Environ.* **28**, 1396–1409.

62. Uhrig, R. G., She, Y. M., Leach, C. A., and Plaxton, W. C. (2008). Regulatory monoubiquitination of phosphoenolpyruvate carboxylase in germinating castor oil seeds. *JBC* **283**, 29650–29657.

63. de Vetten, N. C. and Ferl, R. J. (1994). Two genes encoding GF14 (14-3-3) proteins in *Zea mays*. Structure, expression, and potential regulation by G-box-binding complex. *Plant Physiol.* **106**, 1593–1604.

64. Bihn, E. A., Paul, A. L., Wang, S. W., Erdos, G. W., and Ferl, R. J. (1997). Localization of 14-3-3 proteins in the nuclei of *Arabidopsis* and maize. *Plant J.* **12**, 1439–1445.

65. Roberts, M. R. (2000). Regulatory 14-3-3 protein-protein interactions in plant cells. *Curr. Opin. Plant Biol.* **3**, 400–405.

66. Bachmann, M., Huber, J. L., Athwal, G. S., Wu, K., Ferl, R. J., and Huber, S. C. (1996). 14-3-3 proteins associate with the regulatory phosphorylation site of spinach leaf nitrate reductase in an isoform-specific manner and reduce dephosphorylation of Ser-543 by endogenous protein phosphatases. *FEBS Lett.* **398**, 26–30.

67. Ikeda, Y., Koizumi, N., Kusano, T., and Sano, H. (2000). Specific binding of a 14-3-3 protein to autophosphorylated WPK4, an SNF1-related wheat protein kinase, and to WPK-4-phosphorylated nitrate reductase. *JBC* **275**, 31695–31700.

68. Dickson, R., Weiss, C., Howard, R. J., Alldrick, S. P., Ellis, R. J., Lorimer, G., Azem, A., and Viitenen, P. V. (2000). Reconstitution of higher plant chloroplast chaperonin 60 tetradecamers active in protein folding. *JBC* **275**, 11829–11835.

69. Averill, R. H., Bailey-Serres, J., and Kruger, N. J. (1998). Co-operation between cytosolic and plastidic oxidative pentose phosphate pathways revealed by 6-phosphogluconate dehydrogenase-deficient genotypes of maize. *Plant J.* **14**, 449–457.

70. Malkin, R. and Niyogi, K. (2000). Photosynthesis. In *Biochemistry and Molecular Biology of Plants*. B. Buchanan, W. Gruissem, and R. Jones (Eds.). *Am. Soc. Plant Physiol.* Rockville, pp. 568–628.

71. Rapala-Kozik, M., Kowalaska, E., and Ostrowska, K. (2008). Modulation of thiamine metabolism in *Zea mays* seedlings under conditions of abiotic stress. *J. Exp. Bot.* **59**, 4133–4143.

72. Edwards, G. E., Franceschi, V. R., and Voznesenskaya, E. V. (2004). Single-cell C_4 phothosynthesis versus the dual-cell (Kranz) paradigm. *Annu. Rev. Plant Biol.* **55**, 173–196.

73. Ueno, Y., Imanari, E., Emura, J., Yoshizawa-Kumagaye, K., Nakajiama, K., Inami, K., Shiba, T., Sakakibara, H., Sugiyama, T., and Izui, K. (2000). Immunological analysis of the phosphorylation state of maize C_4-form phosphoenolpyruvate carboxylase with specific antibodies raised against a synthetic phosphorylated peptide. *Plant J.* **21**, 17–26.

74. Izui, K., Matsumura, H., Furumoto, T., and Kai, Y. (2004). Phosphoenolpyruvate carboxylase: A new era of structural biology. *Annu. Rev. Plant. Biol.* **55**, 69–84.

75. Nemchenko, A., Kunze, S., Feussner, I., and Kolomietes, M. (2006). Duplicate maize 13-lipoxygenase genes are differentially regulated by circadian rhythm, cold stress, wounding, pathogen infection, and hormonal treatments. *J. Exp. Bot.* **57**, 3767–3779.

76. Feussner, I., Bachmann, A., Höhne, M., and Kindl, H. (1998). All three acyl moieties of trilinolein are efficiently oxygenated by recombinant His-tagged lipid body lipoxygenase *in vitro*. *FEBS Lett.* **431**, 433–436.

77. James, H. E. and Robinson, C. (1991). Nucleotide sequence of cDNA encoding the precursor of the 23 kDa protein of the photosynthetic oxygen-evolving complex from wheat. *Plant Mol. Biol.* **17**, 179–182.

78. Yoshiba, Y., Yamaguchi-Shinozaki, K., Shinozaki, K., and Harada, Y. (1995). Characterization of a cDNA clone encoding 23 kDa polypeptide of the oxygen-evolving

complex of photosystem II in rice. *Plant Cell Physiol.* **36**, 1677–1682.

79. Sourosa, M. and Aro, E. M. (2007). Expression, assembly and auxiliary functions of photosystem II oxygen-evolving proteins in higher plants. *Photosynth. Res.* **93**, 89–100.
80. Ifuku, K., Yamamoto, Y., Ono, T., Ishihara, S., and Sato, F. (2005). PsbP protein, but not PsbQ protein, is essential for the regulation and stabilization of photosystem II in higher plants. *Plant Physiol.* **139**, 1175–1184.
81. Gaude, N., Bréhélin, C., Tischendorf, G., Kessler, F., and Dörmann, P. (2007). Nitrogen deficiency in *Arabidopsis* affects galactolipid composition and gene expression and results in accumulation of fatty acid phytyl esters. *Plant J.* **49**, 729–739.
82. Sakurai, I., Mizusawa, N., Wada, H., and Sato, N. (2007). Digalactosyldiacylglycerol is required for stabilization of the oxygen-evolving complex in photosystem II. *Plant Physiol.* **145**, 1361–1370.
83. Cataldo, D. A., Haroon, M., Schrader, L. E., and Youngs, V. L. (1975). Rapid colorimetric determination of nitrate in plant tissue by nitration of salicylic acid. *Commun. Soil Sci. Plant Anal.* **6**, 71–80.
84. Ferrario-Méry, S., Valadier, M. H., and Foyer, C. H. (1998). Overexpression of nitrate reductase in tobacco delays drought-induced decreases in nitrate reductase activity and mRNA. *Plant Physiol.* **117**, 293–302.
85. Nelson, N. A. (1944). A photometric adaptation of the Somogy method for the determination of glucose. *JBC*, **153**, 375–384.
86. Moore, S. and Stein, W. H. (1954). A modified ninhydrin reagent for the photometric determination of amino acids and related compounds. *JBC* **211**, 907–913.
87. Martínez-Garcia, J. F., Monte, E., and Quall, P. H. (1999). A simple, rapid and quantitative method for preparing *Arabidopsis* protein extracts for immunoblot analysis. *Plant J.* **20**, 251–257.
88. Lichtenthaler, H. K. (1987). Chlorophylls and carotenoids: Pigments of photosynthetic biomembranes. *Met. Enzymol.* **148**, 350–382.
89. Heat, R. L. and Packer, K. (1968). Photoperoxidation in isolated chloroplasts. I. Kinetics and stoichiometry of fatty acid peroxidation. *Arch. Biochem. Biophys.* **125**, 189–198.
90. Krause, G. H. and Weis, E. (1991). Chlorophyll fluorescence and photosynthesis: The basics. *Annu. Rev. Plant Physiol. Plant Mol. Biol.* **42**, 313–349.
91. Genty, B., Briantais, J. M., and Baker, N. R. (1989). The relationship between the quantum yield of photosynthetic electron transport and quenching of chlorophyll fluorescence. *BBA* **990**, 87–92.
92. Hurkman, W. J. and Tanaka, C. K. (1986). Solubilization of plant membrane proteins for analysis by two-dimensional gel electrophoresis. *Plant Physiol.* **81**, 802–806.
93. Laemmli, U. K. (1970). Cleavage of structural proteins during the assembly of the head of bacteriophage. T4. *Nature* **227**, 680–685.
94. Neuhoff, V., Arold, N., Taube, D., and Ehrhardt, W. (1988). Improved staining of proteins in polyacrylamide gels including isoelectric focusing gels with clear background at nanogram sensitivity using Coomassie Brilliant Blue G-250 and R-250. *Electrophoresis* **9**, 255–262.
95. Magni, C., Scarafoni, A., Herndl, A., Sessa, F., Prinsi, B., Espen, L., and Duranti, M. (2007). Combined electrophoretic approaches for the study of white lupin mature seed storage proteome. *Phytochemistry* **68**, 997–1007.
96. National Center for Biotechnology Information. Retrieved from [http://www.ncbi.nlm.nih.gov/].
97. Eng, J. K., McCormack, A. L., and Yates, J. R. III (1994). An approach to correlate tandem mass spectral data of peptides with amino acid sequences in a protein database. *J. Am. Soc. Mass Spectrom.* **5**, 976–989.
98. Mackey, A. J., Haystead, T. A. J., and Pearson, W. R. (2002). Getting more from less: Algorithms for rapid protein identification with multiple short peptide sequences. *Mol. Cell Proteomics* **1**, 139–147.
99. ExPASy Proteomics Server. Retrieved from [http://www.expasy.org/].

2

1. Sanborn, M., Kerr, K. J., Sanin, L. H., Cole, D. C., Bassil, K. L., and Vakil, C. (2007). Non-cancer health effects of pesticides: Systematic review and implications for family doctors. *Can. Fam. Physician.* **53**(10), 1712–1720.
2. Alavanja, M. C. R., Hoppin, J. A., and Kamel, F. (2004). Health effects of chronic pesticde exposure: Cancer and neurotoxicity. *Ann. Rev. Public Health* **25**, 155–197.
3. Costa, L. G. (2006). Current issues in organophosphate toxicology. *Clinica Chimica Acta* **366**, 1–13.
4. Tahmaz, N., Soutar, A., and Cherrie, J. W. (2003). Chronic fatigue and organophosphate pesticides in sheep farming: A retrospective study amongst people reporting to a UK pharmacovigilance scheme. *Ann. Occup. Hyg.* **47**(4), 261–267.
5. Hospital Episode Statistics. Retrieved from [http://www.hesonline.nhs.uk].
6. Rushton, L. and Mann, V. (2008). Retrieved from [http://www.hse.gov.uk/research/rrhtm/rr608.htm]. *Estimating the Prevalence and Incidence of Pesticide-Related Illness Presented to General Practitioners in Great Britain. Research Report 608, HSE.*
7. General Practice Research Framework. Retrieved from [http://www.gprf.mrc.ac.uk].
8. General Practitioner Workload (2004). Retrieved from [http://www.rcgp.org.uk/pdf/ISS_INFO_03_APRIL04.pdf].
9. Thundyil, J. G., Stober, J., Besbelli, N., and Pronczuk, J. (2008). Acute pesticide poisoning: A proposed classification tool. *Bull. WHO* **86**(3), 206–211.
10. Keifer, M., McConnell, R., Pachero, F., Daniel, W., and Rosenstock, L. (1996). Estimating underreported pesticide poisonings in Nicaragua. *Am. J. Ind. Med.* **30**, 195–201.
11. Das, R., Steege, A., Baron, S., Beckman, J., and Harrison, R. (2001). Pesticide-related illness among Migrant Farm workers in the United States. *Int. J. Occup. Environ. Health* **7**, 303–312.
12. Casey, P. and Vale, J. A. (1994). Deaths from pesticide poisoning in England and Wales: 1945–1989. *Hum. Exper. Toxicol.* **13**, 95–101.
13. Profile of UK General Practitioners (2006). Retrieved from [https://www.rcgp.org.uk/pdf/ISS_INFO_01_JUL06.pdf].
14. Rudent, J., Manegaux, F., Leverger, G., et al. (2007). Household exposure to pesticides and risk of childhood haematopoietic malignancies: The ESCALE study (SFCE). *Env. Health Perspect* **115**, 1787–1793.
15. Ma, X., Buffler, P. A., Gurvier, R. B., et al. (2002). Critical windows of exposure to household pesticides and risk of childhood leukaemia. *Env. Health Perspect.* **110**, 955–960.
16. Grey, C. N. B., Nieuwenhuijsen, M. J., and Golding, J. (2006). Usage and storage of domestic pesticides in the UK. *Sci. Tot. Env.* **368**, 465–470.
17. Weale, V. P. and Goddard, H. (1998). *The Effectiveness of Non-agricultural Pesticide Labelling (Contract Research Report 161/1998).* HSE Books, Sudbury, UK.
18. Calvert, G. M., Plate, D. K., Das, R., Rosales, R., Sahfey, O., Thomsen, C., et al. (2004). Acute occupational pesticide-related illness in the US, 1998–1999: Surveillance findings from the SENSOR-pesticides program. *Am. J. Industrial. Med.* **45**, 14–23.
19. Health and Safety Executive (HSE) (2004). *Pesticide Incidents Report 1 April 2003–31 March 2004.*
20. Deckers, J. G. M., Paget, W. J., Schellevis, F. G., and Fleming, D. M. (2006). European primary care surveillance networks: Their structure and operation. *Family Prac.* **23**, 151–158.
21. Green, L. A. and Hickner, J. (2006). A short history of primary care practice-based research networks: From concept to essential research laboratories. *JABFM* **19**(1), 1–10.
22. Freedman, D. O., Weld, L. H., Kozarsky, P. E., Fisk, T., Robins, R., von Sonnenburg, F., et al. (2006). The GeoSentinel Surveillance Network: Spectrum of disease and relationship to place of exposure among ill returned travellers. *New Engl. J. Med.* **354**, 119–130.
23. Jelinek, T. and Muhlberger, N. (2005). Surveillance of imported diseases as a window to travel health risks. *Infect. Dis. Clin. North Am.* **19**, 1–13.

24. London, L., Bourne, D., Sayed, R., and Eastman, R. (2004). Guillain-Barre syndrome in a rural farming district in South Africa: A possible relationship to environmental organophosphate exposure. *Arch. Environ. Health* **59**(11), 575–80.
25. Soutar, C. S. (2001). *Frequencies of Disease Presenting to General Practitioners According to Patients' Occupation. Contract Research Report 340*. HSE Books, Sudbury, UK.
26. Kass, D. E., Their, A. L., Leighton, J., Cone, J. E., and Jeffery, N. L. (2004). Developing a comprehensive pesticide health effects tracking system for an urban setting: New York City's approach. *Env. Health Perspec.* **112**(4), 1419–1423.
27. Wakefield, J. (2003). Pesticides initiative: basic training for health care providers. *Env. Health Perspec.* **111**(10), A520–522.
28. Sandiford, P. (1992). What can information systems do for Primary Health Care? An international perspective. *Soc. Sci. Med.* **34**, 1077–1087.
29. London, L. and Bailie, R. (2001). Challenges for improving surveillance for pesticide poisoning: Policy implications for developing countries. *Int. J. Epid.* **30**, 564–570.

3

1. DeLucia, E. H., Hamilton, J. G., Naidu, S. L., Thomas, R. B., Andrews, J. A., et al. (1999). Net primary production of a forest ecosystem with experimental CO_2 enrichment. *Science* **284**, 1177–1179.
2. Allen, A. S., Andrews, J. A., Finzi, A. C., Matamala, R., Richter, D. D., and Schlesinger, W. H. (2000). Effects of free-air CO_2 enrichment (FACE) on belowground processes in a Pinus taeda forest. *Ecol. Appl.* **10**, 437–448.
3. Schlesinger, W. H. and Lichter, J. (2001). Limited carbon storage in soil and litter of experimental forest plots under increased atmospheric CO_2. *Nature* **411**, 466–469.
4. Finzi, A. C., Allen, A. S., DeLucia, E. H., Ellsworth, D. S., and Schlesinger, W. H. (2001). Forest litter production, chemistry, and decomposition following two years of free-air CO_2 enrichment. *Ecology* **82**, 470–484.
5. Zak, D. R., Holmes, W. E., Finzi, A. C., Norby, R. J., and Schlesinger, W. H. (2003). Soil nitrogen cycling under elevated CO_2: A synthesis of forest FACE experiments. *Ecol. Appl.* **13**, 1508–1514.
6. Zhang, X., Zwiers, F. W., Hegerl, G. C., Lambert, H., Gillett, N. P., et al. (2007). Detection of human influence on twentieth-century precipitation trends. *Nature* **448**, 462–465.
7. Raich, J. W., Russell, A. E., Kitayama, K., Parton, W. J., and Vitousek, P. M. (2006). Temperature influences carbon accumulation in moist tropical forests. *Ecology* **87**, 76–87.
8. Bardgett, R. D. and Wardle, D. A. (2003). Herbivore mediated linkages between aboveground and belowground communities. *Ecology* **84**, 2258–2268.
9. Sayer, E. J. (2006). Using experimental manipulation to assess the roles of leaf litter in the functioning of forest ecosystems. *Biol. Rev.* **81**, 1–31.
10. Dixon, R. K., Brown, S., Houghton, R. A., Solomon, A. M., Trexler, M. C., and Wisniewski, J. (1994). Carbon pools and flux of global forest ecosystems. *Science* **263**, 185–190.
11. Bernoux, M., Carvalho, M. S., Volkoff, B., and Cerri, C. C. (2002). Brazil's soil carbon stocks. *Soil. Sci. Soc. Am. J.* **66**, 888–896.
12. Clark, D. A. (2004). Sources or sinks? The responses of tropical forests to current and future climate and atmospheric composition. *Philos. T. Roy. Soc. B.* **359**, 477–491.
13. Jobbagy, E. G. and Jackson, R. B. (2000). The vertical distribution of soil organic carbon and its relation to climate and vegetation. *Ecol. Appl.* **10**, 423–436.
14. Raich, J. W., Potter, C. S., and Bhagawati, D. (2002). Interannual variability in global soil respiration, 1980–1994. *Glob. Change. Biol.* **8**, 800–812.
15. Palmroth, S., Oren, R., McCarthy, H. R., Johnsen, K. H., Finzi, A. C., et al. (2006). Aboveground sink strength in forests controls the allocation of carbon below ground and its [CO_2]-induced enhancement. *Proc. Nat. Acad. Sci. USA* **103**, 19362–19367.

16. Kirschbaum, M. U. F. (2000). Will changes in soil organic carbon act as a positive or negative feedback on global warming? *Biogeochemistry* **48**, 21–51.
17. Davidson, E. A. and Janssens, I. A. (2006). Temperature sensitivity of soil carbon decomposition and feedbacks to climate change. *Nature* **440**, 165–173.
18. Neff, J. C., Townsend, A. R., Gleixner, G., Lehman, S. J., Turnbull, J., and Bowman, W. D. (2002). Variable effects of nitrogen additions on the stability and turnover of soil carbon. *Nature* **419**, 915–917.
19. Reich, P. B., Hungate, B. A., and Luo, Y. (2006). Carbon-nitrogen interactions in terrestrial ecosystems in response to rising atmospheric carbon dioxid. *Annu. Rev. Ecol. Evol. Syst.* **37**, 611–636.
20. Cleveland, C. C. and Townsend, A. R. (2006). Nutrient additions to a tropical rain forest drive substantial soil carbon dioxide losses to the atmosphere. *Proc. Natl. Acad. Sci. USA* **103**, 10316–10321.
21. Raich, J. W. and Nadelhoffer, K. J. (1989). Belowground carbon allocation in forest ecosystems: Global trends. *Ecology* **70**, 1346–1354.
22. Davidson, E. A., Savage, K., Bolstad, P., Clark, D. A., Curtis, P. S., et al. (2002). Belowground carbon allocation in forests estimated from litterfall and IRGA-based soil respiration measurements. *Agr. Forest. Meteorol.* **113**, 39–51.
23. Cornejo, F. H., Varela, A., and Wright, S. J. (1994). Tropical forest litter decomposition under seasonal drought: Nutrient release, fungi and bacteria. *Oikos* **70**, 183–190.
24. Vasconcelos, S. S., Zarin, D. J., Capanu, M., Littell, R., Davidson, E. A., et al. (2004). Moisture and substrate availability constrain soil trace gas fluxes in an eastern Amazonian regrowth forest. *Global. Biogeochem. Cy.* **18**, GB2009.
25. Li, Y., Xu, M., Sun, O. J., and Cui, W. (2004). Effects of root and litter exclusion on soil CO_2 efflux and microbial biomass in wet tropical forests. *Soil. Biol. Biochem.* **36**, 2111–2114.
26. Priess, J. A. and Folster, H. (2001). Microbial properties and soil respiration in submontane forests of Venezuelan Guyana: Characteristics and response to fertilizer treatments. *Soil. Biol. Biochem.* **33**, 503–509.
27. Sayer, E. J., Tanner, E. V. J., and Lacey, A. L. (2006a). Effects of litter manipulation on early-stage decomposition and meso-arthropod abundance in a tropical moist forest. *Forest. Ecol. Manage.* **229**, 285–293.
28. Sayer, E. J., Tanner, E. V. J., and Cheesman, A. W. (2006b). Increased litterfall changes fine root distribution in a moist tropical forest. *Plant. Soil.* **281**, 5–13.
29. Pregitzer, K. S., Laskowski, M. J., and Burton, A. J. (1998). Variation in sugar maple root respiration with root diameter and soil depth. *Tree Physiol.* **18**, 665–670.
30. Desrochers, A., Landhausser, S. M., and Lieffers, V. J. (2002). Coarse and fine root respiration in aspen *(Populus tremuloides). Tree Physiol.* **22**, 725–732.
31. Singh, J. S. and Gupta, S. R. (1977). Plant decomposition and soil respiration in terrestrial ecosystems. *Bot. Rev.* **43**, 449–528.
32. Anderson, J. M., Proctor, J., and Vallack, H. W. (1983). Ecological studies in four contrasting lowland rain forests in Gunung-Mulu National Park, Sarawak. 3. Decomposition processes and nutrient losses from leaf litter. *J. Ecol.* **71**, 503–527.
33. Rout, S. K. and Gupta, S. R. (1989). Soil respiration in relation to abiotic factors, forest floor litter, root biomass and litter quality in forest ecosystems of Siwaliks in northern India. *Acta. Oecol.* **10**, 229–244.
34. Metcalfe, D. B., Meir, P., Aragão, L. E. O. C., Mahli, Y., da Costa, A. C. L., et al. (2007). Factors controlling spatio-temporal variation in carbon dioxide efflux from surface litter, roots, and soil organic matter at four rain forest sites in eastern Amazon. *J. Geophys. Res.* **112**, G04001.
35. Bingeman, C. W., Varner, J. E., and Martin, W. P. (1953). The effect of the addition of organic materials on the decomposition of an organic soil. *Soil. Sci. Soc. Am. Proc.* **29**, 692–696.
36. Kuzyakov, Y., Friedel, J. K., and Stahr, K. (2000). Review of mechanisms and quantification of priming effects. *Soil Biol. Biochem.* **32**, 1485–1498.

37. DeNobili, M., Contin, M., Mondini, C., and Brookes, P. C. (2001). Soil microbial biomass is triggered into activity by trace amounts of substrate. *Soil Biol. Biochem.* **33**, 1163–1170.

38. Hamer, U. and Marschner, B. (2005). Priming effects in soils after combined and repeated substrate additions. *Geoderma* **128**, 38–51.

39. Fontaine, S., Bardoux, G., Abbadie, L., and Mariotti, A. (2004a). Carbon input to soil may decrease soil carbon content. *Ecol. Lett.* **7**, 314–320.

40. Cavalier, J. (1992). Fine-root biomass and soil properties in a semi-deciduous and a lower montane rain forest in Panama. *Plant Soil* **142**, 187–201.

41. Powers, J. S., Treseder, K. K., and Lerdau, M. T. (2005). Fine roots, arbuscular mycorrhizal hyphae and soil nutrients in four neotropical rain forests: Patterns across large geographic distances. *New Phytol.* **165**, 913–921.

42. Leigh, E. G. (1999). *Tropical Forest Ecology*. Oxford University Press, Oxford.

43. Leigh, E. G. and Wright, S. J. (1990). Barro Colorado Island and tropical biology. In. *Four Neotropical Rainforests*. A. H. Gentry (Ed.). Yale University Press, New Haven, pp. 28–47.

44. Brookes, P. C., Landman, A., Pruden, G., and Jenkinson, D. S. (1985). Chloroform fumigation and the release of soil nitrogen: A rapid direct extraction method to measure microbial biomass nitrogen in the soil. *Soil Biol. Biochem.* **17**, 837–842.

45. Beck, T., Joergensen, R. G., Kandeler, E., Makeschin, F., Nuss, E., et al. (1997). An inter-laboratory comparison of ten different ways of measuring soil microbial biomass C. *Soil Biol. Biochem.* **29**, 1023–1032.

4

1. Wallace, A. R. (1867). *Proc. Entomol. Soc. Lond.* (4 March).

2. Merilaita, S. and Ruxton, G. D. (2007). Aposematic signals and the relationship between conspicuousness and distinctiveness. *J. Theor. Biol.* **245**, 268–277.

3. Fisher, R. A. (1930). The genetical theory of natural selection. Clarendon, Oxford.

4. Harvey, P. H., Bull, J. J., Pemberton, M., and Paxton, R. J. (1982). The evolution of aposematic coloration in distasteful prey: A family model. *Am. Nat.* **119**, 710–718.

5. Harvey, P. H. and Paxton, R. J. (1981). The evolution of aposematic coloration. *Oikos* **37**, 391–396.

6. Eisner, T. and Grant, R. P. (1981). Toxicity, odor aversion, and olfactory aposematism. *Science* **213**, 476–476.

7. Cott, H. B. (1940). *Adaptive Coloration in Animals*. Methuen, London.

8. Rothschild, M. (1961). Defensive odours and Müllerian mimicry among insects. *Trans. R. Entomol. Soc. Lond.* **113**, 101–121.

9. Prudic, K. L., Noge, K., and Becerra, J. X. (2008). Adults and nymphs do not smell the same: The different defensive compounds of the giant mesquite bug (*Thasus neocalifornicus*: Coreidae). *J. Chem. Ecol.* **34**, 734–741.

10. Eisner, T., Eisner, M., and Seigler, M. (2005). *Secret Weapons: Defenses of Insects, Spiders, Scorpions, and Other Many-Legged Creatures*. Belknap Press of Harvard University Press, Cambridge, Mass.

11. Speed, M. P. and Ruxton, G. D. (2005). Warning displays in spiny animals: One (more) evolutionary route to aposematism. *Evolution* **59**, 2499–2508.

12. Gamberale-Stille, G. (2000). Decision time and prey gregariousness influence attack probability in naïve and experienced predators. *Anim. Behav.* **60**, 95–99.

13. Merilaita, S. and Ruxton, G. D. (2007). Aposematic signals and the relationship between conspicuousness and distinctiveness. *J. Theor. Biol.* **245**, 268–277.

14. Skelhorn, J. and Rowe, C. (2006). Avian predators taste–reject aposematic prey on the basis of their chemical defence. *Biol. Let.* **2**(3), 348–350.

15. Rowe, C. (1999). Receiver psychology and the evolution of multicomponent signals. *Anim. Behav.* **58**, 921–931.

16. Rowe, C. and Guilford, T. (1999). The evolution of multimodal warning displays. *Evol. Ecol.* **13**, 655–671.
17. Malakoff, D. (1999). Olfaction: Following the scent of avian olfaction. *Science* **286**(5440), 704–705.
18. Steiger, S. S., Fidler, A. E., Valcu, M., and Kempenaers, B. (2008). Avian olfactory receptor gene repertoires: Evidence for a well-developed sense of smell in birds? *Proc. R. Soc. B.* **275**, 2309–2317.

5

1. Yonaha, M., Saito, M., and Sagai, M. (1983). Stimulation of lipid peroxidation by methyl mercury in rats. *Life Sci.* **32**, 1507–1514.
2. Sarafian, T. and Verity, M. A. (1991). Oxidative mechanisms underlying methyl mercury neurotoxicity. *Int. J. Dev. Neurosci.* **9**, 147–153.
3. Shanker, G. and Aschner, M. (2003). Methylmercury-induced reactive oxygen species formation in neonatal cerebral astrocytic cultures is attenuated by antioxidants. *Mol. Brain. Res.* **110**, 85–91.
4. Shanker, G., Aschner, J. L., Syversen, T., and Aschner, M. (2004). Free radical formation in cerebral cortical astrocytes in culture induced by methylmercury. *Mol. Brain. Res.* **128**, 48–57.
5. Ali, S. F., LeBel, C. P., and Bondy, S. C. (1992). Reactive oxygen species formation as a biomarker of methylmercury and trimethyltin neurotoxicity. *Neurotoxicology* **13**, 637–648.
6. Thompson, S. A., White, C. C., Krejsa, C. M., Eaton, D. L., and Kavanagh, T. J. (2000). Modulation of glutathione and glutamate-L-cysteine ligase by methylmercury during mouse development. *Toxicol. Sci.* **57**, 141–146.
7. Ding, Y., Gonick, H. C., and Vaziri, N. D. (2000). Lead promotes hydroxyl radical generation and lipid peroxidation in cultured aortic endothelial cells. *Am. J. Hypertens.* **13**, 552–555.
8. Hsu, P., Liu, M., Hsu, C., Chen, L., and Guo, Y. (1997). Lead exposure causes generation of reactive oxygen species and functional impairment in rat sperm. *Toxicology* **122**, 133–143.
9. Ercal, N., Treratphan, P., Hammond, T. C., Mathews, R. H., Grannemann, N. H., et al. (1996). *In vivo* indices of oxidative stress in lead exposed C57BL/6 mice are reduced by treatment with meso-2,3-dimercaptosuccinic acid or N-acetyl cysteine. *Free Radic Biol. Med.* **21**, 157–161.
10. Stahnke, T. and Richter-Landsberg, C. (2004). Triethyltin-induced stress responses and apoptotic cell death in cultured oligodendrocytes. *Glia* **46**, 334–344.
11. Jenkins, S. M. and Barone, S. (2004). The neurotoxicant trimethyltin induces apoptosis via caspase activation, p38 protein kinase, and oxidative stress in PC12 cells. *Toxicol. Lett.* **147**, 63–72.
12. Fowler, B. A., Whittaker, M. H., Lipsky, M., Wang, G., and Chen, X. Q. (2004). Oxidative stress induced by lead, cadmium and arsenic mixtures: 30-day, 90-day, and 180-day drinking water studies in rats: An overview. *Biometals* **17**, 567–568.
13. Souza, V., Escobar Mdel, C., Bucio, L., Hernandez, E., and Gutierrez-Ruiz, M. C. (2004). Zinc pretreatment prevents hepatic stellate cells from cadmium-produced oxidative damage. *Cell Biol. Toxicol.* **20**, 241–251.
14. Hei, T. K. and Filipic, M. (2004). Role of oxidative damage in the genotoxicity of arsenic. *Free Radic Biol. Med.* **37**, 574–581.
15. McDonough, K. H. (2003). Antioxidant nutrients and alcohol. *Toxicology* **189**, 89–97.
16. Abdollahi, M., Ranjbar, A., Shadnia, S., Nikfar, S., and Rezale, A. (2004). Pesticides and oxidative stress: A review. *Med. Sci. Monit.* **10**, RA141–147.
17. Suntres, Z. E. (2002). Role of antioxidants in paraquat toxicity. *Toxicology* **180**, 65–77.
18. Smith, L. L., Rose, M. S., and Wyatt, I. (1978). The pathology and biochemistry of paraquat. *Ciba. Found. Symp.* **65**, 321–341.
19. Giray, B. (2001). Cypermethrin-induced oxidative stress in rat brain and liver is prevented by vitamin E or allopurinol. *Toxicol. Lett.* **118**, 139–146.

20. Gupta, A. (1999). Effect of pyrethroid-based liquid mosquito repellent inhalation on the blood-brain barrier function and oxidative damage in selected organs of developing rats. *J. Appl. Toxicol.* **19**, 67–72.

21. Kale, M., Rathore, N., John, S., and Bhathagar, D. (1999). Lipid peroxidative damage on pyrethroid exposure and alteration in antioxidant status in rat erythrocytes. A possible involvement of reactive oxygen species. *Toxicol. Lett.* **105**, 197–205.

22. Gultekin, F. (2000). The effect of organophosphate insecticide chlorpyrifos-ethyl on lipid peroxidation and antioxidant enzymes (*in vitro*). *Arch. Toxicol.* **74**, 533–538.

23. Gupta, R. C. (2001). Depletion of energy metabolites following acetylcholinesterase inhibitor-induced status epilepticus: Protection by antioxidants. *Neurotoxicology* **22**, 271–282.

24. Akhgari, M., Abdollahi, M., and Kebryaeezadeh, A., (2003). Biochemical evidence for free radical-induced lipid peroxidation as a mechanism for subchronic toxicity of malathion in blood and liver of rats. *Hum. Exp. Toxicol.* **22**, 205–211.

25. Banerjee, B. D., Seth, V., Bhattacharya, A., Pasha, S. T., and Chakraborty, A. K. (1999). Biochemical effects of some pesticides on lipid peroxidation and freeradical scavengers. *Toxicol. Lett.* **107**, 33–47.

26. Ranjbar, A., Pasalar, P., and Abdollahi, M. (2002). Induction of oxidative stress and acetylcholinesterase inhibition in organophosphorous pesticide manufacturing workers. *Hum. Exp. Toxicol.* **21**, 179–182.

27. Noble, M., Mayer-Proschel, M., and Proschel, C. (2005). Redox regulation of precursor cell function: Insights and paradoxes. *Antioxid. Redox. Signal.* **7**, 1456–1467.

28. Nathan, C. (2003). Specificity of a third kind: Reactive oxygen and nitrogen intermediates in cell signaling. *J. Clin. Invest.* **111**, 769–778.

29. Droge, W. (2006). Redox regulation in anabolic and catabolic processes. *Curr. Opin. Clin. Nutr. Metab. Care* **9**, 190–195.

30. Cerdan, S., Rodrigues, T. B., Sierra, A., Benito, M., Fonseca, L. L., et al. (2006). The redox switch/redox coupling hypothesis. *Neurochem. Int.* **48**, 523–530.

31. Squier, T. C. (2006). Redox modulation of cellular metabolism through targeted degradation of signaling proteins by the proteasome. *Antioxid. Redox. Signal* **8**, 217–228.

32. Sager, P. R., Doherty, R. A., and Olmsted, J. B. (1983). Interaction of methylmercury with microtubules in cultured cells and *in vitro*. *Exp. Cell. Res.* **146**, 127–137.

33. Lopachin, R. M. and Barber, D. S. (2006). Synaptic cysteine sulfhydryl groups as targets of electrophilic neurotoxicants. *Toxicol. Sci.* **94**, 240–255.

34. Denny, M. F. and Atchison, W. D. (1996). Mercurial-induced alterations in neuronal divalent cation homeostasis. *Neurotoxicology* **17**, 47–61.

35. Goldstein, G. W. (1993). Evidence that lead acts as a calcium substitute in second messenger metabolism. *Neurotoxicology* **14**, 97–102.

36. Simons, T. J. B. (1993). Lead-calcium interactions in cellular lead toxicity. *Neurotoxicology* **14**, 77–86.

37. Costa, L. G., Guizzetti, M., Lu, H., Bordi, F., Vitalone, A., et al. (2001). Intracellular signal transduction pathways as targets for neurotoxicants. *Toxicology* **160**, 19–26.

38. Deng, W. and Poretz, R. D. (2002). Protein kinase C activation is required for the lead-induced inhibition of proliferation and differentiation of cultured oligodendroglial progenitor cells. *Brain. Res.* **929**, 87–95.

39. Choi, B. H., Yee, S., and Robles, M. (1996). The effects of glutathione glycoside in methylmercury poisoning. *Toxicol. Appl. Pharmacol.* **141**, 357–364.

40. Shenker, B. J., Guo, T. L. O. I., and Shapiro, I. M. (1999). Induction of apoptosis in human T-cells by methyl mercury: Temporal relationship between mitochondrial dysfunction and loss of reductive reserve. *Toxicol. Appl. Pharmacol.* **157**, 23–35.

41. Anderson, A. C., Puerschel, S. M., and Linakis, J. G. (1996). Pathophysiology of lead poisoning. In *Lead Poisoning In Children*. S. M. Pueschel, J. G. Linakis, and A. C. Anderson (Eds.). P.H. Brookes, Baltimore, pp. 75–96.

42. He, L., Poblenz, A. T., Medrano, C. J., and Fox, D. A. (2000). Lead and calcium

produce rod photoreceptor cell apoptosis by opening the mitochondrial permeability transition pore. *J. Biol. Chem.* **275**, 12175–12184.

43. Tiffany-Castiglioni, E., Sierra, E. M., Wu, J. N, and Rowles, T. K. (1989). Lead toxicity in neuroglia. *Neurotoxicol* **10**, 417–443.
44. Bressler, J. P. and Goldstein, G. W. (1991). Mechanisms of lead neurotoxicity. *Biochem. Pharmacol.* **41**, 479–484.
45. Pounds, J. G. (1984). Effect of lead intoxication on calcium homeostasis and calcium-mediated cell function: A review. *Neuro.Toxicology.* **5**, 295–332.
46. Trasande, L., Landrigan, P. J., and Schechter, C. (2005). Public health and economic consequences of methyl mercury toxicity to the developing brain. *Environ. Health Perspect.* **113**, 590–596.
47. Raff, M. C., Miller, R. H., and Noble, M. (1983). A glial progenitor cell that develops *in vitro* into an astrocyte or an oligodendrocyte depending on the culture medium. *Nature* **303**, 390–396.
48. Barres, B. A., Hart, I. K., Coles, H. S., Burne, J. F., Voyvodic, J. T., et al. (1992). Cell death in the oligodendrocyte lineage. *J. Neurobiol.* **23**, 1221–1230.
49. Noble, M., Mayer-Proschel, M., and Miller, R. H. (2005). The oligodendrocyte. In *Developmental Neurobiology*. M. S. Rao and M. Jacobson (Eds.). Kluwer Academic/Plenum, New York.
50. Noble, M., Pröschel, C., and Mayer-Proschel, M. (2004). Getting a GR(i)P on oligodendrocyte development. *Dev. Biol.* **265**, 33–52.
51. Levine, J. M., Reynolds, R., and Fawcett, J. W. (2001). The oligodendrocyte precursor cell in health and disease. *TINS* **24**, 39–47.
52. Miller, R. H. (2002). Regulation of oligodendrocyte development in the vertebrate CNS. *Prog. Neurobiol.* **67**, 451–467.
53. Deng, W., McKinnon, R. D., and Poretz, R. D. (2001). Lead exposure delays the differentiation of oligodendroglial progenitors *in vitro*, and at higher doses induces cell death. *Toxicol. Appl. Pharmacol.* **174**, 235–244.
54. Bichenkov, E. and Ellingson, J. S. (2001). Ethanol exerts different effects on myelin basic protein and 2′,3′-cyclic nucleotide 3′-phosphodiesterase expression in differentiating CG-4 oligodendrocytes. *Brain. Res. Dev. Brain. Res.* **128**, 9–16.
55. Zoeller, R. T., Butnariu, O. V., Fletcher, D. L., and Riley, E. P. (1994). Limited postnatal ethanol exposure permanently alters the expression of mRNAS encoding myelin basic protein and myelin-associated glycoprotein in cerebellum. *Alcohol. Clin. Exp. Res.* **18**, 909–916.
56. Harris, S. J., Wilce, P., and Bedi, K. S. (2000). Exposure of rats to a high but not low dose of ethanol during early postnatal life increases the rate of loss of optic nerve axons and decreases the rate of myelination. *J. Anat.* **197**(Prt. 3), 477–485.
57. Özer, E., Saraioglu, S., and Güre, A. (2000). Effect of prenatal ethanol exposure on neuronal migration, neurogenesis and brain myelination in the mice brain. *Clin. Neuropathol.* **19**, 21–25.
58. O'Callaghan, J. P. and Miller, D. B. (1983). Acute postnatal exposure to triethyltin in the rat: Effects on specific protein composition of subcellular fractions from developing and adult brain. *J. Pharmacol. Exp. Ther.* **224**, 466–472.
59. Smith, J., Ladi, E., Mayer-Pröschel, M., and Noble, M. (2000). Redox state is a central modulator of the balance between self-renewal and differentiation in a dividing glial precursor cell. *Proc. Natl. Acad. Sci. USA* **97**, 10032–10037.
60. Noble, M., Murray, K., Stroobant, P., Waterfield, M. D., and Riddle, P. (1988). Platelet-derived growth factor promotes division and motility and inhibits premature differentiation of the oligodendrocyte/type-2 astrocyte progenitor cell. *Nature* **333**, 560–562.
61. Richardson, W. D., Pringle, N., Mosley, M., Westermark, B., and Dubois-Dalcq, M. (1988). A role for platelet-derived growth factor in normal gliogenesis in the central nervous system. *Cell* **53**, 309–319.
62. Calver, A., Hall, A., Yu, W., Walsh, F., Heath, J., et al. (1998). Oligodendrocyte population dynamics and the role of PDGF *in vivo*. *Neuron* **20**, 869–882.

63. Barres, B. A., Lazar, M. A., and Raff, M. C. (1994). A novel role for thyroid hormone, glucocorticoids and retinoic acid in timing oligodendrocyte development. *Development* **120**, 1097–1108.
64. Ibarrola, N., Mayer-Proschel, M., Rodriguez-Pena, A., and Noble, M. (1996). Evidence for the existence of at least two timing mechanisms that contribute to oligodendrocyte generation *in vitro*. *Dev. Biol.* **180**, 1–21.
65. Grinspan, J. B., Edell, E., Carpio, D. F., Beesley, J. S., Lavy, L., et al. (2000). Stage-specific effects of bone morphogenetic proteins on the oligodendrocyte lineage. *J. Neurobiol.* **43**, 1–17.
66. Mabie, P., Mehler, M., Marmur, R., Papavasiliou, A., Song, Q., et al. (1997). Bone morphogenetic proteins induce astroglial differentiation of oligodendroglial-astroglial progenitor cells. *Neurosci* **17**, 4112–4120.
67. Castoldi, A. F., Barni, S., Turin, I., Gandini, C., Manzo, L. (2000). Early acute necrosis, delayed apoptosis and cytoskeletal breakdown in cultured cerebellar granule neurons exposed to methylmercury. *J. Neurosci. Res.* **60**, 775–787.
68. Park, S. T., Lim, K. T., Chung, Y. T., and Kim, S. U. (1996). Methylmercury induced neurotoxicity in cerebral neuron culture is blocked by antioxidants and NMDA receptor antagonists. *Neurotoxicology* **17**, 37–46.
69. Aschner, M., Yao, C. P., Allen, J. W., and Tan, K. H. (2000). Methylmercury alters glutamate transport in astrocytes. *Neurochem. Int.* **37**, 199–206.
70. Markowski, V. P., Flaugher, C. B., Baggs, R. B., Rawleigh, R. C., Cox, C., et al. (1998). Prenatal and lactational exposure to methylmercury affects select parameters of mouse cerebellar development. *Neurotoxicology* **19**, 879–892.
71. Peckham, N. H. and Choi, B. H. (1988). Abnormal neuronal distribution within the cerebral cortex after prenatal methylmercury intoxication. *Acta. Neuropathol.* **76**, 222–226.
72. Kakita, A., Inenaga, C., Sakamoto, M., and Takahashi, H. (2002). Neuronal migration disturbance and consequent cytoarchitecture in the cerebral cortex following transplacental administration of methylmercury. *Acta. Neuropathol. (Berl)* **104**, 409–417.
73. Faustman, E. M., Ponce, R. A., Ou, Y. C., Mendoza, M. A., Lewandowski, T., et al. (2002). Investigations of methylmercury-induced alterations in neurogenesis. *Environ. Health Perspect.* **110**, 859–864.
74. Choi, B. H. (1986). Methylmercury poisoning of the developing nervous system: I. Pattern of neuronal migration in the cerebral cortex. *Neurotoxicology* **7**, 591–600.
75. Murata, K., Budtz-Jorgensen, E., and Grandjean, P. (2002). Benchmark dose calculations for methylmercury-associated delays on evoked potential latencies in two cohorts of children. *Risk. Anal.* **22**, 465–474.
76. Murata, K., Weihe, P., Araki, S., Budtz-Jorgensen, E., and Grandjean, P. (1999). Evoked potentials in Faroese children prenatally exposed to methylmercury. *Neurotoxicol. Teratol.* 471–472.
77. Murata, K., Weihe, P., Budtz-Jorgensen, E., Jorgensen, P. J., and Grandjean, P. (2004). Delayed brainstem auditory evoked potential latencies in 14-year-old children exposed to methylmercury. *J. Pediatr.* **144**, 177–183.
78. Hamada, R., Yoshida, Y., Kuwano, A., Mishima, I., and Igata, A. (1982). Auditory brainstem responses in fetal organic mercury poisoning (in Japanese). *Shinkei-Naika* **16**, 282–285.
79. Nakamura, K., Houzawa, J., and Uemura, T. (1986). Auditory brainstem responses in rats with methylmercury poisoning. *Audiol. Jpn.* **29**, 445–446.
80. Algarin, C., Peirano, P., Garrido, M., Pizarro. F., and Lozoff, B. (2003). Iron deficiency anemia in infancy: Long–lasting effects on auditory and visual system functioning. *Pediatr. Res.* **53**, 217–223.
81. Roncagliolo, M., Garrido, M., Walter, T., Peirano, P., and Lozoff, B. (1998). Evidence of altered central nervous system development in infants with iron deficiency anemia at 6 mo: Delayed maturation of auditory brainstem responses. *Am. J. Clin. Nutr.* **68**, 683–690.

82. Heldin, C. H., Ostman, A., and Ronnstrand, L. (1998). Signal transduction via platelet-derived growth factor receptors. *Biochim. Biophys. Acta.* **1378**, F79–113.

83. Rupprecht, H. D., Sukhatme, V. P., Lacy, J., Sterzel, R. B., and Coleman, D. L. (1993). PDGF-induced Egr-1 expression in rat mesangial cells is mediated through upstream serum response elements. *Am. J. Physiol.* **265**, F351–360.

84. Franke, T. F., Yang, S. I., Chan, T. O., Datta, K., Kazlauskas, A., et al. (1995). The protein kinase encoded by the Akt proto-oncogene is a target of the PDGF-activated phosphatidylinositol 3-kinase. *Cell* **81**, 727–736.

85. Choudhury, G. G. (2001). Akt serine threonine kinase regulates platelet-derived growth factor-induced DNA synthesis in glomerular mesangial cells: Regulation of c-fos AND p27(kip1) gene expression. *J. Biol. Chem.* **276**, 35636–35643.

86. Raff, M. C., Lillien, L. E., Richardson, W. D., Burne, J. F., and Noble, M. D. (1988). Platelet-derived growth factor from astrocytes drives the clock that times oligodendrocyte development in culture. *Nature* **333**, 562–565.

87. Lamballe, F., Klein, R., Barbacid, M. (1991). trkC, a new member of the trk family of tyrosine protein kinases, is a receptor for neurotrophin-3. *Cell* **66**, 967–979.

88. Miyake, S., Mullane-Robinson, K. P., Lill, N. L., Douillard, P., and Band, H. (1999). Cbl-mediated negative regulation of platelet-derived growth factor receptor-dependent cell proliferation. A critical role for Cbl tyrosine kinase-binding domain. *J. Biol. Chem.* **274**, 16619–16628.

89. Miyake, S., Lupher, M. L. J., Druke, B., and Band, H. (1998). The tyrosine kinase regulator Cbl enhances the ubiquitination and degradation of the platelet-derived growth factor receptor alpha. *Proc. Natl. Acad. Sci. USA* **95**, 7927–7932.

90. Duan, L., Miura, Y., Dimri, M., Majumder, B., Dodge, I. L., et al. (2003). Cbl-mediated ubiquitinylation is required for lysosomal sorting of epidermal growth factor receptor but is dispensable for endocytosis. *J. Biol. Chem.* **278**, 28950–28960.

91. Rosenkranz, S., Ikuno, Y., Leong, F. L., Klinghoffer, R. A., Miyake, S., et al. (2000). Src family kinases negatively regulate platelet-derived growth factor alpha receptor-dependent signaling and disease progression. *J. Biol. Chem.* **275**, 9620–9627.

92. Schmidt, M. H. and Dikic, I. (2005). The Cbl interactome and its functions. *Nat. Rev. Mol. Cell. Biol.* **6**, 907–919.

93. Tsygankov, A. Y., Mahajan, S., Fincke, J. E., and Bolen, J. B. (1996). Specific association of tyrosine-phosphorylated c-Cbl with Fyn tyrosine kinase in T cells. *J. Biol. Chem.* **271**, 27130–27137.

94. Hunter, S., Burton, E. A., Wu, S. C., and Anderson, S. M. (1999). Fyn associates with Cbl and phosphorylates tyrosine 731 in Cbl, a binding site for phosphatidylinositol 3-kinase. *J. Biol. Chem.* **274**, 2097–2106.

95. Feshchenko, E. A., Langdon, W. Y., and Tsygankov, A. Y. (1998). Fyn, Yes, and Syk phosphorylation sites in c-Cbl map to the same tyrosine residues that become phosphorylated in activated T cells. *J. Biol. Chem.* **273**, 8223–8331.

96. Kassenbrock, C. K., Hunter, S. F., Garl, P., Johnson, G. L., and Anderson, S. M. (2002). Inhibition of Src family kinases blocks epidermal growth factor (EGF)-induced activation of Akt, phosphorylation of c-Cbl, and ubiquitination of the EGF receptor. *J. Biol. Chem.* **277**, 24967–24975.

97. Abe, J. and Berk, B. C. (1999). Fyn and JAK2 mediate ras activation by reactive oxygen species. *J. Biol. Chem.* **274**, 21003–21010.

98. Abe, J., Okuda, M., Huang, Q., Yoshizumi, M., and Berk, B. C. (2000). Reactive oxygen species activate p90 ribosomal S6 kinase via Fyn and Ras. *J. Biol. Chem.* **275** 1739–1748.

99. Sanguinetti, A. R., Cao, H., and Corley Mastick, C. (2003). Fyn is required for oxidative- and hyperosmotic-stress-induced tyrosine phosphorylation of caveolin-1. *Biochem. J.* **376**, 159–168.

100. Hehner, S. P., Breitfreutz, R., Shubinsky, G., Unsoeld, H., Schulze-Osthoff, K., et al. (2000) Enhancement of T cell receptor

signaling by a mild oxidative shift in the intracellular thiol pool. *J. Immunol.* **165**, 4319–4328.

101. Osterhout, D. J., Wolven, A., Wolf, R. M., Resh, M. D., and Chao, M. V. (1999). Morphological differentiation of oligodendrocytes requires activation of Fyn tyrosine kinase. *J. Cell. Biol.* **145**, 1209–1218.

102. Wolf, R. M., Wilkes, J. J., Chao, M. V., and Resh, M. D. (2001). Tyrosine phosphorylation of p190 RhoGAP by Fyn regulates oligodendrocyte differentiation. *J. Neurobiol.* **49**, 62–78.

103. Poole, B. and Ohkuma, S. (1981). Effect of weak bases on the intralysosomal pH in mouse peritoneal macrophages. *J. Cell. Biol.* **90**, 665–669.

104. Brown, W. J., Goodhouse, J., and Farquhar, M. G. (1986). Mannose-6-phosphate receptors for lysosomal enzymes cycle between the Golgi complex and endosomes. *J. Cell. Biol.* **103**, 1235–1247.

105. Laing, J. G., Tadros, P. N., Green, K., Saffitz, J. E., and Beyer, E. C. (1998). Proteolysis of connexin43-containing gap junctions in normal and heat-stressed cardiac myocytes. *Cardiovasc. Res.* **38**, 711–718.

106. Taher, T. E., Tjin, E. P., Beuling, E. A., Borst, J., Spaargaren, M., et al. (2002). c-Cbl is involved in Met signaling in B cells and mediates hepatocyte growth factor-induced receptor ubiquitination. *J. Immunol.* **169**, 3793–3780.

107. Thien, C. B. and Langdon, W. Y. (2005). Negative regulation of PTK signalling by Cbl proteins. *Growth Factors* **23**, 161–167.

108. van Leeuwen, J. E., Paik, P. K., and Samelson, L. E. (1999). The oncogenic 70Z Cbl mutation blocks the phosphotyrosine binding domain-dependent negative regulation of ZAP-70 by c-Cbl in Jurkat T cells. *Mol. Cell Biol.* **19**, 6652–6664.

109. Deng, W. and Poretz, R. D. (2003). Oliogodendroglia in developmental neurotoxicity. *Neurotoxicol.* **24**, 161–178.

110. Hausburg, M. A., Dekrey, G. K., Salmen, J. J., Palic, M. R., and Gardiner, C. S. (2005). Effects of paraquat on development of preimplantation embryos *in vivo* and *in vitro*. *Reprod. Toxicol.* **20**, 239–246.

111. McCarthy, S., Somayajulu, M., Sikorska, M., Borowy-Borowski, H., and Pandey, S. (2004). Paraquat induces oxidative stress and neuronal death; neuroprotection by water-soluble coenzyme Q10. *Toxicol. Appl. Pharmacol.* **201**, 21–31.

112. Matsuda, S., Gomi, F., Katayama, T., Koyama, Y., Tohyama, M., et al. (2006). Induction of connective tissue growth factor in retinal pigment epithelium cells by oxidative stress. *Jpn. J. Ophthalmol.* **50**, 229–234.

113. Kim, S. J., Kim, J. E., and Moon, I. S. (2004). Paraquat induces apoptosis of cultured rat cortical cells. *Mol. Cells* 17

114. Shimizu, K., Matsubara, K., Ohtaki, K., and Shiono, H. (2003). Paraquat leads to dopaminergic neural vulnerability in organotypic midbrain culture. *Neurosci. Res.* **46**, 523–532.

115. Aruoma, O. I., Halliwell, B., Hoey, B. M., and Butler, J. (1989). The antioxidant action of N-acetylcysteine: Its reaction with hydrogen peroxide, hydroxyl radical, superoxide and hypochlorous acid. *Free Radic Biol. Med.* **6**, 593–597.

116. Meister, A., Anderson, M. E., and Hwang, O. (1986). Intracellular cysteine and glutathione delivery systems. *J. Am. Coll. Nutr.* **5**, 137–151.

117. Hoffer, E., Avidor, I., Benjaminov, O., Shenker, L., Tabak, A., et al. (1993). N-acetylcysteine delays the infiltration of inflammatory cells into the lungs of paraquat-intoxicated rats. *Toxicol. Appl. Pharmacol.* **120**, 8–12.

118. Mayer, M. and Noble, M. (1994). N-acetyl-L-cysteine is a pluripotent protector against cell death and enhancer of trophic factor-mediated cell survival *in vitro*. *Proc. Natl. Acad. Sci. USA* **91**, 7496–7500.

119. Chen, Y. W., Huang, C. F., Tsai, K. S., Yang, R. S., Yen, C. C., et al. (2006). The role of phosphoinositide 3-kinase/Akt signaling in low-dose mercury-induced mouse pancreatic {beta}-cell dysfunction *in vitro* and *in vivo*. *Diabetes* **55**, 1614–16124.

120. Ballatori, N., Lieberman, M. W., and Wang, W. (1998). N-acetylcysteine as an antidote in methylmercury poisoning. *Environ. Health Perspect.* **106**, 267–271.

121. Shanker, G., Syversen, T., and Aschner, M. (2005). Modulatory effect of glutathione status and antioxidants on methylmercury-induced free radical formation in primary cultures of cerebral astrocytes. *Brain Res. Mol. Brain Res.* **137**, 11–22.

122. Nehru, B. and Kanwar, S. S. (2004). N-acetylcysteine exposure on lead-induced lipid peroxidative damage and oxidative defense system in brain regions of rats. *Biol. Trace. Elem. Res.* **101**, 257–264.

123. Neal, R., Copper, K., Gurer, H., and Ercal, N. (1998). Effects of N-acetyl cysteine and 2,3-dimercaptosuccinic acid on lead induced oxidative stress in rat lenses. *Toxicology* **130**, 167–174.

124. Yeh, S. T., Guo, H. R., Su, Y. S., Lin, H. J., Hou, C. C., et al. (2006). Protective effects of N-acetylcysteine treatment post acute paraquat intoxication in rats and in human lung epithelial cells. *Toxicology* **223**, 181–190.

125. Satoh, E., Okada, M., Takadera, T., and Ohyashiki, T. (2005). Glutathione depletion promotes aluminum-mediated cell death of PC12 cells. *Biol. Pharm. Bull.* **28**, 941–946.

126. Tandon, S. K., Singh, S., Prasad, S., Khandekar, K., Dwivedi, V. K., et al. (2003). Reversal of cadmium induced oxidative stress by chelating agent, antioxidant or their combination in rat. *Toxicol. Lett.* **145**, 211–217.

127. Flora, S. J. (1999). Arsenic-induced oxidative stress and its reversibility following combined administration of N-acetylcysteine and meso 2,3-dimercaptosuccinic acid in rats. *Clin. Exp. Pharmacol. Physiol.* **26**, 865–869.

128. Zaragoza, A., Diez-Fernandez, C., Alvarez, A. M., Andres, D., and Cascales, M. (2001). Mitochondrial involvement in cocaine-treated rat hepatocytes: Effect of N-acetylcysteine and deferoxamine. *Br. J. Pharmacol.* **132**, 1063–1070.

129. Roberts, J., Nagasawa, H., Zera, R., Fricke, R., and Goon, D. (1987). Prodrugs of L-cysteine as protective agents against acetaminophen-induced hepatotoxicity. 2-(Polyhydroxyalkyl)- and 2-(polyacetoxyalkyl) thiazolidine-4(R)-carboxylic acids. *J. Med. Chem.* **30**, 1891–1896.

130. Yan, H. and Rivkees, S. A. (2002). Hepatocyte growth factor stimulates the proliferation and migration of oligodendrocyte progenitor cells. *J. Neurosci. Res.* **69**, 597–606.

131. Bottaro, D. P., Rubin, J. S., Faletto, D. L., Chan, A. M., Kmiecik, T. E., et al. (1991). Identification of the hepatocyte growth factor receptor as the c-met proto-oncogene product. *Science* **251**, 802–804.

132. Naldini, L., Vigna, E., Narsimhan, R. P., Gaudino, G., Zarnegar, R., et al. (1991). Hepatocyte growth factor (HGF) stimulates the tyrosine kinase activity of the receptor encoded by the proto-oncogene c-MET. *Oncogene* **6**, 501–504.

133. Knapp, P. E. and Adams, M. H. (2004). Epidermal growth factor promotes oligodendrocyte process formation and regrowth after injury. *Exp. Cell Res.* **296**, 135–144.

134. Levkowitz, G., Klapper, L. N., Tzahar, E., Freywald, A., Sela, M., et al. (1996). Coupling of the c-Cbl protooncogene product to ErbB-1/EGF-receptor but not to other ErbB proteins. *Oncogene* **12**, 1117–1125.

135. Rubin, C., Gur, G., and Yarden, Y. (2005). Negative regulation of receptor tyrosine kinases: Unexpected links to c-Cbl and receptor ubiquitylation. *Cell Res.* **15**, 66–71.

136. de Melker, A. A., van der Horst, G., and Borst, J. (2004). c-Cbl directs EGF receptors into an endocytic pathway that involves the ubiquitin-interacting motif of Eps15. *J. Cell Sci.* **117**, 5001–5012.

137. Ravid, T., Heidinger, J. M., Gee, P., Khan, E. M., and Goldkorn, T. (2004). c-Cbl-mediated ubiquitinylation is required for epidermal growth factor receptor exit from the early endosomes. *J. Biol. Chem.* **279**, 37153–37162.

138. Garcia-Guzman, M., Larsen, E., and Vuori, K. (2000). The proto-oncogene c-Cbl is a positive regulator of Met-induced MAP kinase activation: A role for the adaptor protein Crk. *J. Immunol.* **19**, 4058–4065.

139. Tiffany-Castiglioni, E. (1993). Cell culture models for lead toxicity in neuronal and glial cells. *Neurotoxicol.* **14**, 513–536.

140. Krigman, M. R., Druse, M. J, Traylor, T. D., Wilson, M. H., Newell, L. R., et al. (1974). Lead encephalopathy in the developing rat: Effect on myelination. *J. Neuropathol. Exp. Neurol.* **33**, 58–73.

141. Dabrowska-Bouta, B., Sulkowski. G., Bartosz. G., Walski. M., and Rafalowska. U. (1999). Chronic lead intoxication affects the myelin membrane status in the central nervous system of adult rats. *J. Mol. Neurosci.* **13**, 127–139.

142. Deng, W. and Poretz, R. D. (2001). Chronic dietary lead exposure affects galactolipid metabolic enzymes in the developing rat brain. *Toxicol. Appl. Pharmacol.* **172**, 98–107.

143. Weiss, B., Stern, S., Cox, C., and Balys, M. (2005). Perinatal and lifetime exposure to methylmercury in the mouse: Behavioral effects. *Neurotoxicology* **26**, 675–690.

144. Stern, S., Cox, C., Cernichiari, E., Balys, M., and Weiss, B. (2001). Perinatal and lifetime exposure to methylmercury in the mouse: Blood and brain concentrations of mercury to 26 months of age. *Neurotoxicology* **22**, 467–477.

145. Goulet, S., Dore, F. Y., and Mirault, M. E. (2003). Neurobehavioral changes in mice chronically exposed to methylmercury during fetal and early post-natal development. *Neurotoxicol. Teratol.* **25**, 335–347.

146. Sakamoto, M., Kakita, A., de Oliveira, R. B., Pan, H. S., and Takahashi, H. (2004). Dose-dependent effects of methylmercury administered during neonatal brain spurt in rats. *Dev. Brain. Res.* **152**, 171–176.

147. Barone, S. Jr., Haykal-Coates, N., Parran, D. K., and Tilson, H. A. (1998). Gestational exposure to methylmercury alters the developmental pattern of trk-like immunoreactivity in the rat brain and results in cortical dysmorphology. *Brain. Res. Dev. Brain. Res.* **109**, 13–31.

148. Dietrich, J., Han, R., Yang, Y., Mayer-Pröschel, M., and Noble, M. (2006). CNS progenitor cells and oligodendrocytes are targets of chemotherapeutic agents *in vitro* and *in vivo*. *J. Biol.* **5**, 22.

149. Rowitch, D. H., Lu, R. Q., Kessaris, N., and Richardson, W. D. (2002). An "oligarchy" rules neural development. *Trends Neurosci.* **25**, 417–422.

150. Takebayashi, H., Nabeshima, Y., Yoshida, S., Chisaka, O., Ikenaka, K., et al. (2002). The basic helix-loop-helix factor olig2 is essential for the development of motoneuron and oligodendrocyte lineages. *Curr. Biol.* **12**, 1157–1163.

151. Zhou, Q., Choi, G., and Anderson, D. J. (2001). The bHLH transcription factor Olig2 promotes oligodendrocyte differentiation in collaboration with nkx2.2. *Neuron* **31**, 791–807.

152. Mukouyama, Y. S., Deneen, B., Lukaszewicz, A., Novitch, B. G., Wichterle, H., et al. (2006). $Olig2^+$ neuroepithelial motoneuron progenitors are not multipotent stem cells *in vivo*. *Proc. Natl. Acad. Sci. USA* **103**, 1551–1556.

153. Fancy, S. P., Zhao, C., and Franklin, R. J. (2004). Increased expression of Nkx2.2 and Olig2 identifies reactive oligodendrocyte progenitor cells responding to demyelination in the adult CNS. *Mol. Cell Neurosci.* **27**, 247–254.

154. Talbott, J. F., Loy, D. N., Liu, Y., Qiu, M. S., Bunge, M. B., et al. (2005). Endogenous $Nkx2.2+/Olig2^+$ oligodendrocyte precursor cells fail to remyelinate the demyelinated adult rat spinal cord in the absence of astrocytes. *Exp. Neurol.* **192**, 11–24.

155. Homolya, L., Varadi, A., and Sarkadi, B. (2003). Multidrug resistance-associated proteins: Export pumps for conjugates with glutathione, glucuronate or sulfate. *Biofactors* **17**, 103–114.

156. Leslie, E. M., Deeley, R. G., and Cole, S. P. (2001). Toxicological relevance of the multidrug resistance protein 1, MRP1 (ABCC1) and related transporters. *Toxicology* **167**, 3–23.

157. Liang, X., Draghi, N. A., and Resh, M. D. (2004). Signaling from integrins to Fyn to Rho Family GTPases regulates morphologic differentiation of oligodendrocytes. *J. Neurosci.* **24**, 7140–7149.

158. Tsatmali, M., Walcott, E. C., and Crossin, K. L. (2005). Newborn neurons acquire high levels of reactive oxygen species and increased mitochondrial proteins upon

differentiation from progenitors. *Brain Res.* **1040**, 137–150.

159. Goldsmit. Y., Erlich, S., and Pinkas-Kramarski, R. (2001). Neuregulin induces sustained reactive oxygen species generation to mediate neuronal differentiation. *Cell Mol. Neurobiol.* **211**, 753–769.

160. Puceat, M. (2005). Role of Rac-GTPase and reactive oxygen species in cardiac differentiation of stem cells. *Antioxid Redox Signal* **7**, 1435–1439.

161. McGrath, S. A. (1998). Induction of p21WAF/CIP1 during hyperoxia. *Am. J. Respir. Cell Mol. Biol.* **18**, 179–187.

162. McGrath-Morrow, S. A., Cho, C., Soutiere, S., Mitzner, W., and Tuder, R. (2004). The effect of neonatal hyperoxia on the lung of p21Waf1/Cip1/Sdi1-deficient mice. *Am. J. Respir. Cell Mol. Biol.* **30**, 635–640.

163. Seomun, Y., Kim, J. T., Kim, H. S., Park, J. Y., Joo, and C. K. (2005). Induction of p21Cip1-mediated G2/M arrest in H2O2-treated lens epithelial cells. *Mol. Vis.* **11**, 764–774.

164. Esposito, F., Russo, L., Chirico, G., Ammendola, R., Russo, T., et al. (2001). Regulation of p21waf1/cip1 expression by intracellular redox conditions. *IUBMB Life* **52**, 67–70.

165. Barnouin, K., Dubuisson, M. L., Child, E. S., Fernandez De Mattos, S., Glassford, J., et al. (2002). H2O2 induces a transient multi-phase cell cycle arrest in mouse fibroblasts through modulating cyclin D and p21Cip1 expression. *J. Biol. Chem.* **277**, 13761–13770.

166. Hu, Y., Wang, X., Zeng, L., Cai, D. Y., Sabapathy, K., et al. (2005). ERK phosphorylates p66shcA on Ser36 and subsequently regulates p27kip1 expression via the Akt-FOXO3a pathway: Implication of p27kip1 in cell response to oxidative stress. *Mol. Biol. Cell* **16**, 3705–3718.

167. WHO (1990). Environmental health criteria 101: *Methylmercury*. World Health Organization, Geneva. Retrieved from [http://www.inchem.org/documents/ehc/ehc/ehc101.htm]. Accessed December 15, 2006.

168. Cernichiari, E., Brewer, R., Myers, G. J., Marsh, D. O., Lapham, L. W., et al. (1995). Monitoring methylmercury during pregnancy: Maternal hair predicts fetal brain exposure. *Neurotoxicology* **16**, 705–710.

169. Goyer, R. A. (1993). Lead toxicity: Current concerns. *Environ. Health Perspect* **100**, 177–187.

170. Banks, E. C., Ferretti, L. E., and Shucard, D. W. (1997). Effects of low level lead exposure on cognitive function in children: A review of behavioral, neuropsychological and biological evidence. *Neurotoxicology* **18**, 237–282.

171. Lidsky, T. I. and Schneider, J. S. (2003). Lead neurotoxicity in children: Basic mechanisms and clinical correlates. *Brain* **126**, 5–19.

172. Needleman, H. L. and Gatsonis, C. A. (1990). Low level lead exposure and the IQ of children. A meta-analysis of modern studies. *JAMA* **263**, 673–678.

173. Finkelstein, Y., Markowitz, M. E., and Rosen, J. F. (1998). Low-level lead-induced neurotoxicity in children: An update on central nervous system effects. *Brain Res. Brain Res. Rev.* **27**, 168–176.

174. Ballinger, D., Leviton, A., Waternoux, C., Needleman, H., and Rabinowitz, P. (1987). Longitudinal analysis of prenatal and postnatal lead exposure and early cognitive development. *N. Engl. J. Med.* **316**, 1037–1043.

175. Winneke, G., Brockhaus, A., Ewers, U., Krämer, U., and Neuf, M. (1990). Results from the European Multicenter Study on lead neurotoxicity in children: Implications for risk assessment. *Neurotox. Teratol.* **12**, 553–559.

176. Bellinger, D. and Needleman, H. L. (1992). Neurodevelopmental effects of low-level lead exposure in children. In *Human Lead Exposure*. H. Needleman (Ed.). CRC Press, Boca Raton (Florida), pp. 191–208.

177. WHO (1995) Environmental health criteria 165: Inorganic lead. Geneva: World Health Organization. Available: [http://www.inchem.org/documents/ehc/ehc/ehc165.htm.] Accessed December 15, 2006.

178. Kaiser, R., Henderson, A. K., Daley, W. R., Naughton, M., Khan, M. H., et al. (2001). Blood lead levels of primary school children

in Dhaka, Bangladesh. *Environ. Health Perspect* **109**, 563–566.

179. Yakovlev, A. Y., Boucher, K., Mayer-Pröschel, M., and Noble, M. (1998). Quantitative insight into proliferation and differentiation of O-2A progenitor cells *in vitro*: The clock model revisited. *Proc. Natl. Acad. Sci. USA* **95**, 14164–14167.

180. Hyrien, O., Mayer-Proschel, M., Noble, M., and Yakovlev, A. (2005). Estimating the life-span of oligodendrocytes from clonal data on their development in cell culture. *Math. Biosci.* **193**, 255–274.

181. Hyrien, O., Mayer-Proschel, M., Noble, M., and Yakovlev, A. (2005). A stochastic model to analyze clonal data on multi-type cell populations. *Biometrics* **61**, 199–207.

182. Tamm, C., Duckworth, J., Hermanson, O., and Ceccatelli, S. (2006). High susceptibility of neural stem cells to methylmercury toxicity: Effects on cell survival and neuronal differentiation. *J. Neurochem.* **97**, 69–78.

183. Rothenberg, S. J., Poblano, A., and Schnaas, L. (2000). Brainstem auditory evoked response at five years and prenatal and postnatal blood lead. *Neurotoxicol. Teratol.* **22**, 503–510.

184. Bleecker, M. L., Ford, D. P., Lindgren, K. N., Scheetz, K., and Tiburzi, M. J. (2003). Association of chronic and current measures of lead exposure with different components of brainstem auditory evoked potentials. *Neurotoxicology* **24**, 625–631.

185. Lester, B. M., Lagasse, L., Seifer, R., Tronick, E. Z., Bauer, C. R., et al. (2003). The Maternal Lifestyle Study (MLS): Effects of prenatal cocaine and/or opiate exposure on auditory brain response at one month. *J. Pediatr.* **142**, 279–285.

186. Tan-Laxa, M. A., Sison-Switala, C., Rintelman, W., and Ostrea, E. M. J. (2004). Abnormal auditory brainstem response among infants with prenatal cocaine exposure. *Pediatrics* **113**, 357–360.

187. Poblano, A., Belmont, A., Sosa, J., Ibarra, J., Rosas, Y., et al. (2002). Effects of prenatal exposure to carbamazepine on brainstem auditory evoked potentials in infants of epileptic mothers. *J. Child Neurol.* **17**, 364–368.

188. Fruttiger, M., Karlsson, L., Hall, A., Abramsson, A., Calver, A., et al. (1999). Defective oligodendrocyte development and severe hypomyelination in PDGF-A knockout mice. *Development* **126**, 457–467.

189. Hoch, R. V. and Soriano, P. (2003). Roles of PDGF in animal development. *Development* **130**, 4769–4784.

190. Betsholtz, C. (2004). Insight into the physiological functions of PDGF through genetic studies in mice. *Cytokine. Growth Factor Rev.* **15**, 215–228.

191. Wong, R. W. C. and Guillaud, L. (2004). The role of epidermal growth factor and its receptors in mammalian CNS. *Cytokine Growth Factor Rev.* **15**, 147–156.

192. Xian, C. J. and Zhou, X. F. (2004). EGF family of growth factors: Essential roles and functional redundancy in the nerve system. *Front. Biosci.* **9**, 85–92.

193. Holbro, T. and Hynes, N. E. (2004). ErbB receptors: Directing key signaling networks throughout life. *Annu. Rev. Pharmacol. Toxicol.* **44**, 195–217.

194. Gutierrez, H., Dolcet, C., Tolcos, M., and Davies, A. (2004). HGF regulates the development of cortical pyramidal dendrites. *Development* **131**, 3717–3726.

195. Birchmeier, C. and Gherardi, E. (1998). Developmental role of HGF/SF and its receptor, the c-Met tyrosine kinase. *Trends Cell Biol.* **8**, 404–410.

196. Morita, A., Yamashita, N., Sasaki, Y., Uchida, Y., Nakajima, O., et al. (2006). Regulation of dendritic branching and spine maturation by semaphorin3A-Fyn signaling. *J. Neurosci.* **26**, 2971–2980.

197. He, J., Nixon, K., Shetty, A. K., and Crews, F. T. (2005). Chronic alcohol exposure reduces hippocampal neurogenesis and dendritic growth of newborn neurons. *Eur. J. Neurosci.* **21**, 2711–2720.

198. Newey, S. E., Velamoor, V., Govek, E. E., and Van Aeist, L. (2005). Rho GTPases, dendritic structure, and mental retardation. *J. Neurobiol.* **64**, 58–74.

199. Power, J., Mayer-Proschel, M., Smith, J., and Noble, M. (2002). Oligodendrocyte precursor cells from different brain regions express divergent properties consistent with

the differing time courses of myelination in these regions. *Dev. Biol.* **245**, 362–375.

200. Sakamoto, M., Kakita, A., Wakabayashi, K., Takahashi, H., Nakano, A., et al. (2002). Evaluation of changes in methylmercury accumulation in the developing rat brain and its effects: A study with consecutive and moderate dose exposure throughout gestation and lactation periods. *Brain Res.* **949**, 51–59.

6

1. Garner, C. E., Jefferson, W. N., Burka, L. T., Matthews, H. B., and Newbold, R. R. (1999). *In vitro* estrogenicity of the catechol metabolites of selected polychlorinated biphenyls. *Toxicol. Appl. Pharmacol.* **154**, 188–197.
2. Moore, M., Mustain, M., Daniel, K., Chen, I., Safe, S., Zacharewski, T., Gillesby, B., Joyeux, A., and Balaguer, P. (1997). Antiestrogenic activity of hydroxylated polychlorinated biphenyl congeners identified in human serum. *Toxicol. Appl. Pharmacol.* **142**, 160–168.
3. Salama, J., Chakraborty, T. R., Ng, L., and Gore, A. C. (2003). Effects of polychlorinated biphenyls on estrogen receptor-beta expression in the anteroventral periventricular nucleus. *Environ. Health Perspect.* **111**, 1278–1282.
4. United States Environmental Protection Agency, Office of Water Update (2003). National Listing of Fish and Wildlife Advisories. EPA-823-F-03-003.
5. Fernandez, M. A., Gomara, B., Bordajandi, L. R., Herrero, L., Abad, E., Abalos, M., Rivera, J., and Gonzalez, M. J. (2004). Dietary intakes of polychlorinated dibenzo-p-dioxins, dibenzofurans and dioxin-like polychlorinated biphenyls in Spain. *Food Addit. Contam.* **21**, 983–991.
6. Smith, A. G. and Gangolli, S. D. (2002). Organochlorine chemicals in seafood: Occurrence and health concerns. *Food Chem. Toxicol.* **40**, 767–779.
7. Antunes, P. and Gil, O. (2004). PCB and DDT contamination in cultivated and wild Sea bass from Ria de Aveiro, Portugal. *Chemosphere* **54**, 1503–1507.
8. Svensson, B. G., Hallberg, T., Schultz, A., and Hagmar, L. (1994). Parameters of immunological competence in subjects with high consumption of fish contaminated with persistent organochlorine compounds. *Arch. Occup. Environ. Health* **65**, 351–358.
9. Faroon, O., Keith, M. S., Jones, D., and de Rosa, C. (2001). Effects of polychlorinated biphenyls on development and reproduction. *Toxicol. Ind. Health* **17**, 63–93.
10. Brouwer, A., Longnecker, M. P., Birnbaum, L. S., Cogliano, J., Kostyniak, P., Moore, J., Schantz, S., and Winneke, G. (1999). Characterization of potential endocrine-related health effects at low dose levels of exposure to PCBs. *Environ. Health Perspect.* **107**(Suppl. 4), 639–649.
11. Judd, N., Griffith, W. C., and Faustman, E. M. (2004). Contribution of PCB exposure from fish consumption to total dioxin-like dietary exposure. *Regul. Toxicol. Pharmacol.* **40**, 125–135.
12. Schantz, S. L., Gasior, D. M., Polverejan, E., McCaffrey, R. J., Sweeney, A. M., Humphrey, H. E., and Gardiner, J. C. (2001). Impairments of memory and learning in older adults exposed to polychlorinated biphenyls via consumption of Great Lakes fish. *Environ. Health Perspect.* **109**, 605–611.
13. Istituto Superiore di Sanità (2002). Istisan Report 02/38 ISSN 1123-3117, ISS, Roma.
14. Liu, J. W. and Picard, D. (1998). Bioactive steroids as contaminants of the common carbon source galactose. *FEMS Microbiol. Lett.* **159**, 167–171.
15. Liu, J. W., Jeannin, E., and Picard, D. (1999). The anti-estrogen hydroxytamoxifen is a potent antagonist in a novel yeast system. *Biol. Chem.* **380**, 1341–1345.
16. Pinto, B., Picard, D., and Reali, D. (2004). A recombinant yeast strain as a short term bioassay to assess estrogen-like activity of xenobiotics. *Ann. Ig* **16**, 579–585.
17. Pinto, B., Garritano, S., and Reali, D. (2005). Occurrence of estrogen-like substances in the marine environment of the Northern Mediterranean Sea. *Mar. Poll. Bull.* **50**, 1681–1685.
18. Miller, J. H. (1972). *Experiments in Molecular Genetics*. Cold Spring Harbour Laboratory Press, New York.

19. Gong, Y., Chin, H. S., Lim, L. S., Loy, C. J., Obbard, J. P., and Yong, E. L. (2003). Clustering of sex hormone disruptors in Singapore's marine environment. *Environ. Health Perspect.* **111**, 1448–1453.

20. Jackson, J. E. (1991). *A User's Guide to Principal Components.* John Wiley & Sons, Inc, New York, NY.

21. Jolliffe, I. T. (1986). *Principal Components Analysis*. Springer-Verlag, New York Inc., NY.

22. Martens, H. and Naes, T. (1989). *Multivariate Calibration.* John Wiley & Sons, Chichester, UK.

23. Storelli, M. M., Giacominelli-Stuffler, R., D'Addabbo, R., and Marcotrigiano, G. O. (2003). Health risk of coplanar polychlorinated biphenyl congeners in edible fish from the Mediterranean Sea. *J. Food Prot.* **66**, 2176–2179.

24. Perugini, M., Cavaliere, M., Giammarino, A., Mazzone, P., Olivieri, V., and Amorena, M. (2004). Levels of polychlorinated biphenyls and organochlorine pesticides in some edible marine organisms from the Central Adriatic Sea. *Chemosphere* **57**, 391–400.

25. Layton, A. C., Sanseverino, J., Gregory, B. W., Easter, J. P., Sayler, G. S., and Schultz, T. W. (2002). *In vitro* estrogen receptor binding of PCBs: Measured activity and detection of hydroxylated metabolites in a recombinant yeast assay. *Toxicol. Appl. Pharmacol.* **180**, 157–163.

26. Bonefeld-Jørgensen, E. C., Andersen, H. R., Rasmussen, T. H., and Vinggaard, A. M. (2001). Effect of highly bioaccumulated polychlorinated biphenyl congeners on estrogen and androgen receptor activity. *Toxicology* **158**, 141–153.

27. Rivas, A., Fernandez, M. F., Cerrillo, I., Ibarluzea, J., Olea-Serrano, M. F., Pedraza, V., and Olea, N. (2001). Human exposure to endocrine disrupters: Standardisation of a marker of estrogenic exposure in adipose tissue. *APMIS* **109**, 185–197.

28. Miao, X. -S., Swenson, C., Woodward, L. A., and Li, Q. X. (2000). Distribution of polychlorinated biphenyls in marine species from French Frigate Shoals, North Pacific Ocean. *Sci. Tot. Environ.* **257**, 17–28.

29. Bayarri, S., Baldassarri, L. T., Iacovella, N., Ferrara, F., and di Domenico, A. (2001). PCDDs, PCDFs, PCBs and DDE in edible marine species from the Adriatic Sea. *Chemosphere* **43**, 601–610.

30. Llobet, J. M., Bocio, A., Domingo, J. L., Teixido, A., Casas, C., and Muller, L. (2003). Levels of polychlorinated biphenyls in food from Catalonia, Spain: Estimated dietary intake. *J. Food Prot.* **66**, 479–484.

31. Oberdörster, E. and Cheek, A. O. (2000). Gender benders at the beach: Endocrine disruption in marine and estuarine organisms. *Environ. Toxicol. Chem.* **20**, 23–36.

32. De Metrio, G., Corriero, A., Desantis, S., Zubani, D., Cirillo, F., Deflorio, M., Bridges, C. R., Eicker, J., de la Serna, J. M., Megalofonou, P., and Kime, D. E. (2003). Evidence of a high percentage of intersex in the Mediterranean swordfish (*Xiphias gladius* L.). *Mar. Poll. Bull.* **46**, 358–361.

33. Jobling, S., Casey, D., Rogers-Gray, T., Oehlmann, J., Schulte-Oehlmann, U., Pawlowski, S., Baunbeck, T., Turner, A. P., and Tyler, C. R. (2004). Comparative responses of molluscs and fish to environmental estrogens and an estrogenic effluent. *Aquat. Toxicol.* **66**, 207–222.

34. Nash, J. P., Kime, D. E., Van der Ven, L. T., Wester, P. W., Brion. F., Maack, G., Stahlschmidt-Allner, P., and Tyler, C. R. (2004). Long-term exposure to environmental concentrations of the pharmaceutical ethynylestradiol causes reproductive failure in fish. *Environ. Health Perspect.* **112**, 1725–1733.

35. Tyler, C. R., Spary, C., Gibson, R., Santos, E. M., Shears, J., and Hill, E. M. (2005). Accounting for differences in estrogenic responses in rainbow trout (*Oncorhynchus mykiss*: salmonidae) and roach (*Rutilus rutilus*: Cyprinidae) exposed to effluents from wastewater treatment works. *Environ. Sci. Technol.* **39**, 2599–2607.

36. Grimvall, E., Rylander, L., Nilsson-Ehle, P., Nilsson, U., Strömberg, U., Hagmar, L., and Östman, C. (1997). Monitoring of polychlorinated biphenyls in human blood plasma: Methodological developments and influence of age, lactation, and fish consumption. *Arch. Environ. Contam. Toxicol.* **32**, 329–336.

37. Grandjean, P., Weihe, P., Burse, V. W., Needham, L. L., Storr-Hansen, E., Heinzow, B., Debes, F., Murata, K., Simonsen, H., Ellefsen, P., Budtz-Jorgensen, E., Keiding, N., and White, R. F. (2001). Neurobehavioural deficits associated with PCB in 7-year-old children prenatally exposed to seafood neurotoxicants. *Neurotoxicol. Teratol.* **23**, 305–317.

38. Fossi, M. C., Casini, S., Marsili, L., Neri, G., Mori, G., Ancora, S., Moscatelli, A., Ausili, A., and Notarbartolo-di-Sciara, G. (2002). Biomarkers for endocrine disruptors in three species of Mediterranean large pelagic fish. *Marine Environ. Res.* **54**, 667–671.

39. Dewailly, E. and Weihe, P. (2003). The effect of Arctic pollution on population health. [http://www.amap.no] AMAP 2003. *AMAP Assessment 2002: Human health in the Arctic, Chapter 9* Arctic Monitoring and Assessment Programme (AMAP), Oslo, Norway, xiv+137. HH_CO9.pdf ISBN 82-7971-016-7.

40. Bonefeld-Jørgensen, E. C. and Ayotte, P. (2003). Toxicological properties of persistent organic pollutants and related health effects of concern for the Arctic populations. [http://www.amap.no] *AMAP 2003. AMAP Assessment 2002: Human health in the Arctic, Chapter 6* Arctic Monitoring and Assessment Programme (AMAP), Oslo, Norway, xiv+137. HH_CO6.pdf ISBN 82-7971-016-7.

41. Dallinga, J. W., Moonen, E. J., Dumoulin, J. C., Evers, J. L., Geraedts, J. P., and Kleinjans, J. C. (2002). Decreased human semen quality and organochlorine compounds in blood. *Hum. Reprod.* **17**, 1973–1979.

7

1. International Boundary and Water Commission (1998). *Second phase of the binational study regarding the presence of toxic substances in the Rio Grande/Rio Bravo and its tributaries along the boundary portion between the United States and Mexico.* Final Report, United States and Mexico, I.

2. Texas Natural Resource Conservation Commission, Watershed Management Division (1994). *Regional assessment of water quality in the Rio Grande Basin. Austin, TX.*

3. Singh, A. (1992) Detection methods for waterborne pathogens. In *Environmental Microbiology*, R. Mitchell (Ed.). A John Wiley & Sons, Inc, New York, pp. 125–156.

4. Craun, G. F., Berger, P. S., and Calderon, R. L. (1997). Coliform bacteria and waterborne disease outbreaks. *J. Am. Water Works Ass.* **89**(3), 96–104.

5. International Boundary and Water Commission (1997). *Second phase of the binational study regarding the presence of toxic substances in the Rio Grande/Rio Bravo and its tributaries along the boundary portion between the United States and Mexico. Final Report. United States and Mexico*, II.

6. Owen, R. J. (1995). Bacteriology of Helicobacter pylori. *Baillieres Clin. Gastroenterol.* **9**(3), 415–446.

7. Peura, D. A. (1997). The report of the international update conference on Helicobacter pylori. *Digestive Disease Week.* Washington, DC May 14, 1997.

8. Megraud, F. (1995). Transmission of Helicobacter pylori: Fecal-oral versus oraloral route. *Aliment. Pharmacol. Ther.* **9**(2), 85–91.

9. Hulten, K., Han, S. W., Enroth, H., Klein, P. D., Opejun, A. R., Gilman, R. H., Evans, D. G., Engstrand, L., Graham, D. Y., and El-Zaatari, F. A. (1996). Helicobacter pylori in the drinking water in Peru. *Gastroenterology* **110**(4), 1031–1035.

10. Redlinger, T., O'Rourke, K., and Goodman, K. J. (1999). Age distribution of Helicobacter pylori seroprevalence among young children in a United States/Mexico border community: Evidence for transitory infection. *Am. J. Epidemiol.* **150**, 225–230.

11. Rozak, D. B. and Colwell, R. R. (1987). Survival strategies of bacteria in the natural environment. *Microbiol. Rev.* **51**, 365–379.

12. Shahamat, M., u Mai, Paszko-Kova, C., Sessel, M., and Colwell, R. R. (1993). Use of autoradiography to assess viability of Helicobacter pylori in water. *Appl. Environ. Microbiol.* **59**(4), 1231–1235.

13. Alvarez, M. E., Aguilar, M., Fountain, A., Gonzalez, N.,. Rascon, O., and Saenz, D.

(2000). Inactivation of MS2 phage and poliovirus in groundwater. *Can. J. Microbiol.* **46**, 159–165.

14. Botsford, J. L. (1998). A simple assay for toxic chemicals using a bacterial indicator. *World J. Microb. Biot.* **14**, 369–376.
15. American Public Health Association (1995). Standard Methods for the Examination of Water and Wastewater 19 edition, Washington DC.
16. Elmund, G. K., Allen, M. J., and Rice, E. W. (1999). Comparison of *Escherichia coli*, total coliform, and fecal coliform populations as indicators of wastewater treatment efficiency. *Water Environ. Res.* **71**(3), 332–339.

8

1. Berg, G., Knaape, C., Ballin, G., and Seidel, D. (1994). Biological control of Verticillium dahliae KLEB by naturally occurring rhizosphere bacteria. *Arch. Phytopathol. Dis. Prot.* **29**, 249–262.
2. Debette, J. and Blondeau, R. (1980). Presence of *Pseudomonas maltophilia* in the rhizosphere of several cultivated plants. *Can. J. Microbiol.* **26**, 460–463.
3. Heuer, H. and Smalla, K. (1999). Bacterial phyllosphere communities of Solanum tuberosum L and T4-lysozyme producing genetic variants. *FEMS Microbiol. Ecol.* **28**, 357–371.
4. Lambert, B. and Joos, H. (1989). Fundamental aspects of rhizobacterial plant growth promotion research. *Trends Biotechnol.* **7**, 215–219.
5. Whipps, J. (2001). Microbial interactions and biocontrol in the rhizosphere. *J. Exp. Bot.* **52**, 487–511.
6. Binks, P. R., Nicklin, S., and Bruce, N. C. (1995). Degradation of RDX by *Stenotrophomonas maltophilia* PB1. *Appl. Environ. Microbiol.* **61**, 1813–1822.
7. Lee, E. Y., Jun, Y. S., Cho, K. S., and Ryu, H. W. (2002). Degradation characteristics of toluene, benzene, ethylbenzene, and xylene by *Stenotrophomonas maltophilia* T3-c. *J. Air Waste Manag. Assoc.* **52**, 400–406.
8. Juhasz, A. L., Stanley, G. A., and Britz, M. L. (2000). Microbial degradation and detoxification of high molecular weight polycyclic aromatic hydrocarbons by *Stenotrophomonas maltophilia* strain VUN 10,003. *Lett. Appl. Microbiol.* **30**, 396–401.
9. Quinn, J. P. (1998). Clinical problems posed by multiresistant nonfermenting gram-negative pathogens. *Clin. Infect. Dis.* **27**, 117–124.
10. Valdezate, S., Vindel, A., Loza, E., Baquero, F., and Canton, R. (2001). Antimicrobial susceptibilities of unique *Stenotrophomonas maltophilia* clinical strains. *Antimicrob. Agents Chemother.* **45**, 1581–1584.
11. Dignani, M. C., Grazziutti, M., and Anaissie, E. (2003). *Stenotrophomonas maltophilia* infections. *Semin. Respir. Crit. Care Med.* **24**, 89–98.
12. Yamazaki, E., Ishii, J., Sato, K., and Nakae, T. (1989). The barrier function of the outer membrane of Pseudomonas maltophilia in the diffusion of saccharides and beta-lactam antibiotics. *FEMS Microbiol. Lett.* **51**, 85–88.
13. Li, X. Z., Zhang, L., and Poole, K. (2002). SmeC, an outer membrane multidrug efflux protein of Stenotrophomonas maltophilia. *Antimicrob. Agents Chemother.* **46**, 333–343.
14. Alonso, A. and Martinez, J. L. (2000). Cloning and characterization of SmeDEF, a novel multidrug efflux pump from Stenotrophomonas maltophilia. *Antimicrob. Agents Chemother.* **44**, 3079–3086.
15. Zhang, L., Li, X. Z., and Poole, K. (2001). SmeDEF multidrug efflux pump contributes to intrinsic multidrug resistance in Stenotrophomonas maltophilia. *Antimicrob. Agents Chemother.* **45**, 3497–3503.
16. Berg, G., Eberl, L., and Hartmann, A. (2005). The rhizosphere as a reservoir for opportunistic human pathogenic bacteria. *Environ. Microbiol.* **7**, 1673–85.
17. Knudsen, G. R., Walter, M. V., Porteous, L. A., Prince, V. J., Amstrong, J. L., et al. (1988). Predictive model of conjugated plasmid transfer in the rhizosphere and phyllosphere. *Appl. Environ. Microbiol.* **54**, 343–347.
18. Dungan, R. S., Yates, S. R., and Frankenberger, W. T. Jr. (2003). Transformations of selenate and selenite by *Stenotrophomonas maltophilia* isolated from a seleniferous

agricultural drainage pond sediment. *Environ. Microbiol.* **5**, 287–295.

19. Sauge-Merle, S., Cuine, S., Carrier, P., Lecomte-Pradines, C., Luu, D. T., et al. (2003). Enhanced toxic metal accumulation in engineered bacterial cells expressing Arabidopsis thaliana phytochelatin synthase. *Appl. Environ. Microbiol.* **69**, 490–494.
20. Alonso, A., Sanchez, P., and Martinez, J. L. (2000). *Stenotrophomonas maltophilia* D457R contains a cluster of genes from gram-positive bacteria involved in antibiotic and heavy metal resistance. *Antimicrob. Agents Chemother.* **44**, 1778–1782.
21. Fauchon, M., Lagniel, G., Aude, J. C., Lombardia, L., Soularue, P., et al. (2002). Sulfur sparing in the yeast proteome in response to sulfur demand. *Mol. Cell* **9**, 713–723.
22. Park, S. and Imlay, J. A. (2003). High levels of intracellular cysteine promote oxidative DNA damage by driving the Fenton reaction. *J. Bacteriol.* **185**, 1942–50.
23. Holmes, J. D., Richardson, D. J., Saed, S., Evans-Gowing, R., Russell, D. A., et al. (1997). Cadmium-specific formation of metal sulfide "Q-particles" by Klebsiella pneumoniae. *Microbiology* **143**, 2521–2530.
24. Sharma, P. K., Balkwill, D. L., Frenkel, A., and Vairavamurthy, M. A. (2000). A new Klebsiella planticola strain (Cd-1) grows anaerobically at high cadmium concentrations and precipitates cadmium sulfide. *Appl. Environ. Microbiol.* **66**, 3083–3087.
25. Kredich, N. M., Foote, L. J., and Keenan, B. S. (1973). The stoichiometry and kinetics of the inducible cysteine desulfhydrase from Salmonella typhimurium. *J. Biol. Chem.* **218**, 6187–6196.
26. Wang, C. L., Lum, A. M., Ozuna, S. C., Clark, D. S., and Keasling, J. D. (2001). Aerobic sulfide production and cadmium precipitation by Escherichia coli expressing the Treponema denticola cysteine desulfhydrase gene. *Appl. Microbiol. Biotechnol.* **56**, 425–430.
27. Pagès, D., Sanchez, L., Conrod, S., Gidrol, X., Fekete, A., et al. (2007). Exploration of intraclonal strategies of Pseudomonas brassicacearum facing Cd toxicity. *Environ. Microbiol.* **9**, 2820–2835.
28. Rijstenbil, J. W. and Wijnholds, J. A. (1996). HPLC analysis of nonprotein thiols in planktonic diatoms: Pool size, redox state and response to copper and cadmium exposure. *Mar. Biol.* **127**, 45–54.
29. Michalowicz, A. (1991). *Logiciels pour la Chimie*. Société Francaise de Chimie, Paris.
30. Teo, B. K. (1986). *Inorganic Chemistry Concepts*. Springer-Verlag, Berlin.
31. Rehr, J. J. and Albers, R. C. (1990). Scattering-matrix formulation of curved-wave multiple-scattering theory: Application to x-ray-absorption fine structure. *Phys. Rev. B Condens. Matter* **41**, 8139–8149.
32. Zabinsky, S. I., Rehr, J. J., Ankudinov, A., Albers, R. C., and Eller, M. J. (1995). Multiple-scattering calculations of x-ray-absorption spectra. *Phys. Rev. B* **52**, 2995–3009.

9

1. Berenbaum, M. R. (1995). The chemistry of defense—Theory and practice. *Proc. Natl. Acad. Sci. USA* **92**, 2–8.
2. Hay, M. E. and Fenical, W. (1988). Marine plant-herbivore interactions: The ecology of chemical defense. *Ann. Rev. Ecol. Syst.* **19**, 111–145.
3. Pawlik, J. R. (1993). Marine invertebrate chemical defenses. *Chem. Rev.* **93**, 1911–1922.
4. Rosenthal, G. A. and Janzen, D. H. (1979). *Herbivores: Their Interaction with Secondary Plant Metabolites*. Academic Press, Orlando, Florida.
5. Jensen, P. R. and Fenical, W. (1994). Strategies for the discovery of secondary metabolites from marine bacteria: Ecological perspectives. *Ann. Rev. Microbiol.* **48**, 559–584.
6. Costerton, J. W., Stewart, P. S., and Greenberg, E. P. (1999). Bacterial biofilms: A common cause of persistent infections. *Science* **284**, 1318–1322.
7. Hall-Stoodley, L., Costerton, J. W., and Stoodley, P. (2004). Bacterial biofilms: From the natural environment to infectious diseases. *Nat. Rev. Microbiol.* **2**, 95–108.
8. Darby, C., Hsu, J. W., Ghori, N., and Falkow, S. (2002). *Caenorhabditis elegans*:

Plague bacteria biofilm blocks food intake. *Nature* **417**, 243–244.

9. Matz, C. and Kjelleberg, S. (2005). Off the hook—How bacteria survive protozoan grazing. *Trends Microbiol.* **13**, 302–307.
10. Fenchel, T. (1987). *Ecology of Protozoa: The Biology of Free-living Phagotrophic Protists.* Science Tech Publishers, Madison, WI.
11. Sherr, E. B. and Sherr, B. F. (2002). Significance of predation by protists in aquatic microbial food webs. *Antonie Leeuwenhoek* **81**, 293–308.
12. Parry, J. D. (2004). Protozoan grazing of freshwater biofilms. *Adv. Appl. Microbiol.* **54**, 167–196.
13. Fenchel, T. and Blackburn, N. (1999). Motile chemosensory behaviour of phagotrophic protists: Mechanisms for and efficiency in congregating at food patches. *Protist* **150**, 325–336.
14. Kiørboe, T., Tang, K., Grossart, H. P., and Ploug, H. (2003). Dynamics of microbial communities on marine snow aggregates: Colonization, growth, detachment, and grazing mortality of attached bacteria. *Appl. Environ. Microbiol.* **69**, 3036–3047.
15. Jürgens, K. and Matz, C. (2002) Predation as a shaping force for the phenotypic and genotypic composition of planktonic bacteria. *Antonie Leeuwenhoek* **81**, 413–434.
16. Chrzanowski, T. H. and Šimek, K. (1990). Prey-size selection by freshwater flagellated protozoa. *Limnol. Oceanogr.* **35**, 1429–1436.
17. Hahn, M. W., Moore, E. R. B, and Höfle, M. G. (1999). Bacterial filament formation, a defense mechanism against flagellate grazing, is growth rate controlled in bacteria of different phyla. *Appl. Environ. Microbiol.* **65**, 25–35.
18. Pernthaler, J., Sattler, B., Šimek, K., Schwarzenbacher, A., and Psenner, R. (1996). Top-down effects on the size-biomass distribution of a freshwater bacterioplankton community. *Aquat. Microb. Ecol.* **10**, 255–263.
19. Matz, C. and Jürgens, K. (2005). High motilityreduces grazing mortality of planktonic bacteria. *Appl. Environ. Microbiol.* **71**, 921–929.
20. Hay, M. E. and Kubanek, J. (2002). Community and ecosystem level consequences of chemical cues in the plankton. *J. Chem. Ecol.* **28**, 2001–2016.
21. Egan, S., Thomas, T., Holmström, C., and Kjelleberg, S. (2000). Phylogenetic relationship and antifouling activity of bacterial epiphytes from the marine alga *Ulva lactuca. Environ. Microbiol.* **2**, 343–347.
22. Burmølle, M., Webb, J. S., Rao, D., Hansen, L. H., Sørensen, S. J., et al. (2006). Enhanced biofilm formation and increased resistance to antimicrobial agents and bacterial invasion are caused by synergistic interactions in multispecies biofilms. *Appl. Environ. Microbiol.* **72**, 3916–3923.
23. Rao, D., Webb, J. S., and Kjelleberg, S. (2006). Microbial colonization and competition on the marine alga *Ulva australis. Appl. Environ. Microbiol.* **72**, 5547–5555.
24. Rao, D., Webb, J. S., Holmström, C., Case, R., Low, A., et al. (2007). Low densities of epiphytic bacteria from the marine alga *Ulva australis* inhibit settlement of fouling organisms. *Appl. Environ. Microbiol.* **73**, 7844–7852.
25. Longford, S. R., Tujula, N. A., Crocetti, G. R., Holmes, A. J., Holmström, C., et al. (2007). Comparison of diversity of bacterial communities associated with three sessile marine eukaryotes. *Aquat. Microb. Ecol.* **48**, 217–229.
26. Matz, C., McDougald, D., Moreno, A. M., Yung, P. Y., Yildiz, F. H., et al. (2005). Biofilm formation and phenotypic variation enhance predation-driven persistence of *Vibrio cholerae. Proc. Natl. Acad. Sci. USA* **102**, 16819–16824.
27. Patterson, D. J. and Lee, W. Y. (2000). *The Flagellates—Unity, Diversity and Evolution.* B. S. C. Leadbeater and J. C. Green (Eds.). Taylor & Francis, London, pp. 269–287.
28. Boenigk, J. and Arndt, H. (2000). Comparative studies on the feeding behavior of two heterotrophic nanoflagellates: The filter-feeding choanoflagellate *Monosiga ovata* and the raptorial-feeding kinetoplastid *Rhynchomonas nasuta. Aquat. Microb. Ecol.* **22**, 243–249.
29. Matz, C., Bergfeld, T., Rice, S. A., and Kjelleberg, S. (2004). Microcolonies, quorum

sensing and cytotoxicity determine the survival of *Pseudomonas aeruginosa* biofilms exposed to protozoan grazing. *Environ. Microbiol.* **6**, 218–226.

30. Weitere, M., Bergfeld, T., Rice, S. A., Matz, C., and Kjelleberg, S. (2005). Grazing resistance of *Pseudomonas aeruginosa* biofilms depends on type of protective mechanism, developmental stage and protozoan feeding mode. *Environ. Microbiol.* **7**, 1593–1601.
31. Riveros, R., Haun, M., Campos, V., and Duran, N. (1988). Bacterial chemistry IV: Complete characterization of violacein. *Arq. Biol. Technol.* **31**, 475–487.
32. August, P. R., Grossman, T. H., Minor, C., Draper, M. P., MacNeil, I. A., et al. (2000). Sequence analysis and functional characterization of the violacein biosynthetic pathway from *Chromobacterium violaceum*. *J. Mol. Microbiol. Biotechnol.* **2**, 513–519.
33. Sanchez, C., Brana, A. F., Mendez, C., and Salas, J. A. (2006). Reevaluation of the violacein biosynthetic pathway and its relationship to indolocarbazole biosynthesis. *Chem. Biochem.* **7**, 1231–1240.
34. Huang, S. Y. and Hadfield, M. G. (2003). Composition and density of bacterial biofilms determine larval settlement of the polychaete *Hydroides elegans*. *Mar. Ecol. Progr. Ser.* **260**, 161–172.
35. Halda-Alija, L. and Johnston, T. C. (1999). Diversity of culturable heterotrophic aerobic bacteria in pristine stream bed sediments. *Can. J. Microbiol.* **45**, 879–84.
36. Corpe, W. (1953). Variation in pigmentation and morphology of colonies of gelatinous strains of *Chromobacterium* species from soil. *J. Bacteriol.* **66**, 470–477.
37. Rusch, D. B., Halpern, A. L., Sutton, G., Heidelberg, K. B., Williamson, S., et al. (2007). The sorcerer II global ocean sampling expedition: Northwest Atlantic through Eastern Tropical Pacific. *PLoS Biol.* **5**, e77.
38. Seshadri, R., Kravitz, S. A., Smarr, L., Gilna, P., and Frazier, M. (2007). CAMERA: A community resource for metagenomics. *PLoS Biol.* **5**, e75.
39. McClean, K. H., Winson, M. K., Fish, L., Taylor, A., Chhabra, S. R., et al. (1997). Quorum sensing and *Chromobacterium violaceum*: Exploitation of violacein production and inhibition for the detection of N-acylhomoserine lactones. *Microbiology* **143**, 3703–3711.
40. Nakamura, Y., Sawada, T., Morita, Y., and Tamiya, E. (2002). Isolation of a psychrotrophic bacterium from the organic residue of a water tank keeping rainbow trout and antibacterial effect of violet pigment produced from the strain. *Biochem. Engineer.* **12**, 79–86.
41. Matz, C., Deines, P., Boenigk, J., Arndt, H., Eberl, L., et al. (2004). Impact of violacein-producing bacteria on survival and feeding of bacterivorous nanoflagellates. *Appl. Environ. Microbiol.* **70**, 1593–1599.
42. Bidle, K. D. and Falkowski, P. G. (2004). Cell death in planktonic, photosynthetic microorganisms. *Nat. Rev. Microbiol.* **2**, 643–655.
43. Ferreira, C. V., Bos, C. L., Versteeg, H. H., Justo, G. Z., Duran, N., et al. (2004). Molecular mechanism of violacein-mediated human leukemia cell death. *Blood* **104**, 1459–1464.
44. Kodach, L. L., Bos, C. L., Duran, N., Peppelenbosch, M. P., Ferreira, C. V., et al. (2006). Violacein synergistically increases 5-fluorouracil cytotoxicity, induces apoptosis and inhibits Akt-mediated signal transduction in human colorectal cancer cells. *Carcinogenesis* **27**, 508–516.
45. Väätänen, P. (1976). Microbiological studies on coastal waters of the Northern Baltic Sea, I. Distribution and abundance of bacteria and yeasts in the Tvarminne area. *Walter Andre Nottback Found Sci. Rep.* **1**, 1–58.
46. Leist, M. and Jaattela, M. (2001). Four deaths and a funeral: From caspases to alternative mechanisms. *Nat. Rev. Mol. Cell. Biol.* **2**, 589–598.
47. Egan, S., James, S., Holmström, C., and Kjelleberg, S. (2002). Correlation between pigmentation and antifouling compounds produced by *Pseudoalteromonas tunicata*. *Environ. Microbiol.* **4**, 433–442.
48. Markowitz, V. M., Korzeniewski, F., Palaniappan, K., Szeto, E., Werner, G., et al. (2006). The integrated microbial genomes (IMG) system. *Nucleic Acids Res.* **34**, D344–348.

49. Blosser, R. S. and Gray, K. M. (2000). Extraction of violacein from *Chromobacterium violaceum* provides a new quantitative bioassay for N-acyl homoserine lactone autoinducers. *J. Microbiol. Methods* **40**, 47–55.

50. Hild, E., Takayama, K., Olsson, R. M., and Kjelleberg, S. (2000). Evidence for a role of rpoE in stressed and unstressed cells of marine *Vibrio angustum* strain S14. *J. Bacteriol.* **182**, 6964–6974.

51. Gao, L. Y. and Abu Kwaik, Y. (2000). The mechanism of killing and exiting the protozoan host *Acanthamoeba polyphaga* by *Legionella pneumophila. Environ. Microbiol.* **2**, 79–90.

10

1. Nasci, R. S. and Miller, B. R. (1996). Culicine mosquitoes and the agents they transmit. In *The Biology of Disease Vectors*. B. J. Beaty and W. C. Marquardt (Eds.). University Press of Colorado, Niwot, pp. 85–97.

2. Anonymous (2008). California mosquito-borne virus surveillance and response plan. *California Dept. Public Health, Mosq. Vector Cont. Assoc. Calif., and University of California.*

3. Reisen, W. K. and Pfuntner, A. R. (1987). Effectiveness of five methods for urban sampling adult *Culex* mosquitoes in rural and urban habitats in San Bernadino County, California. *J. Am. Mosq. Contr. Assoc.* **3**, 601–606.

4. Allan, S. A. and Kline, D. (2004). Evaluation of various attributes of gravid female traps for collection of *Culex* in Florida. *J. Vector Ecol.* **29**, 285–294.

5. Hardy, J. L., Houk, E. J., Kramer, L. D., and Reeves, W. C. (1983). Intrinsic factors affecting vector competence of mosquitoes for arboviruses. *Ann. Rev. Entomol.* **28**, 229–262.

6. Ishida, Y., Cornel, A. J., and Leal, W. S. (2002). Identification and cloning of a female antenna-specific odorant-binding protein in the mosquito *Culex quinquefasciatus. J. Chem. Ecol.* **28**, 867–871.

7. Wojtasek, H. and Leal, W. S. (1999). Conformational change in the pheromone-binding protein from *Bombyx mori* induced by pH and by interaction with membranes. *J. Biol. Chem.* **274**, 30950–30956.

8. Wogulis, M., Morgan, T., Ishida, Y., Leal, W. S., and Wilson, D. K. (2006). The crystal structure of an odorant binding protein from *Anopheles gambiae*: Evidence for a common ligand release mechanism. *Biochem. Biophys. Res. Commun.* **339**, 157–164.

9. Biessmann, H., Walter, M. F., Dimitratos, S., and Woods, D. (2002). Isolation of cDNA clones encoding putative odourant binding proteins from the antennae of the malaria-transmitting mosquito, *Anopheles gambiae. Insect. Mol. Biol.* **11**, 123–132.

10. Ishida, Y., Cornel, A. J., and Leal, W. S. (2003). Odorant-binding protein from *Culex tarsalis*, the most competent vector of West Nile virus in California. *J. Asia-Pacific Entomol.* **6**, 45–48.

11. Ishida, Y., Chen, A. M., Tsuruda, J. M., Cornel, A. J., Debboun, M., et al. (2004). Intriguing olfactory proteins from the yellow fever mosquito, *Aedes aegypti. Naturwissenschaften* **91**, 426–431.

12. Zhou, J. -J, He, X. -L., Pickett, J. A., and Field, L. M. (2008). Identification of odorant-binding proteins of the yellow fever mosquito *Aedes aegypti*: Genome annotation and comparative analyses. *Insect. Mol. Biol.* **17**, 147–163.

13. Clements, A. N. (1999). *The Biology of Mosquitoes: Sensory Reception and Behaviour*. CAB International, New York.

14. Hwang, Y. -H., Mulla, M. S., Chaney, J. D., Lin, G. -G., and Xu, H. -J. (1987). Attractancy and species specificity of 6-acetoxy-5-hexadecanolide, a mosquito oviposition attractant pheromone. *J. Chem. Ecol.* **13**, 245–252.

15. Laurence, B. R., Mori, K., Otsuka, T., Pickett, J. A., and Wadhams, L. J. (1985). Absolute configuration of mosquito oviposition attractant pheromone, 6-acetoxy-5-hexadecanolide. *J. Chem. Ecol.* **11**, 643–648.

16. Laurence, B. R. and Pickett, J. A. (1982). erythro-6-Acetoxy-5-hexadecanolide, the major component of a mosquito attractant pheromone. *J. Chem. Soc., Chem. Commun.* 59–60.

17. Du, Y. -J. and Millar, J. G. (1999). Electroantennogram and oviposition bioassay responses of *Culex quinquefasciatus* and *Culex tarsalis* (Diptera: Culicidae) to chemicals in odors from Bermuda grass infusions. *J. Med. Entomol.* **36**, 158–166.

18. Mboera, L. E. G., Takken, W., Mdira, K. Y., Chuwa, G. J., and Pickett, J. A. (2000). Oviposition and behavioral responses of *Culex quinquefasciatus* to skatole and synthetic oviposition pheromone in Tanzania. *J. Chem. Ecol.* **26**, 1193–1203.

19. Braks, M. A. H., Leal, W. S., and Cardé, R. T. (2007). Oviposition responses of gravid female *Culex quinquefasciatus* to egg rafts and low doses of oviposition pheromone under semifield conditions. *J. Chem. Ecol.* **33**, 567–578.

20. Dawson, G. W., Mudd, A., Pickett, J. A., Pile, M. M., and Wadhams, L. J. (1990). Convenient synthsesis of mosquito oviposition pheromone and a highly fluorinated analog retaining biological activity. *J. Chem. Ecol.* **16**, 1779–1789.

21. Olagbemiro, T. O., Birkett, M. A., Mordue, A. J., and Pickett, J. A. (1999). Production of (5R,6S)-6-acetoxy-5-hexadecanolide, the mosquito oviposition pheromone, from the seed oil of the summer cypress plant, *Kochia scoparia* (Chenopodiaceae). *J. Agric. Food Chem.* **47**, 3411–3415.

22. Olagbemiro, T. O., Birkett, M. A., Mordue Luntz, A. J., and Pickett, J. A. (2004). Laboratory and field responses of the mosquito, *Culex quinquefasciatus*, to plant-derived *Culex* spp. oviposition pheromone and the oviposition cue skatole. *J. Chem. Ecol.* **30**, 965–976.

23. Syed, Z. and Leal, W. S. (2007). Maxillary palps are broad spectrum odorant detectors in *Culex quinquefasciatus*. *Chem. Senses* **32**, 727–738.

24. Damberger, F., Nikonova, L., Horst, R., Peng, G., Leal, W. S., et al. (2000). NMR characterization of a pH-dependent equilibrium between two folded solution conformations of the pheromone-binding protein from *Bombyx mori*. *Protein Sci.* **9**, 1038–1041.

25. Leal, W. S., Chen, A. M., and Erickson, M. L. (2005). Selective and pH-dependent binding of a moth pheromone to a pheromone-binding protein. *J. Chem. Ecol.* **31**, 2493–2499.

26. Leal, W. S., Chen, A. M., Ishida, Y., Chiang, V. P., Erickson, M. L., et al. (2005). Kinetics and molecular properties of pheromone binding and release. *Proc. Natl. Acad. Sci. USA* **102**, 5386–5391.

27. Ban, L., Zhang, L., Yan, Y., and Pelosi, P. (2002). Binding properties of a locust's chemosensory protein. *Biochem. Biophys. Res. Commun.* **293**, 50–54.

28. Horst, R., Damberger, F., Luginbuhl, P., Guntert, P., Peng, G., et al. (2001). NMR structure reveals intramolecular regulation mechanism for pheromone binding and release. *Proc. Natl. Acad. Sci. USA* **98**, 14374–14379.

29. Lautenschlager, C., Leal, W. S., and Clardy, J. (2005). Coil-to-helix transition and ligand release of *Bombyx mori* pheromone-binding protein. *Biochem. Biophys. Res. Commun.* **335**, 1044–1050.

30. Sandler, B. H., Nikonova, L., Leal, W. S., and Clardy, J. (2000). Sexual attraction in the silkworm moth: Structure of the pheromone-binding-protein-bombykol complex. *Chem. Biol.* **7**, 143–151.

31. Xu, W. and Leal, W. S. (2008). Molecular switches for pheromone release from a moth pheromone-binding protein. *Biochem. Biophys. Res. Commun.* **372**, 559–564.

32. Wojtasek, H., Hansson, B. S., and Leal, W. S. (1998). Attracted or repelled? A matter of two neurons, one pheromone binding protein, and a chiral center. *Biochem. Biophys. Res. Commun.* **250**, 217–222.

33. Kline, D. L., Allan, S. A., Bernier, U. R., and Welch, C. H. (2007). Evaluation of the enantiomers of 1-octen-3-ol and 1-octyn-3-ol as attractants for mosquitoes associated with a freshwater swamp in Florida, U.S.A. *Med. Vet. Entomol.* **21**, 323–331.

34. Christiansen, J. A., Smith, C., Madon, M. B., Albright, J., Hazeleur, W., et al. (2005). Use of gravid traps for collection of California West Nile Virus vectors. *Proc. Papers Calif. Mosq. Control Assoc. Ann. Conf.* pp. 89–93.

35. Trexler, J. D., Apperson, C. S., Gemeno, C., Perich, M. J., Carlson, D., et al. (2003).

Field and laboratory evaluations of potential oviposition attractants for *Aedes albopictus* (Diptera: Culicidae). *J. Am. Mosq. Control. Assoc.* **19**, 228–234.

36. Millar, J. G., Chaney, J. D., Beehler, J. W., and Mulla, M. S. (1994). Interaction of the *Culex quinquefasciatus* egg raft pheromone with a natural chemical associated with oviposition sites. *J. Am. Mosq. Contr. Assoc.* **10**, 374–379.
37. Leal, W. S. (2000). Duality monomer-dimer of the pheromone-binding protein from *Bombyx mori*. Biochem. *Biophys. Res. Commun.* **268**, 521–529.
38. Dobritsa, A. A., van der Goes van Naters, W., Warr, C. G., Steinbrecht, R. A., and Carlson, J. R. (2003). Integrating molecular and cellular basis of odor coding in the *Drosophila* antenna. *Neurons* **37**, 827–841.
39. Pitts, R. J., Fox, A. N., and Zwiebel, L. J. (2004). A highly conserved candidate chemoreceptor expressed in both olfactory and gustatory tissues in the malaria vector *Anopheles gambiae*. *Proc. Natl. Acad. Sci. USA* **101**, 5058–5063.
40. Kotsuki, H., Kadota, I., and Ochi, M. (1990). A novel carbon-carbon bond-forming reaction of triflates with copper(1)-catalyzed Grignard reagents. A new concise and enantiospecific synthesis of (+)-exo-brevicomin, (5R,6S)-(-)-6-acetoxy-5-hexadecanolidea, and L-factor. *J. Org. Chem.* **55**, 4417–4422.
41. Barbosa, R. M. R., Souto, A., Eiras, A. E., and Regis, L. (2007). Laboratory and field evaluation of an oviposition trap for *Culex quinquefasciatus* (Diptera: Culicidae). *Mem. Inst. Oswaldo. Cruz* **102**, 523–529.
42. McAbee, R. D., Green, E. N., Holeman, J., Christiansen, J., Frye, N., et al. (2008). Identification of *Culex pipiens* complex mosquitoes in a hybrid zone of West Nile Virus transmission in Fresno County, California. *Am. J. Trop. Med. Hyg.* **78**, 303–310.

11

1. Bock, K. W. and Kohle, C. (2006). Ah receptor: Dioxin-mediated toxic responses as hints to deregulated physiologic functions. *Biochem. Pharmacol.* **72**, 393–404.
2. Pelclova, D., Fenclova, Z., Preiss, J., Prochazka, B., Spacil, J., Dubska, Z., Okrouhlik, B., Lukas, E., and Urban, P. (2002). Lipid metabolism and neuropsychological follow-up study of workers exposed to 2,3,7,8-tetrachlordibenzo-p-dioxin. *Int. Arch. Occup. Environ. Health* **75**(Suppl.), S60–66.
3. Baccarelli, A., Mocarelli, P., Patterson, D. G. Jr., Bonzini, M., Pesatori, A. C., Caporaso, N., and Landi, M. T. (2002). Immunologic effects of dioxin: new results from Seveso and comparison with other studies. *Environ. Health Perspect.* **110**, 1169–1173.
4. Pesatori, A. C., Consonni, D., Bachetti, S., Zocchetti, C., Bonzini, M., Baccarelli, A., and Bertazzi, P. A. (2003). Short- and long-term morbidity and mortality in the population exposed to dioxin after the "Seveso accident". *Ind. Health* **41**, 127–138.
5. Birnbaum, L. S. (1995). Developmental effects of dioxins. *Environ. Health Perspect.* **103**(Suppl. 7), 89–94.
6. Peterson, R. E., Theobald, H. M., and Kimmel, G. L. (1993). Developmental and reproductive toxicity of dioxins and related compounds: Cross-species comparisons. *Crit. Rev. Toxicol.* **23**, 283–335.
7. Birnbaum, L. S. and Tuomisto, J. (2000). Non-carcinogenic effects of TCDD in animals. *Food Addit. Contam.* **17**, 275–288.
8. Pocar, P., Fischer, B., Klonisch, T., and Hombach-Klonisch, S. (2005). Molecular interactions of the aryl hydrocarbon receptor and its biological and toxicological relevance for reproduction. *Reproduction* **129**, 379–389.
9. Barker, D. J., Gluckman, P. D., and Robinson, J. S. (1995). Conference report: Fetal origins of adult disease—Report of the First International Study Group, Sydney, October 29–30, 1994. *Placenta* **16**, 317–320.
10. Wynn, M. and Wynn, A. (1988). Nutrition around conception and the prevention of low birthweight. *Nutr. Health* **6**, 37–52.
11. Kwong, W. Y., Wild, A. E., Roberts, P., Willis, A. C., and Fleming, T. P. (2000). Maternal undernutrition during the preimplantation period of rat development causes blastocyst abnormalities and programming

of postnatal hypertension. *Development* **127**, 4195–4202.

12. Hunt, P. A., Koehler, K. E., Susiarjo, M., Hodges, C. A., Ilagan, A., Voigt, R. C., Thomas, S., Thomas, B. F., and Hassold, T. J. (2003). Bisphenol a exposure causes meiotic aneuploidy in the female mouse. *Curr. Biol.* **13**, 546–553.
13. Lo, C. W. and Gilula, N. B. (1979). Gap junctional communication in the post-implantation mouse embryo. *Cell* **18**, 411–422.
14. Reeve, W. J. (1981). The distribution of ingested horseradish peroxidase in the 16-cell mouse embryo. *J. Embryol. Exp. Morphol.* **66**, 191–207.
15. Johnson, M. H. and Maro, B. (1984). The distribution of cytoplasmic actin in mouse 8-cell blastomeres. *J. Embryol. Exp. Morphol.* **82**, 97–117.
16. Houliston, E., Pickering, S. J., and Maro, B. (1987). Redistribution of microtubules and pericentriolar material during the development of polarity in mouse blastomeres. *J. Cell Biol.* **104**, 1299–1308.
17. Ziomek, C. A. and Johnson, M. H. (1980). Cell surface interaction induces polarization of mouse 8-cell blastomeres at compaction. *Cell* **21**, 935–942.
18. Johnson, M. H. and McConnell, J. M. (2004). Lineage allocation and cell polarity during mouse embryogenesis. *Semin. Cell Dev. Biol.* **15**, 583–597.
19. Blankenship, A. L., Suffia, M. C., Matsumura, F., Walsh, K. J., and Wiley, L. M. (1993). 2,3,7,8-Tetrachlorodibenzo-p-dioxin (TCDD) accelerates differentiation of murine preimplantation embryos *in vitro*. *Reprod. Toxicol.* **7**, 255–261.
20. Tsutsumi, O., Uechi, H., Sone, H., Yonemoto, J., Takai, Y., Momoeda, M., Tohyama, C., Hashimoto, S., Morita, M., and Taketani, Y. (1998). Presence of dioxins in human follicular fluid: their possible stage-specific action on the development of preimplantation mouse embryos. *Biochem. Biophys. Res. Commun.* **250**, 498–501.
21. Shi, Z., Valdez, K. E., Ting, A. Y., Franczak, A., Gum, S. L., Petroff, B. K. (2007). Ovarian endocrine disruption underlies premature reproductive senescence following environmentally relevant chronic exposure to the aryl hydrocarbon receptor agonist 2,3,7,8-Tetrachlorodibenzo-p-Dioxin. *Biol. Reprod.* **76**, 198–202.
22. Wu, Q., Ohsako, S., Baba, T., Miyamoto, K., and Tohyama, C. (2002). Effects of 2,3,7,8-tetrachlorodibenzo-p-dioxin (TCDD) on preimplantation mouse embryos. *Toxicology* **174**, 119–129.
23. Moran, F. M., Vande Voort, C. A., Overstreet, J. W., Lasley, B. L., and Conley, A. J. (2003). Molecular target of endocrine disruption in human luteinizing granulosa cells by 2,3,7,8-tetrachlorodibenzo-p-dioxin: Inhibition of estradiol secretion due to decreased 17alpha-hydroxylase/17,20-lyase cytochrome P450 expression. *Endocrinology* **144**, 467–473.
24. Pocar, P., Nestler, D., Risch, M., and Fischer, B. (2005). Apoptosis in bovine cumulus-oocyte complexes after exposure to polychlorinated biphenyl mixtures during *in vitro* maturation. *Reproduction* **130**, 857–868.
25. Li, B., Liu, H. Y., Dai, L. J., Lu, J. C., Yang, Z. M., and Huang, L. (2006). The early embryo loss caused by 2,3,7,8-tetrachlorodibenzo-p-dioxin may be related to the accumulation of this compound in the uterus. *Reprod. Toxicol.* **21**, 301–306.
26. Eichenlaub-Ritter, U., Winterscheidt, U., Vogt, E., Shen, Y., Tinneberg, H. R., and Sorensen, R. (2007). 2-methoxyestradiol induces spindle aberrations, chromosome congression failure, and nondisjunction in mouse oocytes. *Biol. Reprod.* **76**, 784–793.
27. Lattanzi, M. L, Santos, C. B., Mudry, M. D., and Baranao, J. L. (2003). Exposure of bovine oocytes to the endogenous metabolite 2-methoxyestradiol during *in vitro* maturation inhibits early embryonic development. *Biol. Reprod.* **69**, 1793–1800.
28. Brison, D. R. and Schultz, R. M. (1997). Apoptosis during mouse blastocyst formation: Evidence for a role for survival factors including transforming growth factor alpha. *Biol. Reprod.* **56**, 1088–1096.
29. Murray, A. (1994). Cell cycle checkpoints. *Curr. Opin. Cell Biol.* **6**, 872–876.
30. Artus, J., Babinet, C., and Cohen-Tannoudji, M. (2006). The cell cycle of early mammalian embryos: lessons from genetic mouse models. *Cell Cycle* **5**, 499–502.

31. Harrison, R. H., Kuo, H. C., Scriven, P. N., Handyside, A. H., and Ogilvie, C. M. (2000). Lack of cell cycle checkpoints in human cleavage stage embryos revealed by a clonal pattern of chromosomal mosaicism analysed by sequential multicolour FISH. *Zygote* **8**, 217–224.

32. Chatzimeletiou, K., Morrison, E. E., Prapas, N., Prapas, Y., and Handyside, A. H. (2005). Spindle abnormalities in normally developing and arrested human preimplantation embryos *in vitro* identified by confocal laser scanning microscopy. *Hum. Reprod.* **20**, 672–682.

33. Varmuza, S., Prideaux, V., Kothary, R., and Rossant, J. (1988). Polytene chromosomes in mouse trophoblast giant cells. *Development* **102**, 127–134.

34. Burke, B. and Stewart, C. L. (2006). The laminopathies: The functional architecture of the nucleus and its contribution to disease. *Annu. Rev. Genomics Hum. Genet.* **7**, 369–405.

35. Torres-Padilla, M. E., Parfitt, D. E., Kouzarides, T., and Zernicka-Goetz, M. (2007). Histone arginine methylation regulates pluripotency in the early mouse embryo. *Nature* **445**, 214–218.

36. Anway, M. D., Cupp, A. S., Uzumcu, M., and Skinner, M. K. (2005). Epigenetic transgenerational actions of endocrine disruptors and male fertility. *Science* **308**, 1466–1469.

37. Wu, Q., Ohsako, S., Ishimura, R., Suzuki, J. S., and Tohyama, C. (2004). Exposure of mouse preimplantation embryos to 2,3,7,8-tetrachlorodibenzo-p-dioxin (TCDD) alters the methylation status of imprinted genes H19 and Igf2. *Biol. Reprod.* **70**, 1790–1797.

38. Susiarjo, M., Hassold, T. J., Freeman, E., and Hunt, P. A. (2007) Bisphenol A exposure in utero disrupts early oogenesis in the mouse. *PLoS Genet.* **3**(1), e5.

39. Zarrow, M., Yochim, J., and McCarthy, J. (1964). *Experimental Endocrinology: A Source Book of Basic Techniques*. Academic Press, New York.

40. Ataniyazova, O. A., Baumann, R. A., Liem, A. K., Mukhopadhyay, U. A., Vogelaar, E. F., and Boersma, E. R. (2001). Levels of certain metals, organochlorine pesticides and dioxins in cord blood, maternal blood, human milk and some commonly used nutrients in the surroundings of the Aral Sea (Karakalpakstan, Republic of Uzbekistan). *Acta. Paediatr.* **90**, 801–808.

41. Hooper, K., Petreas, M. X., Chuvakova, T., Kazbekova, G., Druz, N., Seminova, G., Sharmanov, T., Hayward, D., She, J., Visita, P., Winkler, J., McKinney, M., Wade, T. J., Grassman, J., and Stephens, R. D. (1998). Analysis of breast milk to assess exposure to chlorinated contaminants in Kazakstan: High levels of 2,3,7, 8-tetrachlorodibenzo-p-dioxin (TCDD) in agricultural villages of southern Kazakstan. *Environ. Health Perspect.* **106**, 797–806.

42. Kociba, R. J., Keyes, D. G., Beyer, J. E., Carreon, R. M., Wade, C. E., Dittenber, D. A., Kalnins, R. P., Frauson, L. E., Park, C. N., Barnard, S. D., Hummel, R. A., and Humiston, C. G. (1978). Results of a two-year chronic toxicity and oncogenicity study of 2,3,7,8-tetrachlorodibenzo-p-dioxin in rats. *Toxicol. Appl. Pharmacol.* **46**, 279–303.

43. Combelles, C. M., Cekleniak, N. A., Racowsky, C., and Albertini, D. F. (2002). Assessment of nuclear and cytoplasmic maturation in *in-vitro* matured human oocytes. *Hum. Reprod.* **17**, 1006–1016.

12

1. Leslie, A. and Cuperus, G. W. (1993). *Successful Implementation of Integrated Pest Management for Agricultural Crops.* Lewis/ CRC Press, Florida.

2. Thrupp, L. A. (1996). *New Partnerships for Sustainable Agriculture*. World Resources Institute, Washington, DC.

3. Premazzi, G. and Ziglio, G. (1995). Regulations and Management. In *Pesticide Risk in Groundwater.* M. Vighi and E. Funari (Eds.). CRC Lewis Publishers, Boca Raton, Chapter 10, pp. 203–240.

4. London, L. (1992). An overview of agrochemical hazards in the South African farming sector. *S. Afr. Med. J.* **81**, 560–564.

5. London, L. (1995). An investigation into the neurological and neurobehavioral effects of long-term agrochemical exposure amongst deciduous fruit farm workers in the Western

Cape, South Africa. Doctoral Thesis. *Cape Town, Department of Community Health, University of Cape Town.*

6. London, L. and Myers, J. E. (1995). General patterns of agrochemical usage in the Southern Region of South Africa. *SA J. Sci.* **91**, 509–514.
7. London, L. and Rother, A. (1998). People, pesticide and the environment: Who bears the brunt of backward policy in South Africa? In *Conference Proceedings: Environmental Justice and the Legal Process*. Environmental Law Unit, University of Cape Town, Cape Town, South Africa and Environmental Law Centre, Macquarie Univeristy, Sydney, Australia.
8. Maroni, M. and Fait, A. (1993). Health Effects in man from long-term exposure to pesticides. A review of the 1975–1991 literature. *Toxicology* **78**, 1–174.
9. Dalvie, M. A., White, N., Raine, R., Myers, J. E., London, L., Thompson, M., and Christiani, D. C. (1999). The long-term respiratory health effects of the herbicide, paraquat, among workers in the Western Cape. *Occup. Environ. Med.* **56**, 391–396.
10. Myers, J. E. (1990). Occupational health of farm workers. *S. Afr. Med. J.* **78**, 562–563.
11. London, L. and Myers, J. E. (1995). Critical issues in agrochemical safety in South Africa. *Am. J. Ind. Med.* **27**, 1–14.
12. Hassett, A. J., Viljoen, P. T., and Liebenberg, J. J. E. (1987). An assessment of chlorinated pesticides in the major surface water resources of the Orange Free State during the period September 1984 to September 1985. *Water SA* **13**(3), 133–136.
13. Schultz, R., Peall, S. K. C., Dabrowski, J. M., and Reinecke, A. J. (2001). Current-use insecticides, phosphates and suspended solids in the Lourens River, Western Cape, during the first rainfall event of the wet season. *Water SA* **27**(1), 65–70.
14. Greichus, Y. A., Greichus, A., Amman, B. D., Call, D. J., Hamman, D. C. D., and Pott, R. M. (1977). Insecticides polychlorinated biphenyl and metals in African ecosystems.1. Hartebeespoort Dam, Transvaal and Voëlvlei Dam, Cape Province, Republic of South Africa. *Arch. Environ. Contam. Toxicol.* **6**, 371–383.
15. Davies, H. (1997). An assessment of the suitability of a series of Western Cape Farm Dams as water bird habitats. *MSc (Conservation Biology) Thesis*. Zoology Department, University of Cape Town.
16. Grobler, D. F. (1994). A note on PCBs and chlorinated hydrocarbon pesticide residues in water, fish and sediment from the Olifants River, Eastern Transvaal, South Africa. *Water SA* **20**(3), 187–194.
17. Weaver, J. M. C. (1993). A preliminary survey of pesticide levels in groundwater from a selected area on intensive agriculture in the Western Cape. *Report to the Water Research Commission*. Pretoria, Division of Water Technology, CSIR.
18. London, L., Dalvie, M. A., Cairncross, E., and Solomon, A. (2001). The quality of surface and groundwater in the rural Western Cape with regard to pesticides. *WRC Report No: K5/795/00*. Pretoria, WRC 2001.
19. World Wild Life Fund (1997). *Known and Suspected Hormone Disruptors List*. World Wild Life, Canada, Toronto.
20. Schettler, T., Solomon, G., Burns, P., and Valenti, M. (1996). Generations at Risk. How Environmental Toxins may affect reproductive health in Massachusetts. *Greater Boston Physicians for Social Responsibility, Massachusetts Public Interest Research Group (MASSPIRG) Education Fund*, Cambridge.
21. EPA (1995). *Solid Phase Extraction Method 3535*. U.S. Environmental Protection Agency, Washington, DC.
22. EPA (1995). *Organochlorine Pesticides by Capillary Column Gas Chromatography Method 8081A*. U.S. Environmental Protection Agency, Washington, DC.
23. McGregor, F. (1999). The mobility of endosulfan and chlorpyrifos in the soil of the Hex River Valley. *Thesis Submitted in Partial Fulfilment of the Requirements for the Degree of Masters of Science in Environmental Geochemistry*. Department of Geological Sciences, University of Cape Town.
24. California Environmental Protection Agency (1997). Sampling for pesticide residues in California well water. 1996 update of the well inventory database. *California Environmental*

Protection Agency. Department of pesticide regulation, California.

25. Council of the European Community (1980). Directive relating to the quality of water intended for human consumption (80/778/EEC). *EEC.*
26. EPA (1992). *Guidelines for Drinking Water Quality. 202, 260–7572.* U.S. Environmental Protection Agency, Washington, DC.
27. WHO (1993). *Guidelines for Drinking Water Quality. Recommendations.* WHO2 Edition, Geneva, p. 1.
28. DWAF (1996). *South African Water Quality Guidelines.* Pretoria, DWAF, pp. 1–7.
29. Dallas, H. F. and Day, J. A (1993). The effect of water quality variables on riverine ecosystems: A review. *Report Prepared for the Water Research Commission.* Rondebosch, Freshwater Research Unit, University of Cape Town.
30. International Programme on Chemical Safety (IPCS) (1993). Summary of Toxicological Evaluations performed by the joint WHO/FAO meeting on pesticide residues (JMPR). WHO, Geneva.
31. Soto, A. M., Chung, K. L., and Sonnenschein, C. (1994). The pesticides endosulfan, toxaphene, and dieldrin have estrogenic effects on human estrogen-sensitive cells. *Environ. Health Perspect.* **102**(4), 380–383.
32. Funari, E., Donati, L., Sandroni, D., and Vighi, M. (1995). Pesticide levels in groundwater: Value and limitations of monitoring. In *Pesticide Risk in Groundwater*. M. Vighi and E. Funari (Eds.). CRC Lewis Publishers, Boca Raton, Chapter 1, pp. 3–44.
33. Oskam, G., Van Genderen, J., Hopman, R., Noij, T. H. M., Noordsij, A., and Piuker, L. M. (1993). A general view of the problem, with special reference to the Dutch situation. *Water Supply* **11**, 1–17.
34. Espigares, M., Coca, C., Fernandez-Crehuet, M. O., Bueno, A., and Galvez, R. (1997). Pesticide concentrations in the waters from a section of the Guadal river basin, Spain. *Environ. Toxicol. Water Qual.* **12**, 249–256.
35. Domagalski, J. (1997). Results of a prototype surface water network design for pesticides developed for the San Joaquin River Basin, California. *J. Hydrology* **192**, 33–50.
36. Gustafson, D. I. (1989). Groundwater ubiquity score: A simple method for assessing pesticide leachability. *Environ. Toxi. Chem.* **8**, 339–357.
37. Jabber, A., Masud, S. Z., Parveen, Z., and Ali, M. (1993). Pesticide residues in cropland soils and shallow groundwater in Punjab Pakistan. *Bull. Environ. Contam. Toxicol.* **51**, 268–273.

13

1. Pratt, G. C., Palmer, K., Wu, C. Y., Oliaei, F., Hollerbach, C., and Fenske, M. J. (2000). An assessment of air toxics in Minnesota. *Environ. Health Perspect.* **108**, 815–825.
2. Woodruff, T. J., Axelrad, D. A., Caldwell, J., Morello-Frosch, R., and Rosenbaum, A. (1998). Public health implications of 1990 air toxics concentrations across the United States. Environ. *Health Perspect.* 106, 245–251.
3. Clean Air Act Amendments: Part A, Section 112. Public Law 1990, pp. 101–549.
4. Leikauf, G. D. (2002). Hazardous Air Pollutants and Asthma. *Environ. Health Perspect.* **110**(S4), 505–526.
5. Rumchev, K., Spickett, J., Bulsara, M., Phillips, M., and Stick, S. (2004). Association of domestic exposure to volatile organic compounds with asthma in young children. *Thorax* **59**, 746–751.
6. International Agency for Research on Cancer (1982). IARC Monograph on the Evaluation of the Carcinogenic Risk of Chemicals: Some Industrial Chemicals and Dyestuffs, Volume 29, IARC, Lyon, France.
7. Lin, M., Chen, Y., Villeneuve, P. J., Burnett, R. T., Lemyre, L., Hertzman, C., Mcgrail, K. M., and Krewski, D. (2004). Gaseous air pollutants and asthma hospitalization of children with low household income in Vancouver, British Columbia. *Am. J. Epidemiol.* **159**, 294–303.
8. U.S. Department of Health and Human Services (2007). The Agency for Toxic Substances and Disease Registry (ATSDR). [http://www.atsdr.cdc.gov/toxprofiles/tp3.pdf] Toxicological Profile for Benzene.
9. Glass, D. C., Gray, C. N., Jolley, D. J., Gibbons, C., Sim, M. R., Fritschi, L.,

Adams, G. G., Bisby, J. A., and Manuell, R. (2003). Leukemia risk associated with low-level benzene exposure. *Epidemiology* **14**, 569–577.

10. U.S. Department of Health and Human Services (2000). The Agency for Toxic Substances and Disease Registry (ATSDR). [http://www.atsdr.cdc.gov/toxprofiles/tp3.pdf] Toxicological Profile for Toluene.
11. Chang, S., Chen, C., Lien, C., and Sung, F. (2006). Hearing Loss in Workers Exposed to Toluene and Noise. *Environ, Health Perspect,* **114**, 1283–1286.
12. Gerin, M., Siemiatychi, J., Desy, M., and Krewski, D. (1998). Associations between several sites of cancer and occupational exposure to benzene, toluene, xylene, and styrene: Results of a case-control study in Montreal. *Am. J. Ind. Med.* **34**, 144–156.
13. Antilla, A., Pukkala, E., Riala, R., Sallmén, M., and Hemminki, K. (1998). Cancer incidence among Finnish workers exposed to aromatic hydrocarbons. *Int. Arch. Occup. Environ. Health* **71**, 187–193.
14. International Agency for Research on Cancer (1999). Monographs on the evaluation of carcinogenic risks to humans. Xylenes. In Part Three. Re-evaluation of some organic chemicals, hydrazine, and hydrogen peroxide. Volume 71. World Health Organization, Lyon, France, pp. 1189–1208.
15. Integrated Risk Information System (2001). Benzene, Toluene, Ethylbenzene, and Xylenes. Integrated Risk Information System, U.S. Environmental Protection Agency.
16. Vyskocil, A., Leroux, T., Truchon, G., Lemay, F., Gendron, M., Gagnon, F., El Majidi, N., and Viau, C. (2008). Ethyl benzene should be considered ototoxic at occupationally relevant exposure concentrations. Toxicol. Ind. **24**, 241–246.
17. U.S. Environmental Protection Agency (1999). *Integrated Risk Information System (IRIS) on Xylenes.* National Center for Environmental Assessment, Office of Research and Development, Washington, DC.
18. U.S. Department of Health and Human Services (2007). The Agency for Toxic Substances and Disease Registry (ATSDR). [http://www.atsdr.cdc.gov/tfacts71.pdf]. Toxicological Profile for Xylene (Update) U.S. Department of Public Health and Human Services, Public Health; Service, Atlanta, GA.
19. Aguilera, I., Sunyer, J., Fernandez-Patier, R., Hoek, G., Aguirre-Alfaro, A., Meliefste, K., Bomboi-Mingarro, M. R., Nieuwenhuijsen, M. J., Herce-Garraleta, D., and Brunekreef, B. (2008). Estimation of outdoor NOx, NO_2, and BTEX exposure in a cohort of pregnant women using land use regression modelling. *Environ. Sci. Technol.* **42**, 815–821.
20. Sexton, K., Adgate, J. L., Ramachandran, G., Pratt, G. C., Mongin, S. J., Stock, T. H., and Morandi, M. T. (2004). Comparison of personal, indoor, and outdoor exposures to hazardous air pollutants in three urban communities. *Environ. Sci. Technol.* **38**, 423–430.
21. Adgate, J. L., Church, T. R., Ryan, A. D., Ramachandran, G., Fredrickson, A. L., Stock, T. H., Morandi, M. T., and Sexton, K. (2004). Outdoor, indoor, and personal exposure to VOCs in Children. *Environ. Health Perspect.* **112**, 1386–1392.
22. Lee, S. (1997). Comparison of indoor and outdoor air quality at two staff quarters in Hong Kong. *Environ. Int.* **23**(6), 791–797.
23. Rava, M., Verlato, G., Bono, R., Ponzio, M., Sartori, S., Blengio, G., Kuenzli, N., Heinrich, J., Götschi, T., and de Marco, R. (2007). A predictive model for the home outdoor exposure to nitrogen dioxide. *Sci, Total Environ.* **384**, 163–170.
24. Brunekreef, B. and Holgate, S. T. (2002). Air pollution and health. *Lancet* **360**, 1233–1242.
25. Pope, C. A. and Dockery, D. W. (2006). Health effects of fine particulate air pollution: Lines that connect. *J, Air Waste Manag, Assoc.* **56**, 709–742.
26. Hoek, G., Beelen, R., de Hooh, K., Vienneau, D., Gulliver, P. F., and Briggs, D. (2008). A review of land use regression models to assess spatial variation of outdoor air pollution. *Atmos. Environ.* **42**(33), 7561–7578.
27. Ryan, P. and LeMasters, G. (2007). A review of land-use regression models for characterising intraurban air pollution exposure. *Inhal. Toxicol.* **19**(Suppl. 1), 127–133.

28. Briggs, D. (2005). The role of GIS: Coping with space (and time) in air pollution exposure assessment. *J. Toxicol. Environ. Health* A **68**, 1243–1261.

29. Dockery, D. W. and Stone, P. H. (2007).Cardiovascular risks from fine particulate air pollution. *N. Eng. J. Med.* **356**, 511–513.

30. Burnett, R. T., Stieb, D., Brook, J. R., Cakmak, S., Dales, R., Raizenne, M., Vincent, R., and Dann, T. (204). Associations between short-term changes in nitrogen dioxide and mortality in Canadian cities. *Arch. Environ. Health* **59**, 228–236.

31. Pope, C. A., Burnett, R. T., Thun, M. J., Calle, E. E., Krewski, D., Ito, K., and Thurston, G. D. (2002). Lung cancer, cardiopulmonary mortality, and long-term exposure to fine particulate air pollution. *J. Am. Med. Assoc.* **287**, 1132–1141.

32. Jerrett, M., Arain, A., Kanaroglou, P., Beckerman, B., Crouse, D., Gilbert, N. L., Brook, J. R., Finkelstein, N., and Finkelstein, M. M. (2007). Modelling the intra-urban variability of ambient traffic pollution in Toronto, Canada. *J. Toxicol. Environ. Health A* **70**, 200–212.

33. Levy, J., Houseman, E. A., Ryan, L., Richardson, D., and Spengler, J. D. (2000). Particle concentration in urban microenvironments. *Environ. Health Persp.* **108**, 1051–1057.

34. Madsen, C., Carlsen, K. C., Hoek, G., Oftedal, B., Nafstad, P., Meliefste, K., Jacobsen, R., Nystad, W., Carlsen, K., and Brunekreef, B. (2007). Modelling the intraurban variability of outdoor traffic pollution in Oslo, Norway–A GA2 LEN project. *Atmos. Environ.* **41**, 7500–7511.

35. Jerrett, M., Arain, A., Kanaroglou, P., Beckerman, B., Potoglou, D., Sahsuvaroglu, T., Morrison, J., and Giovis, C. (2005). A review and evaluation of intra-urban air pollution exposure models. *J. Expo. Anal. Environ. Epidemiol.* **15**, 185–204.

36. Briggs, D., Collins, S., Elliott, P., Kingham, S., Lebret, E., Pryl, K., van Reeuwijk, H., Smallbone, K., and Laan, A. (1997). Mapping urban air pollution using GIS: A regression-based approach. *Int. J. Geogr. Info. Syst.* **11**, 699–718.

37. Ryan, P., LeMasters, G., Biswas, P., Levin, L., Hu, S., Lindsey, M., Bernstein, D., Lockey, J., Villareal, M., Hershey, G. K., and Grinshpun, S. A. (2006). A comparison of proximity and land use regression traffic exposure models and wheezing in infants. *Environ. Health Perspect.* **115**, 278–284.

38. Fischer, P., Hoek, G., van Reeuwijk, H., Briggs, D. J., Lebret, E., van Wijnen, J. H., Kingham, S., and Elliott, P. (2000). Traffic-related differences in outdoor and indoor concentrations of particles and volatile organic compounds in Amsterdam. *Atmos. Environ.* **34**, 3713–3722.

39. Linaker, C., Chauhan, A., Inskip, H., Holgate, S., and Coggon, D. (2000). Personal exposures of children to nitrogen dioxide relative to concentrations in outdoor air. *J. Occup. Environ. Med.* **57**, 472–476.

40. Fung, K., Luginaah, I., and Gorey, K. (2007). Impact of air pollution on hospital admissions in Southwestern Ontario, Canada: Generating hypotheses in sentinel high-exposure places. *Environ. Health* **6**, 1–8.

41. Brauer, M., Hoek, G., van Vliet, P., Meliefste, K., Fischer, P., Gehring, U., Heinrich, J., Cyrys, J., Bellander, T., Lewne, M., and Brunekreef, B. (2003). Estimating long-term average particulate air pollution concentrations: Application of traffic indicators and geographic information systems. *Epidemiology* **14**, 228–239.

42. Statistics Canada (2006). [http://www12.statcan.ca/english/census06/data/profiles/community/Index.cfm?Lang=E] Community Profiles.

43. Gilbertson, M. and Brophy, J. (2001). Community health profile of Windsor, Ontario, Canada: Anatomy of a great lakes area of concern. *Environ. Health Persp.* **109**(Suppl. 6), 827–843.

44. Curren, K. C., Dann, T. F., and Wang, D. K. (2006). Ambient air 1,3-butadiene concentration in Canada (1995–2003): Seasonal, day of week variations, trends, and source influences. *Atmos. Environ.* **40**, 171–181.

45. Health Canada (2000). *Health Data and Statistics Compilations for Great Lakes Areas of Concern.* Health Canada, Ottawa, Ontario.

46. Kanaroglou, P. S., Jerrett, M., Morrison, J., Beckerman, B., Arain, M. A., Gilbert, N. L., and Brook, J. (2005). Establishing an air pollution monitoring network for intra-urban population exposure assessment: A location-allocation approach. *Atmos. Environ.* **39**, 2399–2409.
47. Miller, L., Xu, X., and Luginaah, I. (2009). Spatial variability of VOC concentrations in Sarnia, Ontario, Canada. *J. Toxicol. Environ. Health A* **72**, 1–15.
48. SPSS 15.0 for Windows SPSS Inc. Headquarters, 233 S. Wacker Drive, 11th floor, Chicago, IL 60606 2007.
49. Hamilton, L. (1992). *Regression with Graphics: A Second Course in Applied Statistics*. Duxbury Press, Belmont, California.
50. Odland, J. (1998). *Spatial Autocorrelation*. Sage Publications, New Delhi, India.
51. Griffith, D. A. (1987). *Spatial Autocorrelation: A Primer. Resource publications in geography*. Association of American Geographers, Washington, DC.
52. Isaaks, E. and Srivastava, R. (1989). *An Introduction to Applied Geostatistics*. Oxford University Press, New York, NY.
53. Sahsuvaroglu, T., Arain, A., Kanaroglou, P. S., Finkelstein, N., Newbold, B., Jerrett, M., Beckerman, B., Brook, J. R., Finkelstein, M., and Gilbert, N. L. (2006). A land use regression model for predicting ambient concentrations of nitrogen dioxide in Hamilton, Ontario, Canada. *J. Air Waste Manag. Assoc.* **56**, 1059–1069.
54. Chow, G. (1960). Tests of equality between sets of coefficients in two linear regressions. *Econometrica* **28**, 591–605.
55. Atari, D. O., Luginaah, I., Xu, X., and Fung, K. (2008). Spatial variability of ambient nitrogen dioxide and sulphur dioxide in Sarnia, "Chemical Valley", Ontario, Canada. *J. Toxicol. Environ. Health A* **71**, 1–10.
56. Gilbert, N. L., Goldberg, M. S., Beckerman, B., Brook, J. R., and Jerrett, M. (2005). Assessing spatial variability of ambient nitrogen dioxide in Montreal, Canada, with a Land-Use Regression Model. *J. Air Waste Manag. Assoc.* **55**(8), 1059–1063.
57. Monod, A., Sive, B. C., Avino, P., Chen, T., Blake, D. B., and Rowland, F. S. (2001). Monoaromatic compounds in ambient air of various cities: A focus on correlations between the xylenes and ethylbenzene. *Atmos. Environ.* **35**, 135–149.
58. Environment Canada (2001). National Air Pollution Surveillance (NAPS) Network. [http://www.etc-cte.ec.gc.ca/NAPS/naps_summary_e.html] NAPS Network Summary.
59. Carr, D., von Ehrestein, O., Weiland, S., Wagner, C., Wellie, O., Nicolai, T., and von Mutius, E. (2002). Modelling annual benzene, toluene, NO_2, and soot concentrations on the basis of road traffic characteristics. *Environ. Res. Sect.* **90**, 111–118.
60. Smith, L., Mukerjee, S., Gonzales, M., Stallings, C., Neas, L., and Norris, G. H. (2006). Use of GIS and ancillary variables to predict volatile organic compound and nitrogen dioxide levels at unmonitored locations. *Atmos. Environ.* **40**, 3773–3787.
61. Wheeler, A. J., Smith-Doiron, M., Xu, X., Gilbert, N. L., and Brook, J. R. (2008). Intra-urban variability of air pollution in Windsor, Ontario-measurement and modeling for human exposure assessment. *Environ. Res.* **106**, 7–16.
62. Beckerman, B., Jerrett, M., Brook, J. R., Verma, D. K., Arain, M. A., and Finkelstein, M. M. (2008). Correlation of nitrogen dioxide with other traffic pollutants near a major expressway. *Atmos. Environ.* **42**, 275–290.
63. Ross, Z., English, P. B., Scalf, R., Gunier, R., Smorodinsky, S., Wall, S., and Jerrett, M. (March, 2006). Nitrogen dioxide prediction in Southern California using land use regression modeling: Potential for environmental health analyses. *J. Expo. Sci. Environ. Epidemiol.* **16**(2), 106–114.
64. Pankow, J. F., Luo, W., Bender, D. A., Isabelle, L. M., Hollingsworth, J. S., Chen, C., Asher, W. E., and Zogorski, J. S. (2003). Concentration and co-occurrence correlations of 88 volatile organic compounds (VOC) in the ambient air of 13 semi-rural to urban locations in the United States. *Atmos. Environ.* **37**, 5023–5046.
65. Lebret, E., Briggs, D., van Reeuwijk, H., Fischer, P., Smallbone, K., Harssema, H., Kriz, B., Gorynski, P., and Elliott, P. (2000). Small area variations in ambient NO_2

concentrations in four European areas. *Atmos. Environ.* **34**, 177–185.

14

1. Hammer, U. T. (1978). The saline lakes of Saskatchewan, III. Chemical characterization. *Inter. Revue Der. Ges. Hydrobiol.* **63**, 3311–3335.
2. Cowardin, L. M., Carter, V., Golet, F. C., and LaRoe, E. T. (1979). *Classification of Wetlands and Deepwater Habitats in the United States*. USFW Service.
3. LaBaugh, J. W. (1989). Chemical characteristics of water in northern prairie wetlands. In *Northern Prairie Wetlands*. A. V. Ames Valk (Ed.). Iowa University Press, pp. 57–90.
4. Hammer, U. T. and Haynes, R. C. (1978). The saline lakes of Saskatchewan. II. Locale, hydrology, and other physical aspects. *Inter.Revue Der Ges. Hydrobio.* **63**, 1179–1203.
5. Last, W. and Ginn, F. (2005). Saline systems of the Great Plains of western Canada: An overview of the limnogeology and paleolimnology. *Saline Systems* **1**(1), 10.
6. Wollheim, W. M. and Lovvorn, J. R. (1995). Salinity effects on macroinvertebrate assemblages and waterbird food webs in shallow lakes of the Wyoming high-plains. *Hydrobiologia* **310**(3), 207–223.
7. Donald, D. B. and Syrgiannis, J. (1995). Occurrence of pesticides in prairie lakes is Saskatchewan in relation to drought and salinity. *J. Env. Quality* **24**(2), 266–270.
8. Hammer, U. T. (1978). The saline lakes of Saskatchewan. I Background and rationale for saline lakes research. *Inter. Revue Der Gesamten Hydrobiolgie* **63**, 1173–1177.
9. Lieffers, V. J. and Shay, J. M. (1983). Ephemeral saline lakes in the Canadian prairies—Their classification and management for emergent macrophyte growth. *Hydrobiologia* **105**, 885–894.
10. Redberry Lake Biosphere Rerserve. Retrieved from [http://www.redberrylake.ca/].
11. Hammer, U. T. and Parker, J. (1984). Limnology of a perturbed highly saline Canadian lake. *Verhandlungen Internationale Vereinigung für theoretische und angewandte Limnologie* **102**, 331–342.
12. *Mineral Resource Map of Saskatchewan Regina* (2006). Saskatchewan Department of Industry and Resources.
13. Hammer, U. T., Shamess, J., and Haynes, R. C. (1989). The distribution and abundance of algae in saline lakes of Saskatchewan, Canada. *Hydrobiologia* **105**, 1–26.
14. Haynes, R. C. and Hammer, U. T. (1978). The saline lakes of Saskatchewan. IV Primary production of phytoplankton in selected saline ecosystems. *Inter. Revue. Der. Ges. Hydrobio.* **63**, 3337–3351.
15. Bierhuizen, J. F. H. and Prepas, E. E. (1985). Relationships between nutrients, dominant ions, and phytoplankton standing crop in prairie saline lakes. *Canadian J. Fisher. Aquatic. Sci.* **42**, 11588–11594.
16. Evans, J. C. and Prepas, E. E. (1996). Potential effects of climate change on ion chemistry and phytoplankton communities in prairie saline lakes. *Limnol. Oceanogr.* **41**(5), 1063–1076.
17. Ferreyra, G. A., Demers, S., delGiorgio, P., and Chanut, J. P. (1997). Physiological responses of natural plankton communities to ultraviolet-B radiation in Redberry Lake (Saskatchewan, Canada). *Canadian J. Fisher. Aquatic. Sci.* **54**(3), 705–714.
18. Sorokin, D. Y. and Kuenen, J. G. (2005). *Chemolithotrophic haloalkaliphiles* from soda lakes. *FEMS Microbiol. Eco.* **52**(3), 287–295.
19. Hammer, U. T. (1986). Saline Lake Ecosystems of the World. In *Monographiae Biologicae*, Volume 59. H. J. Dumont and Dr. W. Dordrecht (Eds.). Junk Publishers.
20. Grasby, S. E. and Londry, K. L. (2007). Biogeochemistry of hypersaline springs supporting a mid-continent Marine Ecosystem: An analogue for Martian springs? *Astrobiology* **7**(4), 662–683.
21. Sorensen, K. B., Canfield, D. E., Teske, A. P., and Oren, A. (2005). Community composition of a hypersaline endoevaporitic microbial mat. *Appl. Environ. Microbiol.* **71**(11), 7352–7365.
22. Sorokin, D. Y., Tourova, T. P., Lysenko, A. M., and Muyzer, G. (2006). Diversity of culturable halophilic sulfur-oxidizing

bacteria in hypersaline habitats. *Microbiology* **152**(10), 3013–3023.

23. Hammer, U. T., Sheard, J. S., and Kranabetter, J. (1990). Distribution and abundance of littoral benthic fauna in Canadian prairie saline lakes. *Hydrobiologia* **197**, 173–192.
24. Oren, A. (1999). Microbiology and biogeochemistry of halophilic microorganisms—An overview. In *Microbiolgy and Biogeochemistry of Hypersaline Environments*. A. Oren (Ed.). New York, CRC.
25. Timms, B. V., Hammer, U. T., and Sheard, J. W. (1986). A study of benthic communities in some saline lakes in Saskatchewan and Alberta, Canada. *Inter. Revue. Der. Ges. Hydrobio.* **71**(6), 759–777.
26. Oren, A. (2002). Diversity of halophilic microorganisms: Environments, phylogeny, physiology, and applications. *J. Industrial Microbiol. Biotechnol.* **28**, 56–63.
27. Litchfield, C. D. and Gillevet, P. M. (2002). Microbial diversity and complexity in hypersaline environments: A preliminary assessment. *J. Ind. Microbiol. Biotechnol.* **28**(1), 48–55.
28. Wobeser, G. and Howard, J. (1987). Mortality of waterfowl on a hypersaline wetland as a result of salt encrustation. *J. Wildl. Dis.* **23**(1), 127–134.
29. Schmutz, J. K. (1999). *Community Conservation Plan for the Redberry Lake Important Bird Area*. Center for studies in agriculture, Law, and the Environment, Saskatoon.
30. Traylor, J. J., Alisauskas, R. T., and Kehoe, F. P. (2004). Nesting ecology of white—Winges scoters (*Melanitta fusca* deglandi) at Redberry Lake, Saskatchewan. In *Auk*, Volume 121. American Ornithologists Union, pp. 950–962.
31. Sachs, J. P., Pahnke, K., Smittenberg, R., and Zhang, Z. (2007). Paleoceanography, biological proxies; biomarkers. *Encyclo. Quarter. Sci.* **2**, 1627–1634.
32. Hammer, U. T. (1990). The effects of climate change on the salinity, water levels, and biota of Canadian prairie lakes. *Verhandlungen Internationale Vereinigung Für Theoretische Und Angewandte Limnologie* **24**, 321–326.
33. Last, W. M. (1993). Geolimnology of Freefight Lake: An unusual hypersaline lake in the northern Great Plains of western Canada. *Sedimentology* **40**, 431–4448.
34. Birks, S. J. and Remenda, V. H. (1999). Hydrogeological investigation of Chappice Lake, southeastern Alberta: Groundwater inputs to a saline basin. *J. Paleolimnol.* **21**, 2235–2255.
35. Waiser, M. J. and Robarts, R. D. (1995). Microbial nutrient limitation in prairie saline lakes with high sulfate concentrations. *Limno. Oceano.* **40**(3), 566–574.
36. Conly, F. M. and Van der Kamp, G. (2001). Monitoring the hydrology of Canadian prairie wetlands to detect the effects of climate change and land use changes. *Environ. Monitor. Asses.* **67**(1–2), 195–215.
37. Covich, A. P., Fritz, S. C., Lamb, P. J., Marzolf, R. D., Matthews, W. J., Poiani, K. A., Prepas, E. E., Richman, M. B., and Winter, T. C. (1997). Potential effects of climate change on aquatic ecosystems of the Great Plains of North America. *Hydrol. Proc.* **11**(8), 993–1021.
38. Last, W. M. (1991). Sedimentology, geochemistry, and evolution of saline lakes of the Northern Great Plains. *Saskatoon, Post-conference Fieldtrip Guidebook, Sedimentary and Paleolimnological Records of Saline Lakes.*
39. Environment Canada Monthly Data Report. Retrieved from [http://www.climate.weatheroffice.ec.gc.ca/climatedata/monthlydata_e.html].
40. Mitsch, W. J. (1993). *Wetlands*. Van Nostrand Reinhold, New York.
41. Anati, D. A. (1999). The salinity of hypersaline brines: Concepts and misconceptions. *Inter. J. Salt Lake Res.* **8**(1), 55–70.
42. Development, O. R. (Ed.) (1983). *Methods of Analysis of Water and Wastes*. United States Environmental Protection Agency, Washington, DC.
43. Richards, S. R., Rudd, J. W. M., and Kelley, C. A. (1994). Organic volatile sulfur in lakes ranging in sulfate and dissolved salt concentration over five orders of magnitude. *Limnol. Oceanogr.* **39**, 562–572.

Index

A

Achouak, W., 111–119
Albertini, D. F., 150–162
Alvarez, M., 95–110
Amodio-Cocchieri, R., 82–94
Arabidopsis thaliana seedlings analysis, 2–3
Atari, D. O., 181–196

B

Barbosa, R. M. R., 132–149
BioworksBrowser 3.2 software, 20
Botsford, J., 95–110
Bowman, J. S., 197–214

C

Cairncross, E., 163–180
Calderisi, M., 82–94
Carrier, P., 111–119
Chemical defense in marine biofilm bacteria
  antiprotozoal compounds, 120
  aquatic ecosystems, 120
  bacterial groups, 121–122
  bacteria-protist interactions, 121
  bacterivorous consumers, 120
  bioassay-guided chromatographic fractionation, 123
  biofilm-enhanced production of violacein, 125
  catastrophic decline in population density, 127
  cell density, 125
  cellular features, 122
  chemical defenses, 128
  colorimetric assays, 128
  defensive metabolite violacein, 123–124
  DNA fragmentation, 127
  edible biofilm structures, 123
  eukaryotic organisms, 120
  exopolymeric matrix, 120
  genetic analysis of antiprotozoal activity, 120
    DNA sequence, 130
    fractions and crude extract, 129–130
    integrated microbial genomes system (IMGs), 130
    panhandle PCR method, 130
    P. tunicata, 129–130
  grazing bioassay
    bacterial cultures, 129
    epifluorescence microscopy, 129
    planktonic phase, 129
  grazing resistance and antiprotozoal activity, 122
  grazing-resistant species, 121
  induction of predator death program, 126–127
  mammalian cell lines, 128
  marine macroalga, 121–122
  morphological and biochemical markers, 127
  niche-specific predatory flagellates, 122
  non-food control treatment, 123
  novel therapeutic agent, 128
  nutrient conditions, 126
  PCR-amplification, 124
  plankton
    effects, 122
    environments, 120–121
    populations, 121
  population density, 126
  predator numbers, 122
  predator-prey interactions, 128
  primary ecological function, 126

protozoan fitness analysis
apoptosis inducer actinomycin, 130
eukaryotic cell death programs, 130–131
intracellular activity, 130–131
morphological changes, 130
phosphatidyl serine asymmetry, 130–131
Pseudoalteromonas tunicata and Microbulbifer sp, 123
putative biosynthetic gene cluster, 123
quorum sensing, 126
Rhynchomonas nasuta and Cafeteria roenbergensis, 122
strains and culture conditions, 128–129
surface feeder, 123
transposon mutagenesis, 123
U. australis, 121
violacein
biosynthesis, 124
mediated defense, 126
producing bacteria, 125
production kinetics, 130
Chemical Valley, 181
Chen, A. M., 132–149
Cirillo, T., 82–94
Cocucci, M., 1–21
Conrod, S., 111–119
Cornel, A. J., 132–149
CquiOBP1 expression
olfactory sensilla of Cx. p. Quinquefasciatus
density of labeling, 135
grooved pegs, 135–136
homologous proteins isolation, 134
malaria mosquito and yellow fever mosquito, 134
maxillary palps, 135–136
mosquito oviposition attractant, 135
olfactory sensilla and trichoid sensilla, 134–135
oviposition pheromone reception, 134–135, 136
penetration of antibody, 134–135, 136
perisplamic expression system, 133–134
pH-dependent binding of MOP
binding of test ligands, 136–137
chiral column, 139
Culex pipiens, 139
dichroism data, 137
enantiocompetitive assays, 140
far-UV circular dichroism spectra, 137, 138
fluorescence emission spectra, 136, 138
fluorescent reporter, 136–137
natural stereoisomer, 140
odorant receptor, 139
pheromone-binding proteins, 136–137
racemic mixture, 140
stereoisomers on chiral column, 139–140
Cuine, S., 111–119

D

Dalvie, M. A., 163–180
2-DE analysis and protein identification. See also Zea mays plants from nitrate, protein pattern changes
biochemical and physiological data, 7
dry and fresh weight, 7
electrophoretic analyses, 7–8
maps, 8
spots, 8–10
Digital elevation models (DEMs), 185
Dong, T., 53–81

E

Egan, S., 120–131

Endosulfan farming areas of Western Cape
analytical methodology, 170
bioaccumulation, 163
case study, 179
daily intake, modeling, 177–178
endocrine disruptor, 178
choice of pesticides, analysis
multi-residue screening methods, 169
field
Grabouw/Vyeboom, 175–176
Hex river, 171–173
Piketberg, 174
sampling, 169
quality assurance, 170
quantification limit, 171
quality control, 170
sample extraction
Environmental Protection Agency (EPA) methods, 169
gas chromatography, 170
high-pressure chromatography, 170
study areas identification and sampling sites, 163, 165–167
geohydrologists, 164
sampling points, 168
Environmental toxicant, morphogenesis
abrogate embryogenesis, 150
AhR pathway, 150
aryl hydrocarbon receptor (AhR), 150
compaction
cytoplasmic remodeling, 151
cytoskeletal remodeling, 151
nuclear remodeling, 151
estrogen bisphenol, 151
trophectoderm (TE), 151
Espen, L., 1–21
Ethanol exposure, 54
Evolution of warning coloration
aposematic prey, 46
assessment period, 50
chemicals, 50–51
conspicuousness, 51
defense chemical, 50
discussion of, 46
evolutionary biologists, 46
experimental, 51–52
intensive learning, 50
model
aposematic traits, 47
correlation, 49–50
developement effect, 49
interactions, 47
linear relationship, 47
mechanics of, 47
olfactory signal, 47
OSD, 47
probabilities, 47–48
VC, 47
visual and olfactory components, 48–50
neophobia, 46
perplexing phenomenon, 46
poor smellers, 51
prey, 51
profitable strategy, 52
reliable chemical signal, 46–47
stochastic model, 46

F

Furtado, A., 132–149
Fyn/c-Cbl hypothesis of toxicant convergence, 72
Fyn kinase activation, 53

G

Garritano, S., 82–94
General Practice Research Framework (GPRF), 23
Gohli, J., 46–52
GPRF. See General Practice Research Framework (GPRF)

H

Heavy metal tolerance in Stenotrophomonas maltophilia
antimicrobial agents, 111
cadmium and
analysis of accumulation, 117
cluster of genes, 113–114
content, 113–114
CdS particles formation
aerobic conditions, 115
antibiotic and heavy metal resistance genes, 116
atomic environment, 115–116
EXAFS spectroscopy, 115–116
Klebsiella pneumoniae, 115
nosocomial infections, 116
clinical syndromes, 111–112
cysteine accumulation
homeostasis, 115
HPLC analysis of nonprotein thiols, 114
intracellular cysteine pool, 114
oxidative stress, 115
thiol compounds, 114
efflux mechanisms, 111
energy dispersive X-ray (EDX) analysis, 111
ESEM-EDX AND TEM-EDX observations, 117–118
extended X-ray absorption fine structure (EXAFS) spectroscopy
back-Fourier transformation (BFT), 118
fluorescence detection mode, 118
pseudo-radial distribution functions (RDFs), 118
XANES spectra intensity, 118
gram-negative bacterium, 111
growth conditions, 116–117
growth-promoting activity and antagonistic properties, 111
metals and antibiotics maximum tolerance concentrations, 117
minimal inhibitory concentrations (MICs) of drugs and heavy metals
bacterial resistance, 112
multidrug resistance (MDR) efflux pumps, 112
nosocomial pathogen, 111–112
oxianions reduction
chemical microanalysis, 112–113
EDX pectroscopy, 112
efflux systems, 112
ESEM-EDX and TEM-EDX observations, 112–113
extracellular deposits, 112–113
intracellular concentration, 113
Pseudomonas strains, 112
rhizosphere inhabitant, 111–112
soluble thiols analysis
calibration curves of glutathione, 117
derivatization procedure, 117
high-performance liquid chromatography (HPLC) analysis, 117
transmission electron microscopy (TEM), 111
Hernandez, J., 95–110
Heulin, T., 111–119
Högstedt, G., 46–52
Hutt, K. J., 150–162
Hydroponic system, 4

I

International agency for research cancer (IARC), 182

K

Kjelleberg, S., 120–131

L

Land use regression in Sarnia
approaches, 183
benzene model, 194

BTEX compounds, 182
Chemical Valley, 181
coefficient of ethylbenzene, 194
data collection
BTEX species, 184–185
exposed filters, 185
samplers, 184–185
ethylbenzene, 182
in modeling, 183–184
model selection
benzene surface, 188
BTEX concentrations, 187
Chow test results, 192
cross-validation approach, 186–187
Kriged surfaces, 190
measured BTEX species, 188
modeled concentrations, 188–189
NAPS and sampled BTEX data, comparison, 187–188
Pearson correlation, 188–189
predictive LUR models, variables, 190–192
predictive models, 186
regression, 186, 193
results for BTEX compounds, 191–192
samplers, 187
organic vapor monitors, 181
physical characteristics, 183–184
researchers, 182–183
spatial trends, 185
study area
areas of concern, 184
Chemical Valley, 184
variable generation
meteorological data, 186
traffic counts, 185
Latte, N., 132–149
Leal, W. S., 132–149
Li, Z., 53–81
London, L., 163–180
Luginaah, I. N., 181–196

M

Malondialdehyde (MDA), 17–18
Mann, V., 22–35
Matz, C., 120–131
MeHg exposure, 55
analysis of cell division, 57
associated suppression of ERK1/2 phosphorylation, 62
cell division in clonal assays, 58
Fyn and c-Cbl activation, 60–63
glial progenitor cells, 55
induced reductions, 62
PDGF effects, 58–59
role of redox regulation, 57
Mendoza, J., 95–110
Microbial contamination and chemical toxicity
anthropogenic activities, 96
antibiotic resistance analyses (ARA), 108
bacterial cultures, 108
bacterial suspension, 108–109
biological variables, 108
Botsford's assay, 104–105
chemical contamination, 97
chemical discharges, 104
chemically-defined medium (CDM), 108
chemical toxicity, 95, 99
contamination potential of water, 95
degree of inhibition, 109
drought conditions, 104
E. coli, 95, 99
environmental parameters, 95, 108
fecal coliform bacteria, 100
contamination, 104
counts, 99
indicators, 108
gastro-intestinal illness, 104
genotyping and ribotyping, 108
H. pylori, 95–97, 99

H. pylori HpSa ELISA test, 99, 101–102
indicator organisms, 97
industrial discharges, 104
infectious diseases, 96–97
infectious micro-organisms and toxic pollutants, 96
inhibition capacity (IC), 95
linear relationship, 109–110
MF technique, 109
mode of transmission, 96–97
most probable number (MPN), 99, 101
physical/chemical parameter, 109–110
potential contributors, 103–104
relative toxicity, 105
Rhizobium meliloti, 108
risk factors, 96–97
river flow
    concrete-lined canal, 97–98
    drought-condition, 97–98
    irrigation purposes, 97–98
    microbial population and chemical composition, 97–98
    sample site descriptions, 97, 99
    sampling locations Rio Grande basin, 97–98
    treated wastewater, 98
statistical analyses
    bacterial population, 105
    biological variables and chemical toxicity, 105
    Pearson-Moment correlation procedure, 105
    Pearson Product-Moment coefficients, 105–107
    physiochemical parameters, 105
    regression procedures, 105
    sample temperature, 105
    survival rates, 105
stepwise regression, 109–110
surface water testing, 105
    synergistic effect, 105
    water pollutants, 96–97
Montoya, A., 95–110
Morgan, T. I., 132–149
Myelin-forming oligodendrocytes, 56–58

N

National air pollution surveillance (NAPS), 187–188
Negri, A. S., 1–21
Nitrate content and nitrate reductase (NR) activity, 5, 7
Noble, M., 53–81

O

Olfactory signal/defense (OSD), 47
Oxygen-evolving complex (OEC) of photosystem II, 14

P

Pages, D., 111–119
Penesyan, A., 120–131
PEPCase activity, 12
Pesaresi, P., 1–21
Pesticide-related illness
    challenges in surveillance, 23
    checklist, 24
    computer-based interview, 24–25
    data collection, 23–24
    design, 32
    eligibility for interview invitation
        distributions of, 27
        GP surgery session, 27
    eligible patients, 24
    environmental factors, 34
    ethical approval, 25
    exact method, 25
    frequency, 25
    general practitioners, 22
    GPRF, 23
    in Great Britain, 22

health effects, 22
high proportion, 33
home use of
interviewes reported, 28
majority of, 29
multivariable models, 31
pest control chemicals, 28, 29
proportions of patients, 29
quantity of, 29
questionnaire, 29–30
univariable logistic regression models, 30–31
impractical and cost, 33
incidence and prevalence of symptoms
annual incidence of, 28
consultations, 28
general practitioner's opinion, 27
study, 28
in-depth interview, 24
information pack, 24
monitoring systems, 32
National strategies, 34
numbers of consultations, 31–32
possible case, 33
practices
GB and GP, 25–26
locations of, 26
pregnancy, 33
pre-pilot phase, 25
robust standard, 25
SENSOR system, 33–34
statistical methods, 25
studies, 32–33
study, 22, 23
surgery sessions, 23
surveillance system, 34
World Health Organization, role in, 32
Petroff, B. K., 150–162
Phosphoenolpyruvate carboxylase (PEP-Case)
Arabidopsis thaliana, 2
low bio-availability of nitrogen, 2
nitrate assimilation, 2
nitric oxide (NO) and LOX, role, 1
post-translational modifications, 1
side-effect, 2
Pinto, B., 82–94
Plastidic type 2 linoleate 13-LOX, 14
Powers, J. S., 36–45
Precursor cell function, chemically diverse toxicants effects on, 53
animal treatment, 79
antibodies and immunoblotting, 75
c-Cbl targets, 66–67
cell isolation, culture, and treatment, 74–75
cell transfection and luciferase activity assay, 75
clonal analysis, 78–79
convergence of, 63
cysteine pro-drugs, 65–66
developmental exposure, 67–68
DNA vector-based RNA interference, 77
Fyn and c-Cbl activation, 60–63
Fyn kinase activity, 76–77
images, data processing, and statistics, 80
immunoprecipitation assay, 76
intracellular reactive oxygen species measurement, 76
MeHg exposure, 56–59
pathway-specific disruption of PDGF-mediated signaling, 59–60
Pb and paraquat exposure, 64
Rho kinase activity, 77
RNA isolation and real-time RT-PCR, 78
signaling pathways, 73
tissue preparation, 79–80
viral packaging, cell infection, and selection, 78
in vitro BrdU incorporation assay, 75–76, 80
in vivo analysis, 69

Prinsi, B., 1–21
Pröschel, C., 53–81

R

Rat pre-implantation embryo, morphogenesis
aberrant cytokinesis, 150
aberrant mitotic spindles, 158
AhR ligand, 150
animals, 160
apoptosis, 158
blastocyst, 150
cell pre-implantation
acute periconceptional TCDD, 154–156
chronic maternal TCDD, 152–154
comparison, 158
environmental perturbations, 162
epidemiological studies, 150
experimental design, 160–161
follicular fluid, 158
immunofluorescence, 161
knockout and transgenic studies, 159
laminopathies, 160
2-methoxyestradiol (2-ME), 158
oocyte maturation, 160
pre-implantation embryos,classification, 161–162
pre-implantation embryo survival to blastocyst
acute periconceptional TCDD, 156–157
rat model
cytoskeletal modifications, 151
statistical analysis, 162
TCDD, 150
Reali, D., 82–94
Receptor tyrosine kinase (RTK) signaling, 53
Reverse and conventional chemical ecology in Culex mosquitoes
analytical procedures
circular dichroism (CD) spectra, 147
fluorescence measurements, 147
high pressure liquid chromatography (HPLC) separations, 147
binding assays, 146
chemical analysis
flame ionization detector (FID), 147
gas chromatograph, 147
chemicals, 146–147
CquiOBP1 expression
dimerization of monomer, 133–134
disulfide bridges, 133–134
gel filtration elution profiles, 133–134
ion-exchange chromatography and gel filtration, 133–134
mass spectral data, 133
perisplamic expression system, 133–134
evaluation of Cx. p. Quinquefasciatus attractants
Aedes albopictus, 142
binary/tertiary synthetic mixtures, 142–143
olfaction and chemical ecology, 143–144
skatole-baited traps, 142
statistically reliable data, 141–142
surveillance and management programs, 144
tangible measurement, 143
TMA-baited gravid traps, 141–142
traps baited with synthetic attractant, 143–144
field tests, 148
GC-EAD-based identification
EAD-active peaks, 141
electroantennographic detector, 141
solid-phase micro extraction (SPME), 141
gravid traps, 133

human diseases, 132–133
immunohistochemistry (IHC) studies, 132, 145
infusions, 147–148
monomeric and dimeric forms, 132
odorant-binding protein (OBP), 132
pathogens vectors, 132
populations and mosquito-borne virus activity, 132–133
prospecting for oviposition attractants
olfactory receptor neurons, 140
reverse chemical ecology, 140
protein expression, purification, and antibody production
perisplasmic expression, 144–145
recombinant proteins concentrations, 145
theoretical extinction coefficients, 145
rabbit chow infusion, 133
single sensillum recording, 146
surveillance and control programs, 132
Western blot, 146
Rose, J., 111–119
Rushton, L., 22–35

S

Sachs, J. P., 197–214
Saenz, R., 95–110
Saline Lakes in Alberta and Saskatchewan
Big Quill Lake
predominant anion, 208
biochemical responses, 198–199
Bitter Lake
in situ measurements, 203
Canada's Northern great plains, 197
Chaplin Lakes
chemical and physical properties of, 204–205
microbial mat, 204–205
satellite image of, 203–204
Chappice Lake, 201
Diaptomus and Artemia, 197–198
filtered and unfiltered water samples studies, 213–214
Freefight Lakes
chemical and physical properties, 201–203
Hawkeye digital sonar handheld sonar system, 213
Hypersaline Ingebright Lake
measurements, 201–202
importance of, 197–198
lakes investigated, sampling location, 199
Little Manitou East and West
measurements, 206
turbidity and salinity, 206–207
locations of lakes investigated, 200
Manito Lake
Secchi depth, 207
Muskiki Lake, 206
Patience Lake, 201
Redberry Lake
measurements, 208
pycnocline and chemocline, 208
salinity trends
annual total precipitation, 210
ecological significance of, 212
extension of Hammer time series, 209
function of specific conductivity, 213
parameters, 212–213
Redberry Lake bathymetry, 211
times of high precipitation water, 209–210
water management projects, 210
satellite images of lakes investigated, 200
survey of lakes, 197
Sayer, E. J., 36–45
Schupp, P. J., 120–131
Seafood, chemical contaminants
analytical sample preparation, 83–84
assay
centrifugation, 85

dose-response curve, 86
length of incubation, 85
pellet, 85
volume of culture, 85
β-galactosidase activity
agonistic activity, 91
extracts, 91
M. merluccius, 91
natural hormone, 91
β-galactosidase activity and PCBs correlation
aryl hydrocarbon (Ah) receptor, 92
bioaccumulation of xenoestrogens, 93
congener, 92
endogenous hormones, 93
estrogen level, 93
screening programs, 93
seafood sample, 92
yeast assay, 92
chemical analysis
acetonitrile, 84
calibration curve, 84
column adsorption-chromatography, 84
gas chromatography mass spectrometry (GCMS), 84
PCBs quantification, 84
endocrine disrupting chemicals (EDCs), 82
environmental pollutants, 82
estrogen receptor, 82–83
fat extraction, 84
fish advisory, 83
immunotoxicity and neurobehavioral deficits, 83
male and female reproductive system, 82–83
marine habitat, 82–83
polychlorinated biphenyls (PCBs) assessment, 82–83
concentrations in marine species, 87–88
congeners distribution, 89
differences in content, 87, 89
marine organisms samples, 87
multifactor analysis, 91
principal components analysis loading plot, 89–90
S. scombrus and M. barbatus, 91
sampling
commercial size, 83
consumed seafood, 83
fish, crustaceans, and cephalopods, 83
marine environment, 83
selective medium, 85
statistical analysis
estrogenicity, 87
partial least square regression (PLS), 86–87
PLS regression, 87
principal components analysis (PCA), 86
target tissues, 83
therapeutic agents, 82–83
xenobiotics, 83
yeast strain
assay, 85
expression vector, 84–85
organic extracts, 84
reporter gene, 82
S. cerevisiae, 84–85
sensitivity and specificity, 85
Shi, Z., 150–162
Shui Yen Phang, 120–131
Solomon, A., 163–180
Steinberg, P., 120–131
Syed, Z., 132–149

T

Tanner, E. V. J., 36–45
c-CblTarget platelet-derived growth factor receptor-a, 53
2,3,7,8- Tetrachlorodibenzo-p-dioxin (TCDD), 150–151

Total dissolved solids (TDS), 197
Tropical forests, litterfall effects
  CO_2 efflux, 40
  CO_2 production, 36
  elevated, consequence of, 36
  experimental treatment, 40
  fine root and microbial biomass
    control treatment, 39, 40
    CT and L- plots, 39, 41
    measurement, 44
  global carbon cycle, 36–37
  global climate change, 36
  lack of moisture, 40
  long-term, large-scale, 36
  predicted and observed soil, 42
  site description
    carbon dynamics, 42
    lateral nutrient, 42–43
    long-term, 42
    nutrient cycle of, 42
    phosphorus concentration, 42
  soil carbon losses, 36
  soil respiration
    annual respiration rates, 43–44
    differences in, 38, 41
    litter manipulation treatments, 37, 41
    measurement sites, 43
    seasonal pattern, 37
  soil temperature and water content, 38–39
    gravimetric measurements, 44
    respiration measurements, 44
  statistical analyses, 44
TurboSEQUEST®, 20
Two-dimensional gel electrophoresis (2-DE), 1

V

Valles, A., 95–110
Vazquez, A., 95–110
Vehicle miles traveled (VMT), 186
Visual conspicuousness (VC), 47

W

Webb, J. S., 120–131
Wei Xu, 132–149

Y

Yuko Ishida, 132–149

Z

Zea mays plants from nitrate, protein pattern changes
  amino acids and total proteins, 17
  chlorophyll, 17
  2-DE analysis, 1, 18–19
  2-DE analysis and protein identification, 7–10
  economical relevance, 2
  experimental design and biochemical parameters, 3–7
  extraction of protein samples, 18
  functional role and quantitative change
    in leaves, 13–15
    in roots, 10–12
  in-gel digestion and LC-ESI-MS/MS analysis, 19–20
  LCESI-MS/MS, 1
  levels of nitrate, 16
  LOX, up-accumulation of, 1
  MDA, 17–18
  mineral nutrient, 2
  nitrate assimilation, 2
  nitrate reductase activity, 16–17
  nitrogen nutrition, 1–2
  PEPCase , modification, 1
  plant material and growth conditions, 16
  reducing sugars and sucrose, 17
  statistical analyses, 18
  transcriptomic analyses, 2
ZmLOX10 gene, 14